Study Guide

Neil D. Jespersen
St. John's University

to accompany

CHEMISTRY
The Molecular Nature of Matter

Seventh Edition

Neil D. Jespersen
St. John's University

Alison Hyslop
St. John's University

With significant contributions by

James E. Brady *St. John's University*

WILEY

Table of Contents

Before You Begin...

Before you begin your general chemistry course, read the next several pages. They're designed to tell you how to use this study guide and to give you a few tips on improving your study habits. If your instructor has decided to use this book for active-learning/group-style work in class, be sure to follow any instructions you are given.

How to Use the Study Guide

This book has been written to parallel the topics covered in your text, *Chemistry: The Molecular Nature of Matter*, Seventh Edition. Each chapter of the Study Guide begins with a very brief overview of the chapter contents. For each section in the textbook, you will find a corresponding section in the Study Guide. In the Study Guide, the Sections are divided into **Learning Objectives, Review, Examples,** and a **Self-Test**. Each chapter Section has a learning objective. Read this objective before beginning each Section, and then reread it again after you've finished to be sure you have met the goals described.

After you've read a section in the text, turn to the study guide and read the **Review**. This will point out specific ideas that you should be sure you have learned. Sometimes you will be referred back to the text to review topics or Examples there. In this edition many additional worked-out **Examples** are given. Work with the Study Guide and the text together to be sure you have mastered the material before going on.

In most sections you will also find a short **Self-Test** to enable you to test your knowledge and problem-solving ability. The answers to all of the Self-Test questions are located at the ends of the chapters in the Study Guide. However, you should try to answer the Self-Test questions without looking up the answers. A space is left after each Self-Test question so that you can write in your answers and then check them all after you've finished.

Chemical vocabulary

An important aspect of learning chemistry is becoming familiar with the language. There are many cases where lack of understanding can be traced to a lack of familiarity with some of the terms used in a discussion or a problem. A great deal of effort was made in your textbook to adequately define terms before using them in discussions. Once a term has been defined, however, it is normally used with the assumption that you've learned its meaning. It's important, therefore, to learn new terms as they appear, and for that reason, most of them are set in boldface type in the text. At the end of the textbook there is a glossary which you can use to be sure you understand the meanings of the new terms. The glossary also directs you to sections in the textbook where the terms are discussed.

Study Habits

Do you need an A (or at least a B) in chemistry? That's not as impossible as you may have been led to believe, but it's going to take some work. Chemistry is not an easy subject—it involves a mix of memorizing facts, understanding theory, and solving problems. It requires that you can see the "big picture" while simultaneously paying attention to precise terminology and details. There is a lot of material to be covered, but it won't overwhelm you if you ***stay up to date***. Don't fall behind, because if you do, you are likely to find that you can't catch up. Your key to success, then, is *efficient* study, so your precious study time isn't wasted.

Efficient study requires a regular routine, not hard study one night and nothing the next. At first, it's difficult to train yourself, but after a short time you will be surprised to find that your study routine has become a study habit, and your chances of success in chemistry, or any other subject, will be greatly improved.

To help you get more out of class, try to devote a few minutes the evening before to reading, in the text, the topics that you will cover the next day. Read the material quickly just to get a feel for what the topics are about. Don't worry if you don't understand everything; the idea at this stage is to be aware of what your teacher will be talking about. It is also a good study habit to write down questions you have about the chapter and then see if your teacher answers them in lecture. If not, ask your question at the appropriate time.

Your lecture instructor and your textbook serve to complement one another; they provide you with two views of the same subject. Try to attend lecture regularly and take notes during class. Bring your textbook, or the relevant pages to class since they may help your note taking. Print out, beforehand, any prepared notes your instructor provides in electronic format. Pay particular attention to items your instructor stresses or indicates will "be on the test." If you pay attention carefully to what your teacher presents in class, your notes will probably be somewhat sketchy. They should, however, give an indication of the major ideas. After class, when you have a few minutes, look over your notes and try to fill in the bare spots while the lecture is still fresh in your mind. This will save you a lot of time later when you finally get around to studying your notes in detail.

In the evening (or whatever part of the day you close yourself off from the rest of the world to really study intensely) review your class notes once again. Use the text and study guide as directed above and really try to learn the material presented to you that day. When you find yourself really struggling with a topic, get on your computer and go to our web site, www.wiley.com/college/brady, where you will find additional tutorial help on difficult concepts. If you have prepared before class and briefly reviewed the notes afterward, you'll be surprised at how quickly and how well your concentrated study time will progress. You may even find yourself enjoying chemistry!

As you study, continue to fill in the bare spots in your class notes. Write out the definitions of new terms in your notebook. In this way, when it comes time for an exam you should be able to review for it simply from your notes.

At this point you're probably thinking that there isn't enough time to do all the things described above. Actually, the preparation before class and brief review of the notes shortly after class take very little time and will probably save more time than they consume.

Well, you're on your way to an A. There are a few other things that can help you get there. If you possibly can, spend about 30 minutes to an hour at the end of a week to review the week's work. Psychologists have found that a few brief exposures to a subject are more effective at fixing them in the mind than a "cram" session before an exam. The brief time spent at the end of a week can save you hours just before an exam (efficiency!). Try it (you'll like it); it works.

There are some people (you may be one of them) who still have difficulty with chemistry even though they do follow good study habits. Often this is because of weaknesses in their earlier education. A lot of the material on our web site is designed to help you if you fall into this group. It has been carefully crafted by experienced teachers who know what your problems are, so we urge you to take advantage of the assets that are available to you there.

If, after following intensive study and working with the web-based material, you are still fuzzy about something, speak to your teacher about it. Try to clear up these problems before they get worse.

Group activities and study also are important in the study of chemistry. Many institutions encourage active learning with group work both in and out of class. Join in and actively participate in class. Create, or join, a study group. It is well known that study sessions with fellow classmates help everyone to get over the stumbling blocks. Group study is very effective, especially if you actively participate. When you find that you can explain a chemical concept, or show the solution for a problem, to someone else, you really know the subject. But if you can't explain a topic, then it requires more study.

Problem Solving—Using Chemical Tools

Your course in chemistry provides a unique opportunity for you to develop and sharpen your problem-solving skills. Problem solving is the most effective way to reinforce and even test the fundamental concepts of chemistry. Just as in life outside the classroom, the problems you will encounter in chemistry are not only numerical ones. In chemistry, you will also find problems related to theory and the application of concepts. The techniques that we apply to these various kinds of problems do not differ much, and one of the goals of your textbook and this Study Guide is to provide a framework within which you can learn to solve all sorts of problems effectively.

If you've read the "To the Student" message at the beginning of the textbook, you learned that we view solving a chemistry problem as not much different than solving a plumbing problem. Both involve the application of specific tools that accomplish specific tasks. A plumber uses tools such as screwdrivers and wrenches; you will learn to use a different set of tools—ones that we might call *chemical tools*.

Chemical tools are the simple one-step tasks that you will learn how to do, such as changing units from feet to meters, or degrees Fahrenheit to degrees Celsius. Solving more complex problems just involves combining simple tools in various ways. The secret to solving complex problems, therefore, is learning how to choose the chemical tools that must be used.

Building a chemical toolbox

Our first goal is to clearly identify the tools you will have at your disposal. As you study the text, the concepts you will need to solve problems are marked by an icon in the margin when they are introduced. (To see what the icon looks like, refer to the "To the Student" message at the beginning of the textbook.) The chemical tools are summarized at the end of a textbook chapter in a section titled *Tools for Problem Solving* and they are also collected at the end of each of the chapters in this Study Guide.

Solving problems

In both the text and the Study Guide there are worked examples that illustrate a wide variety of problems and their solutions. You will notice that each begins with a section titled *Analysis*. The Analysis section describes the thinking that goes into solving the problem and identifies the tools needed to do the job. This is the most important step in problem solving, because once you've figured out *how* to solve the problem, the rest is easy. (Be sure to study the Examples thoroughly, both in the Text and Study Guide. Also be sure to work on the Practice Exercises that follow the Examples in the Text.)

The Analysis step is where you determine which of the chemical tools that you've learned will be applied to the specific problem at hand. Let's look at how you might go about this when working on a problem you haven't seen before—one for which the solution is not immediately obvious. To do this, we will look at a problem of the type you will encounter in Chapter 3. If you've had a previous course in chemistry, you will recognize many of the concepts presented. If they are unfamiliar, don't be concerned. The goal at this time is to illustrate *how* the chemical tools approach can be used to help find a solution to a problem.

Problem
Assemble all the information needed to determine the number of grams of Al that will react with 900 molecules of O_2 to form Al_2O_3, and then describe how to find the answer.

The first step in solving the problem is determining what kind of problem it is. In this case, it is a problem dealing with a subject we call *stoichiometry*. (Don't worry, you will learn about all this later.)

Now that we have identified the *kind* of problem, we look over the tools that apply to stoichiometry problems. In this edition the description of each tool has been expanded to include important equations as needed. These are given in the table below.

Tools that apply to stoichiometry:

Atomic mass (Section 3.1)

gram atomic mass of X = molar mass X = 1 mole X

Molecular mass and formula mass (Section 3.1)

Used to form a conversion factor to calculate mass from moles of a compound, or moles from the mass of a compound.

gram molecular mass of X = molar mass X = 1 mole X

gram formula mass of X = molar mass X = 1 mole X

Molar mass (Section 3.1)

This is a general term encompassing atomic, molecular and formula masses. All are the sum of the masses of the elements in the chemical formula.

molar mass of X = 1 mole of X

Avogadro's number (Section 3.1)

1 mole X = 6.02×10^{23} particles of X

Chemical symbols and subscripts in a chemical formula (Section 2.3)

In a chemical formula, the chemical symbol stands for an atom of an element. Subscripts show the number of atoms of each kind, and when a subscript follows a parentheses, it multiplies everything within the parentheses.

This isn't a very large set of tools, and we can be fairly confident that these will be sufficient to solve the problem. Therefore, we next examine the problem to identify the quantities that relate to the tools we have at hand. Notice that we've drawn boxes around the quantities.

Assemble all the information needed to determine the number of $\boxed{\text{grams of Al}}$ that will react with $\boxed{\text{900 molecules of } O_2}$ to form $\boxed{Al_2O_3}$ and then describe how to find the answer.

Now we begin to assign specific numbers to quantities as we assemble the final set of tools we will use to solve the problem. We've collected the information in a table just to make it easier for you to follow. Notice that we have not used all the tools related to stoichiometry. Instead, we have selected just the tools that apply to the quantities in the problem.

Quantity in question	Tool related to it	Relationship
$\boxed{\text{grams of Al}}$	atomic mass	27 g Al = 1 mol Al
$\boxed{\text{900 molecules of } O_2}$	Avogadro's number	6.02×10^{23} molecules O_2 = 1 mol O_2
$\boxed{Al_2O_3}$	chemical formula	2 mol Al = 3 mol O
$\boxed{O_2}$		1 mol O_2 = 2 mol O

The information in the column at the right is what we use to obtain the answer. As you will learn, we can use a method called the *factor label method* or *dimensional analysis* make sure the units of the answer work out correctly. The proper setup of the solution is

$$900 \text{ molecules } O_2 \times \frac{1 \text{ mol } O_2}{6.02 \times 10^{23} \text{ molecules } O_2} \times \frac{2 \text{ mol } O}{1 \text{ mol } O_2} \times \frac{2 \text{ mol } Al}{3 \text{ mol } O_2} \times \frac{27g \text{ Al}}{1 \text{ mol } Al} = \text{answer}$$

As you can see, we have not actually calculated the answer. Nevertheless, we really have *solved* the problem; we just haven't done the dirty work of doing the arithmetic.

At the end of most chapters in the textbook you will find questions that are not mathematical. These questions ask you to figure out what you need to know to solve various problems, but not what the answers are. The goal is to make you *think* about how to solve the problems without having to worry about the answers. They are worthwhile exercises and you should be sure to work on them. As you will see, some are pretty difficult. But as they say, "No pain, no gain!"

We realize, of course, that many problems have more than one path to the answer. After correctly analyzing a problem and after recognizing what tools must be used, intermediate calculations and thought processes can validly follow more than one *sequence*. Therefore, you might choose a path in which the order of the steps is different from ours. Remember that any valid method will give the correct answer. To provide assurance that you are progressing well, answers are provided for all questions in this Study Guide. Answers to all practice exercises and selected questions are given in the Textbook while more detailed solutions are given in the Student Solutions Manual. More detailed help with selected problems is available online with the Interactive Learning Ware (ILW) and Office Hours (OH) problems.

Finally, when there are no answers given, as on a test or exam, you need a method for determining if the correct answer has been reached. To cultivate these methods, each worked Example has a section called *Is the Answer Reasonable?* In this section many of the techniques used by professional chemists to check their answers are shown. These include methods for estimating answers and reasonable ranges of results based on the given data. As you become even more familiar with chemistry, try to cultivate the practice of asking yourself, "Does my answer make sense?" When you are able to answer that question, you will have gone a long way toward mastering the study of chemistry itself.

Time to Begin

As you begin your chemistry classes, we wish you well. Move on to the course now, and good luck on getting that A!

Notes:

Chapter 0

A Very Brief History of Chemistry

In this chapter the main ideas that chemists base their science upon are briefly presented. We begin with current theories of how the elements were created and finally distributed around our planet, Earth. These concepts are part of cosmological theory and fundamental physics. This chapter concludes with the story of how scientists in the late 19th and early 20th centuries used rather simple instruments to deduce the basic structure of the atom.

0.1   Chemistry's Important Concepts

Learning Objective

We will develop a sense of the scope and purpose of the chemical sciences.

Review

This section is very short, yet very important. Here we find the big concepts that help us focus on why the study of chemistry is important. First and foremost is the *atomic theory* proposed by John Dalton. Second is the idea that *careful measurements and observations* will enable scientists to tease fundamental truths of nature out of a mass of data. Third is the importance of *energy changes* that accompany chemical and physical changes. Finally, the recognition that in addition to the atoms that make up a chemical compound, the overall *shape of a molecule* is very important in its function, particularly in biological settings.

Self-Test

1.   Why is Dalton's atomic aheory important? _____

    _____

    _____

2.   Why are careful measurements important? _____

    _____

    _____

3.   Why are energy changes in chemistry important?   _____

    _____

    _____

4.   Why is molecular structure important?   _____

    _____

    _____

0.2   Supernovas and the Elements

Learning Objective

It is important to learn how the elements were formed.

Review

In this section we review the cosmological theory of how the universe started and after that, how the elements formed. The current theory is often called the "big-bang" theory, where the universe started with a tremendous explosion of matter and energy. Fundamental particles eventually combined to form the nuclei of the lightest elements: mostly hydrogen with small amounts of helium and lithium within the first few minutes after the big bang. When stars formed, their mass and energy ignited nuclear reactions that fused hydrogen atoms to make helium and start the process of nucleosynthesis. When enough helium was produced, helium atoms fused together to make even heavier atoms. As heavier atoms were produced, they separated into layers within the star and produced even heavier atoms, until iron started being produced in significant amounts. Fusion of all atoms lighter than iron give off energy that sustains the star. Iron consumed energy, with the result that the star's core collapsed, starting the sequence leading to a supernova. In the few seconds when the supernova experiences its highest temperatures, elements heavier than iron were formed.

Self-Test

5.   Why does nucleosynthesis only occur for a short time after a supernova?_____

_____

_____

6.   Why do nuclei segregate into layers in a star?_____

_____

_____

7.   How can the process of nucleosynthesis be used to support the big-bang theory?_____

_____

_____

0.3   Elements and the Earth

Learning Objective

We will understand that the distribution of substances around the world is not accidental.

Review

Since the origin of the universe it is estimated that stars and the elements have cycled through many supernovas. Although the relative mass of the elements, besides hydrogen and helium, is small, those elements did not tend to become part of stars, but they accreted into objects we call planets, asteroids, comets, and meteors. In our solar system the inner planets, Mercury, Venus, Earth, and Mars are called rocky planets. Our Earth has accumulated all of the naturally occurring elements but they are not evenly mixed throughout the planet. Geologists have been able to explain the uneven distribution based on the physical properties of the substances such as density, melting point, crystal structure, and solubility.

Self-Test

8.   What effect does density have on distributing elements on/in the earth?_____

_____

_____

9.   Why is solubility in water important in the distribution of elements?   ____ _____ ____

_____

_____

10.   Dalton predicted the existence of the then unknown element #32. He also predicted that it would be found in samples that were rich in silicon and/or tin. What physical properties make this a logical conclusion on Dalton's part?   _____

_____

_____

0.4   Dalton's Atomic Theory

Learning Objective

This section helps us appreciate the powerful nature of the atomic theory.

Review

Although the concept of atoms was not new when Dalton presented his atomic theory, the theory nevertheless revolutionized chemistry. This is because it was based on two *experimentally observed* laws relating to the compositions of substances and chemical reactions. We call them laws of chemical combination.

Be sure you study the definitions of the law of conservation of mass and the law of definite proportions given in this section in the text. Notice that the laws refer to experimentally measurable quantities—the masses of substances in compounds and in chemical reactions.

The example below illustrates how we can use the law of definite proportions and some proportional reasoning to work a chemical calculation. (NOTE: You can skip these problems since similar ones are not in the text. However, they will give good practice for answering more difficult questions later.)

Example 0.1   Proportional reasoning and the law of definite proportions

Rust is a compound of iron and oxygen. A particular sample of rust was found to contain 1.25 g of iron and 0.536 g of oxygen. If a 3.56 g sample of iron were to form rust, how many grams of rust would be formed?

Analysis:

According to the law of definite proportions, all samples of rust will have the same mass ratio of iron to oxygen. We see that in one sample, the ratio is 1.25 g iron to 0.536 g oxygen. The second sample of rust will have iron and oxygen in the same ratio, so we will do some proportional reasoning to find how much oxygen must combine with 3.56 g of iron to have the same iron to oxygen ratio. To find the total mass of rust, we just add the mass of oxygen to the mass of iron.

Assembling the Tools:

The only tool we need is the law of definite (constant) proportions.

Solution:

If we let the unknown mass of oxygen in the rust be represented by x, we can set up two ratios of oxygen to iron.

$$\frac{0.536 \text{ g oxygen}}{1.25 \text{ g iron}} \qquad\qquad \frac{x}{3.56 \text{ g iron}}$$

<u>Known rust sample</u>   <u>New rust sample</u>

According to the law of definite proportions, these two ratios must be the same, so we can equate them.

$$\frac{0.536 \text{ g oxygen}}{1.25 \text{ g iron}} = \frac{x}{3.56 \text{ g iron}}$$

Next, we solve for x.

$$x = 3.56 \text{ g iron} \times \frac{0.536 \text{ g oxygen}}{1.25 \text{ g iron}} = 1.53 \text{ g oxygen}$$

Thus, the new rust sample will contain 3.56 g iron and 1.53 g of oxygen. The total mass of the rust will be the sum of these two masses.

$$\text{Total mass of rust} = 3.56 \text{ g} + 1.53 \text{ g} = 5.09 \text{ g}$$

Is the Answer Reasonable?

Let's compare the masses of iron in the two samples. The second mass (3.56 g) is almost three times larger than the first (1.25 g), so we would expect that the amount of oxygen in the second sample would be almost three times that in the first. Three times 0.536 g would be a little larger than 1.5 g, so the mass of oxygen we calculated seems to be about right. For the total mass, it's easy to check the arithmetic to see that it's okay.

Dalton's atomic theory

Study the postulates of Dalton's theory. You should be able to describe how they explain the laws of chemical combination mentioned above. Study the discussion of the scanning tunneling microscope, a device that provides direct experimental evidence for the existence of atoms.

Self-Test

11. State the law of conservation of mass. _____

_____

12. State the law of definite proportions. _____

_____

13. In a 4.00 g sample of table salt, NaCl, there is 1.57 g of Na. Use the law of conservation of mass to calculate the mass of Cl in that sample.

_____

14.   If the ratio by mass of the elements in the carbon dioxide in a sample obtained in New York City is 2.66 g O to 1.00 g C, what ratio would be present in a sample of carbon dioxide compound taken in Los Angeles?

0.5   Internal Structure of the Atom

Learning Objective

Here we develop an understanding of how we came to know about the structure of the atom.

Review

At the center of an atom is a tiny particle called a nucleus, which is composed of subatomic particles called protons and neutrons. Protons are positively charged and neutrons are electrically neutral. Surrounding the nucleus are electrons, which are negatively charged. The amount of charge carried by a proton and an electron is the same; however, the *kind* of charge (positive and negative) is different. Each proton has one unit of positive charge, and each electron has one unit of negative charge. Because of this, in a neutral atom the number of electrons surrounding the nucleus equals the number of protons in the nucleus.

Protons and neutrons (collectively referred to as nucleons) have nearly the same mass (approximately 1 atomic mass unit) and are much heavier than electrons. As a result, nearly all the mass of an atom is contained within its nucleus.

All of the atoms in a given element have the same number of protons, no matter if the element consists of several isotopes. Atoms of different elements have different numbers of protons. This makes the number of protons—called the atomic number (Z)—something unique for each element, and it corresponds to the size of the positive charge on the nuclei of the element.

The isotopes of an element differ only in their numbers of neutrons, so they differ only in mass. The sum of the protons and neutrons—called the mass number (A)—is one way to specify which isotope of an element is being discussed. Thus, the name carbon-12 identifies the one isotope of carbon that has a mass number of 12. (Because carbon's atomic number is 6, meaning six protons, the nucleus of a carbon-12 atom must have 6 neutrons so that the total number of nucleons is 12.)

Isotopes

Contrary to Dalton's theory, not all the atoms of a given element have exactly the same mass. Atoms of an element with slightly different masses are said to be isotopes of the element. The existence of isotopes did not affect the validity of Dalton's theory because of two reasons.

1.   Any sample of an element large enough to see has an enormous number of atoms.

2.   The proportions of the different isotopes in samples of an element are uniform from sample to sample.

As a result, an element behaves as if its atoms have masses corresponding to the *average mass* of the isotopes.

Carbon-12 atomic mass scale

A major feature of Dalton's theory was the notion that atoms of each element have a characteristic atomic mass (sometimes called *atomic weight*). The atomic masses we use today are relative masses, with carbon-12 (an isotope of carbon) being the foundation of the mass scale: one atom of carbon-12 has a mass of *exactly* 12 u (atomic mass units).

$$1 \text{ atom } {}^{12}C = 12 \text{ u (exactly)}$$

For example, scientists have found that magnesium atoms are, on average, a little more than twice as heavy as atoms of carbon-12. Therefore, if we assign one atom of ^{12}C a mass of 12 u, the mass of a magnesium atom must be slightly larger than 24 u. (More precisely, the average atomic mass of magnesium is 24.305 u.) This means that if we have a sample of carbon with a mass of 12 g, and we want a sample of magnesium *with as many atoms*, we have to take a sample of 24.305 g of magnesium.

Average atomic masses are calculated from isotopic abundances and accurate mass of each of the isotopes. Be sure to study Example 1.2 and work Practice Exercises 3 through 5.

Example 0.2 Calculating the average mass of an element

What is the average atomic mass of copper if Cu-63 has a mass of 62.929599 u and an abundance of 69.17% and the only other isotope, Cu-65, has an exact mass of 64.927792 u?

Analysis:

To calculate the weighted average of all isotopes we need their exact masses and the percentage composition. It appears we are missing the percentage of Cu-65. However, since there are only two isotopes and we know the percentage of one of them, we can use the fact that all percentages must add up to 100% to calculate the percentage of the second isotope. The second isotope must have a percentage composition of 30.83%.

Assembling the Tools:

There is no specific tool for this calculation but we follow common mathematical operations to determine the average mass.

Solution:

To calculate the weighted average that is usually called the average atomic mass, we multiply each exact mass by the fraction of each isotope. The fraction of each isotope is the percentage divided by 100 (recall that we multiply a fraction by 100 to obtain the percent, this is the reverse of that process).

$$62.929599 \text{ u} \left(\frac{67.19\%}{100\%} \right) + 64.927792 \text{ u} \left(\frac{30.83\%}{100\%} \right) = 63.5456419 \text{ u}$$

Since the percentages have four significant figures, the mass should be 63.55 u.

Is the Answer Reasonable?

The easiest test is that the weighted average must be between the highest and lowest exact mass, and it is. It is also closer to the lower mass, as expected, because approximately 70% of the atoms are the isotope with the lower mass.

Sometimes it is desirable to specify the mass number and the atomic number when writing the symbol of an isotope. This is illustrated in the text and some other examples are shown below.

${}^{35}_{17}Cl$	${}^{190}_{77}Ir$	${}^{87}_{37}Rb$
17 protons	77 protons	37 protons
18 neutrons	113 neutrons	50 neutrons
17 electrons	77 electrons	37 protons

Self-Test

15. How do isotopes of the same element differ chemically?

_____

16. How do isotopes of the same element differ physically?

_____

17. The atomic mass of seborgium (rounded) is 266. How much heavier are seborgium atoms than carbon-12 atoms?

_____

18. The element iridium is composed of 37.3% of ^{191}Ir, which has a mass of 190.960 u, and 62.7% ^{193}Ir, which has a mass of 192.963 u. What is the average atomic mass of iridium?

_____

19. Name the three subatomic particles.

_____

_____

_____

20. Name the nucleons.   _____

21. Name the particles that are in the nucleus. _____

22. An atom with a mass number of 56 has an atomic number of 26. How many of the following particles does it have?

    (a) electrons _____   (c) neutrons _____

    (b) protons _____   (d) nuclei _____

23. An atom with 48 electrons has 64 neutrons. What is its approximate atomic mass?

_____

24. An isotope of oxygen has 10 neutrons. Write its symbol.   _____

25. Which is the only atom that has an atomic mass that is exactly a whole number?_____

Answers to Self-Test Questions

1. Dalton's atomic theory explains how atoms and molecules interact at the most fundamental level. With all our modern knowledge, his theory still applies.

2. Measurements and observations are the basis of modern science. The better these measurements and observations are, the better we understand the world around us in a scientific sense.

3. Energy changes that accompany chemical and physical processes allow chemists to predict if a reaction will be spontaneous along with the useful energy it may produce.

4. Shapes of molecules are important; for example, shapes of molecules determine the odor they produce. Many biochemical reactions occur because the shape of one molecule fits the shape of another.

5. Nucleosynthesis requires high temperatures that last for a very short time after a supernova or the big bang. Nuclei that can collide in this time period have a chance of fusing together to make a new nucleus.

6. Inside a star the heavier nuclei sink toward the center of the star. These nuclei tend to form layers of similar nuclei since they have the same density.

7. As heavier nuclei are formed, physicists can predict that the probability of certain nuclei forming is greater than other nuclei. This produces a distinctive pattern of abundances. If the observed abundances match the predictions, it provides evidence supporting the theory.

8. Density has an effect similar to the separation of nuclei in stars. The densest materials, thought to be nickel-iron are at the core and lighter elements are closer to the surface. Rocks, being much less dense than metals, compose the Earth's crust.

9. When the oceans developed on the Earth, substances that could be dissolved were moved into solution. They could be deposited elsewhere or remain dissolved. Large salt deposits where table salt is mined are the result of ancient seas evaporating.

10. The main similarity would be crystal structure so the unknown substance would fit easily into the crystals of silicon and tin. Similarity of chemistry is another. All of the oxides are expected to have the same formulas.

11. The law of conservation of mass states that in a chemical or physical change, mass is neither lost nor gained.

12. The law of definite proportions states that a pure chemical substance will always have the same proportions by weight of its constituent elements.

13. 2.43 g of Cl

14. A ratio of 2.66 g O to 1.00 g C. (The same ratio)

15. They don't differ chemically.

16. They differ slightly in their masses.

17. 22.2 times as heavy

18. 192.22 becomes 192 when properly rounded

19. electron, proton, neutron

20. proton and neutron

21. proton and neutron

22. (a) 26, (b) 26, (c) 30, (d) 1

23. 112 u

24. $^{18}_{8}O$

25. ^{12}C

Tools for problem solving

We have learned the following concepts that can be applied as tools in solving problems. Study each one carefully so that you know what each is used for. When faced with solving a problem, recall what each tool does and consider whether it will be helpful in finding a solution. This will aid you in selecting the tools you need. If necessary, refer to this table when working on the exercises in the chapter. Remember that tools from Chapter 1 may be needed at times to solve problems in this chapter.

You may wish to tear out these pages to use while solving problems.

Law of definite proportions (Section 0.4)

If we know the **mass ratio** of the elements in one sample of a compound, we know the ratio will be the same in a different sample of the same compound.

Law of conservation of mass (Section 0.4)

The total mass of chemicals present before a reaction starts equals the total mass after the reaction is finished. We can use this law to check whether we have accounted for all the substances formed in a reaction.

Numbers of subatomic particles in atoms (Section 0.5)

Number of electrons = number of protons (in neutral atoms only)

Atomic number (Z) = number of protons

Mass number (A) = number of protons + number of neutrons

Symbols for the elements (Section 0.5)

Each element has a one- or two-letter abbreviation with the first letter always capitalized and the second letter, if present, in lower case.

Atomic symbols for isotopes (Section 0.5)

The mass number (A) comes before the element symbol as a superscript and the atomic number (Z) also comes before the element as a subscript.

Notes:

Chapter 1

Scientific Measurements

This chapter introduces some basic concepts that you will need in order to understand future discussions in class and in the textbook. You will also need them to function effectively in the laboratory part of your course.

If you've had a prior course in chemistry, much of what we discuss in this chapter will seem familiar. Nevertheless, be sure you really understand all of it fully and can do the assigned homework. In particular, be sure you've learned the meanings of the boldfaced terms in the text as well as equations that are highlighted by a shaded background.

We begin the chapter by reviewing the laws and theories touched on in the previous chapter. Matter is the general term for all substances that have mass and occupy space. We will classify matter in terms of elements, compounds, and mixtures. It will be important to understand physical versus chemical properties of matter and how they are measured. You will also learn about the scientific method, which describes how scientists learn about nature.

When learning about nature, scientists must describe a wide variety of properties. In modern science these descriptions start as numerical measurements, or data. When correctly combined, these measurements yield valid results from which we can draw conclusions. The correct use of data is an essential skill that should be mastered as soon as possible.

1.1   Laws and Theories: The Scientific Method

Learning Objective

We will learn to explain the scientific method.

Review

The sequence of steps described by the scientific method is little more than a formal description of how people logically analyze any problem, scientific or otherwise. Observations are made in order to collect data (empirical facts) from which we often can draw conclusions. Generalizations come from the analysis of data and can lead to laws, which are concise statements about the behavior of chemical or physical systems. Laws, however, offer no explanations about *why* nature behaves the way it does. Tentative explanations are called hypotheses; tested explanations are called theories. The scientific method consists of collecting data in experiments, formulating theories, and testing the theories by more experimentation. Based on the results of new experiments, the theories are refined, tested further, refined again, and so on.

When scientists create theories, they build mental images that help them visualize how nature operates. The atomic theory is one of the most useful models of the intimate structure of matter. According to the theory, all substances are made up of very tiny atoms that combine in various combinations to form molecules. Often, chemists represent molecules by drawings of the kind illustrated in Figure 1.3 in the text.

Self-Test

1.   Identify each of the following as an observation or a conclusion.

   (a)   If you drop a stone, a force called gravity causes the stone to fall to the ground.

   _____

   (b)   Water boils if it is heated to a temperature of 212 °F. _____

2.   Identify each of the following statements as either a law or a theory.

   (a)   In general, what goes up must come down.   _____
   (b)   The ice ages resulted from the tilting of the earth's rotation axis, which was caused by the earth being hit by very large meteors.

   _____

3.   What does *empirical* mean? _____

   _____

4.   What is the difference between a scientific theory and a hypothesis?

   _____

1.2   Matter and Its Classifications

Learning Objective

   In this section we will learn to classify matter.

Review

Substances are classified according to their properties. One such classification is physical state—solid, liquid, or gas. Study Figure 1.12 to see how they differ in the way their atomic-sized particles are organized.

   Chemically, substances are classified as elements, compounds, or mixtures of elements and/or compounds. Elements are the basic building blocks of all substances and cannot be decomposed into simpler substances by chemical reactions. Each element is identified by its chemical symbol. Most symbols are simple abbreviations of the English name of the element. Eleven elements have symbols derived from other languages, mostly Latin (see Table 1.1). It will speed up your work if you are familiar with the symbols for the more common elements and your teacher will also expect you to learn the symbols of a number of the common elements.

   Compounds are formed from elements. They contain elements in fixed (constant) proportions by mass. Stated differently, the proportions of the elements in a compound cannot be changed. Mixtures are formed from elements and/or compounds and can be of variable composition. A mixture composed of two or more phases is said to be heterogeneous; a mixture composed of just a single phase is a solution and is said to be homogeneous.

   Separating a compound into its components (elements) involves a chemical change, just as in the combination of the elements to form the compound. Separation of a mixture usually can be accomplished by a physical change—a change in which the chemical identities of the components aren't altered. Be sure to study Figures 1.7 and 1.8.

Example 1.1 Identifying elements, compounds, and mixtures

A liquid sample is brought into your lab. How can you tell if this is an element, a compound, or a mixture?

Analysis:

We have some definitions of the nature of, and differences between, these three categories of materials. They are (1) an element cannot be decomposed into other elements, (2) a compound must have a fixed ratio of elements present, and (3) a mixture has two or more substances and those substances do not have a fixed ratio.

Assembling the Tools:

Here will be using the definitions of elements, compounds, and mixtures.

Solution:

There are only two elements that are liquids, mercury and bromine, and they have very distinctive properties. If the sample is not an element, we can get a good idea if it is a mixture by boiling it. A compound will have a fixed boiling point and evaporate completely. Mixtures often boil over a wide range of temperatures or leave a residue when fully evaporated.

Is the Answer Reasonable?

The procedures suggested seem correct. They may be difficult to carry out in a lab if the boiling points are very high.

Self-Test

5.   A student placed some sand and salt in a beaker and stirred them with a glass rod.

   (a)   After stirring, does the beaker contain a compound or a mixture?

   _____

   (b)   How many phases are present? _____

   (c)   Can you suggest a method of separating the components in the beaker? Does this method involve a chemical or a physical change?

   _____

   _____

   _____

   _____

6.   What is the chemical symbol of the following elements?

   (a) carbon   _____   (d) nickel   _____

   (b) bromine   _____   (e) silicon   _____

   (c) copper   _____   (f) argon   _____

7.   Write the name of the element with the symbol.

(a) Al   _____   (b) Mg   _____

(c) Li_____   (d) I   _____

1.3   Physical and Chemical Properties

Learning Objective

To develop an understanding of which physical and chemical properties are important to measure.

Review

In the text, we see that the properties of substances can be classified in two ways. One is to divide them into either physical properties or chemical properties. The other is to divide them into intensive or extensive properties.

Physical and chemical properties

Physical properties are ones that can be observed without changing the chemical makeup of a substance. In general, physical properties can be specified without reference to another chemical substance. Examples are an object's color, or the temperature at which it melts, or its volume. Any change that occurs during the observation of a physical property is a physical change, as when a substance changes from solid to liquid when we observe its melting point.

When we describe a chemical property of a substance, we describe how the substance reacts chemically with something else. Such a chemical reaction (chemical change) produces new chemical substances, so after we've observed a chemical property, the substance is no longer the same; it has changed into a different substance. For example, a chemical property of iron is that it rusts when in contact with air and moisture. When we observe this property, the iron changes to rust and no longer has the same appearance.

Intensive and extensive properties

Extensive properties, such as mass or volume, depend on the size of the sample of matter being examined. Although extensive properties are important for a given sample, they are not especially useful for identifying substances. More useful are intensive properties, because all samples of a given substance have identical values for its intensive properties. For example, if we were asked whether a sample of a liquid was water, we would examine its properties and compare them to those of a known sample of water. If we were to find that the mass of the liquid is 12.0 g, we still would not be any closer to knowing whether or not the sample is water. By itself, the mass is of no value, because different samples of water, or any other liquid, have different masses. However, if we further note that the liquid is clear, has no color, has no odor, and freezes at 0 °C, we would strongly suspect the sample to be water. This is because *all* samples of pure water, regardless of size, are clear, colorless, odorless, and freeze at 0 °C.

Self-Test

8.   Identify the following as chemical properties or physical properties.

(a)   White phosphorous spontaneously ignites in air. _____

(b)   Gold is a yellow metal.   _____

9.   Sodium is a soft, silvery metal that melts at 97.8 °C. It burns with a yellow light in the presence of chlorine gas to give the compound sodium chloride (table salt).

(a)   What are some physical properties of sodium? _____

_____

(b)   Give a chemical property of sodium. _____

_____

10.   (a)   Give two examples of extensive properties. _____

_____

(b)   Give two examples of intensive properties. _____

_____

1.4   Measurement of Physical and Chemical Properties

Learning Objective

To understand measurements and the use of SI units.

Review

This section introduces you to the concept of measurement and differentiates between numbers in a mathematical sense and numbers that arise from measurement. A measurement always has two parts, a number and a unit. Units describe the nature of the measured quantity and its size. Units are an absolutely essential part of any measurement, so whenever you report a measured value, you *must* include its units as well.

Before we continue, you may have noticed a small TOOL ICON in the margin of your textbook at about this point. These icons indicate information that can be classified as a tool to help solve problems. Just as a plumber needs a variety of tools to do the job, you will need appropriate tools to solve chemistry problems. Just as the plumber must learn when and how to use the tools in a toolbox, we will be learning when and how to use the variety of tools that chemists work with. Often we will mention the tool being used as examples are given throughout this book. Now we can consider measurements and our first tool, the SI units of measurement.

When we write the value of a measurement, we always leave a space between the number and the unit.

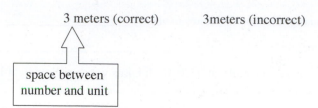

3 meters (correct)   3meters (incorrect)

space between
number and unit

Similarly, a mass of 23.4 kg is correct and 23.4kg is incorrect.

The units used in science, such as the meter above, are based on the precisely defined International System of Units, or SI. Learn the SI base units and their symbols for length, mass, time, and temperature (Table 1.2). The mole, another essential unit, will be defined in Chapter 3.

Quantities other than those given in Table 1.2 are obtained from the base quantities by mathematical operations, and their units (called derived units) are obtained from the base units by the same operations. Old style metric units (not to be confused with SI metric units) are given in Table 1.3. Conversions to the English system, still widely used in the United States, are given in Table 1.4.

Example 1.2 Using derived SI units

You most likely use the units "miles per hour" to describe how fast your car is moving. What is the expression for speed (velocity) using SI base units and the SI abbreviations?

Analysis:
The two essential parts of speed are a distance (miles) divided by (per) time (hour). We need to make substitutions using the SI units.

Assembling the Tools:
Here we will be using the SI units.

Solution:
The table of base SI units gives the meter as the unit of length and the second as the unit of time. We conclude that speed is meters per second using SI units. We can use the abbreviations to write

$$\text{speed} = \text{m/s} = \text{m s}^{-1}$$

Is the Answer Reasonable?
All we can do here is check that the units are representing the same property, in this case meters and miles represent distance and hours and seconds represent time.

The previous example showed us how derived units are made by combining base units. Notice in particular that units undergo the same kinds of mathematical operations that numbers do. For example, volume (of a rectangular solid) is a product of three measurements with distance units.

$$\text{length} \times \text{width} \times \text{height} = \text{volume}$$

The unit for volume is the product of the units for length, width, and height.

$$\text{meter} \times \text{meter} \times \text{meter} = \text{meter}^3$$

$$\text{m} \times \text{m} \times \text{m} = \text{m}^3$$

Often, the base units (or the derived units that come from them) are too large or too small to be used conveniently. For example, if we were to use cubic meters to express the volumes of liquids that we measure in the laboratory, we would find ourselves using very small numbers such as 0.000025 m³ or 0.000050 m³. Because they have so many zeros, these values are difficult to comprehend. To make life easier for us, the SI has a simple way of making larger or smaller units out of the basic ones. This is done with the decimal multipliers and SI prefixes given in Table 1.5. Be sure you learn the ones in the table below. (These are the ones in colored type in Table 1.5.)

SI Prefixes and Decimal Multipliers

Prefix	Symbol	Multiplication Factor
mega	M	10^6
kilo	k	10^3
deci	d	10^{-1}
centi	c	10^{-2}
milli	m	10^{-3}
micro	μ	10^{-6}
nano	n	10^{-9}
pico	p	10^{-12}

The SI prefixes and decimal multipliers are tools we use to scale units to convenient sizes and to translate between differently sized units. You will see how this is done in Section 1.6. Notice that each prefix stands for a particular decimal multiplier. Thus *kilo* means "× 10^3" in exponential form. This lets us translate a quantity with a prefix into the value that it has when expressed in terms of the base units. For example, suppose we wanted to know how many meters are in 25 kilometers (25 km). Since kilo (k) means "× 10^3" we substitute "× 10^3" for "k."

$$25 \text{ km} = 25 \times 10^3 \text{ m}$$

Similarly, a length of 25 millimeters (25 mm) would be

$$25 \text{ mm} = 25 \times 10^{-3} \text{ m}$$

$$= 0.025 \text{ m}$$

Common laboratory units of measurement

In the lab, the most common measurements are those of length, volume, mass, and temperature.

Length is usually measured in units of centimeters (cm) or millimeters (mm). Remember the following:

$$1 \text{ m} = 100 \text{ cm} = 1000 \text{ mm}$$

$$1 \text{ cm} = 10 \text{ mm}$$

Volume is conveniently measured in cubic centimeters (cm³). Frequently we use the non-SI unit called the liter (L), which is slightly larger than a quart. Remember that 1 L = 1000 cm³. Often, glassware is graduated in milliliters (mL): 1000 mL = 1 L. It is also important to remember that 1 cm³ = 1 mL.

Although the SI base unit of mass is the kilogram, in the lab we usually report mass measurements in units of grams (g). Common laboratory balances are graduated in grams.

Temperature is measured with a thermometer, and in the sciences it is measured in units of degrees Celsius (°C). The Celsius and Fahrenheit degree units are of different sizes; five degree units on the Celsius scale correspond to nine degree units on the Fahrenheit scale. Equation 1.2 in the text enables you to make conversions between °C and °F.

The SI unit of temperature is the kelvin (K). Zero on the Kelvin scale corresponds to −273 °C (rounded to the nearest degree) and is called absolute zero, because it is the coldest temperature. (Notice that the name of the SI *temperature scale* is capitalized; the name of the *unit* measured on that scale, the kelvin, is not capitalized.) The kelvin and Celsius degree are the same size, so a *temperature change* of 10 K, for example, is the same as a temperature change of 10 °C. Be sure you can convert from °C to K (Equation 1.2), because when the temperature is needed in a calculation, it nearly always must be expressed in kelvins. In mathematical equations, the capital letter T is used to stand for the Kelvin temperature and the lower case t or t_c is used for Celsius temperatures.

Example 1.3   Temperature scale conversions

Zinc metal is obtained by roasting zinc oxide, ZnO, ore at 1375 K. (a) What temperature is that if the Celsius scale is used? (b) What is this temperature on the Fahrenheit scale?

Analysis:

This question involves only one topic, conversion between temperature scales. We need to find appropriate equations by looking for an equation that has the given units and also has the units we seek.

Assembling the Tools:

We use the equations that relate the temperature scales.

Solution:

The equations that relate the temperature scales are

$$T = (t_C + 273.15 \text{ °C})(1 \text{ K}/1 \text{ °C}) \quad \text{and} \quad t_C = (t_F - 32 \text{ °F})(9 \text{ °F}/5 \text{ °C})$$

All we need to do is make the appropriate substitutions.

(a) $1375 \text{ K} = (t_C + 273.15 \text{ °C})(1 \text{ K}/1 \text{ °C})$

$t_C = 1375 \text{ K} - 273.15 \text{ K}$

$t_C = 1102 \text{ °C}$

(b) $1102 \text{ °C} = (t_F - 32)(5 \text{ °C}/9 \text{ °F})$

$1102 \text{ °C} (9 \text{ °F}/5 \text{ °C}) = (t_F - 32 \text{ °F})$

$1983 \text{ °F} = (t_F - 32 \text{ °F})$

$t_F = 1983 \text{ °F} + 32 \text{ °F}$

$t_F = 2015 \text{ °F}$

Is the Answer Reasonable?

About all we can do is to check our algebra. Also note that the kelvin temperature is always numerically larger than the Celsius temperature. At high temperatures like this, the Fahrenheit temperature is approximately twice the Kelvin or Celsius temperature.

Self-Test

11. Give the SI base unit and its abbreviation for

    (a) mass _____ (b) length _____

    (c) time _____ (d) temperature_____

    (e) amount of substance _____

12. Torque (pronounced "tork") is a quantity that describes a twisting force, such as that applied to a nut or a bolt by a wrench. It is a product of distance × force and in English units is normally given in foot-pounds. The SI derived unit for force is the Newton (symbol, N). What is the SI derived unit for torque?

    Answer _____

13. Fill in the blanks with the correct prefixes.

    (a) 1 _____ gram = 0.01 gram   (e) 1 pm = _____ m

    (b) 1 _____ meter = 10^{-9} meter   (f) 1 μg = _____ g

    (c) 1 _____ g = 0.001 g   (g) 1 dm = _____ m

    (d) 1 _____ m = 1000 m   (h) 1 ns = _____ s

14. Fill in the blanks with the correct numbers.

    (a) 63 dm = _____ m   (c) 2450 nm = _____ m

    (b) 0.023 Mg = _____ g   (d) 2487 cm = _____ m

15. Fill in the blanks with the correct numbers.

    (a) _____ cm = 1.35 m   (g) _____ mL = 0.022 L

    (b) _____ mm = 22.4 cm   (h) _____ L = 346 mL

    (c) _____ cm = 32.6 mm   (i) _____ L = 2.41 mL

    (d) _____ mL = 1.250 L   (j) _____ K = 25 °C

    (e) _____ cm^3 = 246 mL   (k) _____ °C = 265 K

    (f) _____ L = 525 cm^3 (l) _____ °C = 300 K

16. (a) What Celsius temperature corresponds to 23 °F? _____

    (b) What Fahrenheit temperature equals 10 °C? _____

1.5   The Uncertainty of Measurements

Learning Objective

To understand how errors in measurements occur and are reported using significant figures.

Review

Measurements are inexact and contain errors that arise from various sources. When we write the result of a measurement, we use the concept of significant figures to provide an estimate of how certain we are of the measured value. For a given measurement, its significant figures include all the digits known for sure *plus* the first digit that contains some uncertainty.

Expressing a measurement to the correct number of significant figures allows us to convey to someone else who sees the measured value how precise the measurement is. For example, a measured length of 32.47 cm has four significant figures. It tells us that the 3, 2, and 4 are known with certainty and the measuring instrument allowed the hundredths place to be *estimated* to be a 7. Since no digit is reported in the thousandths place, it is assumed that no estimate of that place could be obtained. In general, the larger the number of significant figures in a measurement, the greater is its precision.

Accuracy refers to how close a measured quantity is to the actual correct value. *Precision* refers to how close repeated measurements of the same quantity are to each other. One of the skills you need to learn is counting the number of significant figures in a measured quantity. Usually, this is a simple matter. However, zeros sometimes cause a problem. The general rules for zeros are:

1. Leading zeros (those before any nonzero digits) are never significant.
2. Imbedded zeros (those zeros between two nonzero digits) are always significant.
3. Trailing zeros (those zeros after the last nonzero digit) are always significant if there is a decimal point somewhere in the number. If there is no decimal point, zeros are not counted as significant.

Notice that scientific notation eliminates confusion if trailing zeros appear in a number without a decimal point. Additionally, some numbers are considered as exact numbers. For instance, there are exactly 100 centimeters in one meter. This 100 is considered to have a decimal point and as many digits as needed so it does not have an effect on rounding the final answer.

The rules given in this section are tools we use to correctly express the number of significant figures in a computed quantity.

Multiplication and Division

The answer cannot contain more significant figures than the factor that has the fewest number of significant figures. Study Practice Exercises 1.9 and 1.10.

Addition and Subtraction

The answer is rounded to the same number of decimal places as the quantity with the fewest decimal places.

When using exact numbers in calculations, they can be considered to have as many significant figures as desired. They contain no uncertainty.

Self-Test

17. Which term, *accuracy* or *precision*, is related to the number of significant figures in a measured quantity?

_____

Example 1.4   Calculations using significant figures

Assume that all of the numbers in the following expression come from measurements. Compute the answer to the correct number of significant figures.

$$(3.25 \times 10.46) + 2.44 = ?$$

Analysis:

We need to perform the given calculations and then round off the result shown on the calculator screen (usually too many digits) to a number that has only significant digits.

Assembling the Tools:

We use the rules presented for obtaining the correct number of significant figures. Recall there is one rule for multiplication and division and another for addition and subtraction.

Solution:

When we perform the multiplication, we obtain 33.995. This should be rounded to three significant figures because that's how many there are in 3.25. The result is 34.0, which is then added to 2.44.

34.0	one decimal place
+ 2.44	two decimal places
36.4	rounded to one decimal place

The correct answer, therefore, is 36.4.

Is the Answer Reasonable?

There's no simple check. Keep in mind, however, that the order in which the rules are applied is the same as the order in which the different parts of the calculation are performed.

Self-Test

18. Perform the following arithmetic and express the answers to the proper number of significant figures (assume all numbers come from measurements).

(a) 4.873×3.1 _____

(b) $84.297 \div 2.102$ _____

(c) $78.66 + 0.022$ _____

(d) $105.3 - 0.68$ _____

1.6   Dimensional Analysis

Learning Objective

We will learn how to effectively use dimensional analysis.

Review

The dimensional analysis method is a procedure we use to set up the arithmetic correctly when solving a problem. It is based on the ability of units to cancel from numerator and denominator of a fraction, as described earlier. To apply this method, we view a problem as a conversion from some initial set of units to a final desired set of units. For example, if we need to know how many inches are in 1.5 ft, we begin by restating the problem as

$$1.5 \text{ ft} = ? \text{ in.}$$

To change the units from ft to in., we multiply the given quantity, 1.5 ft, by a conversion factor that will change the units from ft to in.

$$1.5 \text{ ft} \times \left(\frac{\text{conversion}}{\text{factor}} \right) = \text{answer in inches}$$

A **conversion factor** is a fraction that we form from a relationship between units. In this case, the relationship is between feet and inches, which can be expressed by the equation

$$1 \text{ ft} = 12 \text{ in.}$$

We can make a conversion factor by dividing both sides of the equation by 1 ft.

$$\frac{1 \text{ ft}}{1 \text{ ft}} = \frac{12 \text{ in.}}{1 \text{ ft}}$$

Notice that the quantities in the numerator and denominator on the left cancel, so we can write

$$1 = \frac{1 \text{ ft}}{1 \text{ ft}} = \frac{12 \text{ in.}}{1 \text{ ft}}$$

Because the fraction 12 in./1 ft is numerically equivalent to 1, we can multiply something by it and not change its magnitude. (Multiplying something by 1 doesn't change its size.) Using this as our conversion factor, then, gives

$$1.5 \text{ ft} \times \frac{12 \text{ in.}}{1 \text{ ft}} = \text{answer in inches}$$

Notice that the units ft cancel,[1] leaving us with units of inches, which are the units we want for the answer. Performing the arithmetic gives 18 in., which is the correct answer.

$$1.5 \, \text{ft} \times \frac{12 \text{ in.}}{1 \text{ ft}} = 18 \text{ in.}$$

The relationship between feet and inches that we employed above can actually be used to form *two* different conversion factors. One is formed by dividing both sides by 1 ft, as we've done. The other is formed by dividing both sides by 12 in.

$$\frac{1 \text{ ft}}{12 \text{ in.}} = \frac{12 \text{ in.}}{12 \text{ in.}} = 1$$

Notice that if we were to use this second conversion factor in performing the conversion, the units "ft" do not cancel. Instead, the units work out to be ft²/in., which make no sense. The answer, of course, is also incorrect.

$$1.5 \text{ ft} \times \frac{1 \text{ ft}}{12 \text{ in.}} = 0.12 \text{ ft}^2/\text{in.}$$

This illustrates one of the greatest benefits in using the dimensional analysis method — if the problem is set up incorrectly, the units will not cancel properly to give the desired units of the answer. If the units are wrong, we can also be sure that the answer is wrong. Thus, the units have informed us that we've made an error and that we must rework the setup of the problem.

Forming and applying conversion factors

In general, when we can express a relationship between units in the form of an equation, two conversion factors can be derived from it. For example, if we wished to convert between minutes and seconds, we would use the following.

$$60 \text{ s} = 1 \text{ min}$$

Two fractions (conversion factors) can be formed.

$$\frac{60 \text{ s}}{1 \text{ min}} \qquad \frac{1 \text{ min}}{60 \text{ s}}$$

The connection between feet and inches used earlier, and between minutes and seconds are both derived from definitions. Often we will use relationships that come to us in this way. We can also form conversion factors from relationships established by measurements. For instance, in an earlier example we described a situation in which we've determined that a car is able to travel 274 miles using 12.6 gallons of fuel. In making these measurements, we've established a relationship between distance traveled and fuel consumed that we can express in an equation format.

$$274 \text{ mile} \Leftrightarrow 12.6 \text{ gallon}$$

---

[1]A quantity such as 1.5 ft can be viewed as a fraction in which the denominator is 1.

$$1.5 \text{ ft} = \frac{1.5 \text{ ft}}{1}$$

This places the unit ft in the numerator where it can be canceled by the unit ft in the denominator of the conversion factor.

Notice, however, that we've used the symbol ⇔ rather than an equal sign. This is because miles can't actually "equal" gallons; they're two different kinds of units. The symbol ⇔ is read as "is equivalent to." In other words, for this car, "traveling 274 miles *is equivalent to* using 12.6 gallons of fuel." Although the symbol is different, it behaves like an equal sign when we wish to form conversion factors. Thus,

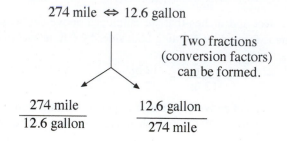

$$274 \text{ mile} \Leftrightarrow 12.6 \text{ gallon}$$

Two fractions (conversion factors) can be formed.

$$\frac{274 \text{ mile}}{12.6 \text{ gallon}} \qquad \frac{12.6 \text{ gallon}}{274 \text{ mile}}$$

Once you've learned how to form conversion factors, the next step is learning how to select the correct one for use in a calculation. Here we let the units be our guide. For instance, suppose we wished to know how many gallons of fuel we would need to travel 850 miles in the car described above. Let's begin by stating the problem in the form of an equation (or an equivalence, to be more exact).[2]

$$850 \text{ miles} \Leftrightarrow ? \text{ gallons}$$

To find the answer by the dimensional analysis method, we will multiply the 850 miles by a conversion factor to convert to gallons.

$$850 \text{ miles} \times \left(\begin{array}{c} \text{conversion} \\ \text{factor} \end{array} \right) \Leftrightarrow ? \text{ gallons}$$

The conversion factor we select has to eliminate the mile units by cancellation. Only one of the two conversion factors does this, so that's the one we select.

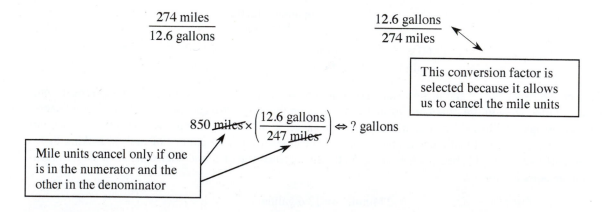

$$\frac{274 \text{ miles}}{12.6 \text{ gallons}} \qquad\qquad \frac{12.6 \text{ gallons}}{274 \text{ miles}}$$

This conversion factor is selected because it allows us to cancel the mile units

$$850 \text{ miles} \times \left(\frac{12.6 \text{ gallons}}{247 \text{ miles}} \right) \Leftrightarrow ? \text{ gallons}$$

Mile units cancel only if one is in the numerator and the other in the denominator

Once we've chosen the conversion factor, we cancel the units and perform the arithmetic.

---

[2] Don't be overly concerned about whether you use an equal sign or the symbol ⇔. The setup of the problem and finding the answer follows the same route.

$$850 \text{ miles} \times \left(\frac{12.6 \text{ gallons}}{247 \text{ miles}} \right) \Leftrightarrow 39.1 \text{ gallons}$$

Notice that if we had selected the other conversion factor, the units would not work out correctly.

> Notice that the miles
> units do not cancel

$$850 \text{ miles} \times \frac{274 \text{ miles}}{12.6 \text{ gallons}} \Leftrightarrow ? \frac{\text{miles}^2}{\text{gallons}}$$

As you see in this example, the cancellation of units serves as a guide in selecting conversion factors as we set up the solution to the problem. In fact, the need to cancel particular units in arriving at the answer to a problem often serves as a clue to the kinds of information and relationships we have to find. For example, when we have arrived at the stage

$$850 \text{ miles} \times \left(\begin{array}{c} \text{conversion} \\ \text{factor} \end{array} \right) \Leftrightarrow ? \text{ gallons}$$

we would recognize that to obtain an answer we must have a relationship between gallons of fuel used and miles traveled. If this information were not already available, we would have to obtain it by some means before we could solve the problem.

The importance of correct relationships between units

In setting up problems using the dimensional analysis method, it's essential that we use proper and correct relationships between the units. In converting between feet and inches, we need to use the relationship

$$1 \text{ ft} = 12 \text{ in.}$$

If we didn't know this relationship, we couldn't just make one up, such as

$$5 \text{ ft} = 24 \text{ in.}$$

Although we could use it to get the units to cancel, the answer is bound to be wrong because the relationship we've used is incorrect. Establishing correct relationships between units is one of the difficulties students often have in applying this approach. Before you form and apply a conversion factor, ask yourself, "Is this relationship between the units correct?"

The dimensional analysis method and the "toolbox" approach

In our earlier discussion of how the toolbox concept can be applied to problem solving, we emphasized the importance of analyzing a problem so that the tools that are needed to obtain a solution can be identified. If the problem involves calculations, the tools are usually relationships between units which can be stated in equation form. Once these statements have been assembled, the dimensional analysis method provides guidance in assembling the arithmetic. By assuring the units cancel correctly, the dimensional analysis method gives us confidence that we are doing the correct math. In a sense, then, the dimensional analysis method is itself a tool that we use to assemble the arithmetic correctly. With this in mind now, let's take a look at an example that illustrates how we apply the dimensional analysis method.

Example 1.5 Calculations using the dimensional analysis method

During a cross-country trip on Interstate I-80, a driver became concerned that her speedometer was inaccurate. To avoid a speeding ticket she decided to check the speedometer. She adjusted the speed of her car until the speedometer read 65 mph and set the car on speed control so the speed would remain constant. She then used the stopwatch feature of her wristwatch to time how long it took to travel exactly five miles, using the mile markers along the side of the road to measure her progress. She found that it took 4 minutes and 24 seconds to travel the five miles. She then pulled into a rest stop to calculate her actual speed over the five-mile stretch. What was her actual speed in miles per hour (mph)?

Analysis:

The first, and most important step in solving a problem is the **analysis**. Here we study the data given, search for any equivalences that might exist, and decide how we will proceed with the solution. We look for unit relationships (our tools), which we will use to solve the problem. After doing all this, we come to the easy part — setting up the **solution** and doing necessary calculations. We encourage you to think of problem solving as involving these two distinct stages: analysis first, followed by solution.

This problem involves a lot of words, so let's begin by extracting the critical information.

It took **4 minutes and 24 seconds**

to travel **5.00 miles**

How fast was she going, in **miles per hour**?

As we've noted, the word "per" can be interpreted to mean "divided by," so our answer must have the units miles "divided by" hour.

$$\text{units of answer} = \frac{\text{miles}}{\text{hour}}$$

Understanding that we need to assemble the data in a way to give us this desired ratio of units is really our basic clue in solving this problem. We see that we need to have distance in the numerator and time in the denominator. Let's use the data to set this up.

$$\text{speed} = \frac{5.00 \text{ miles}}{4 \text{ minutes and } 24 \text{ seconds}}$$

Now we can see that we're not too far from having the answer. We need to change the denominator into units of hour. Let's change 24 seconds to minutes, add the result to the 4 minutes, and then change the total minutes to hours. To do this, we need the relationships (equivalences) between seconds and minutes and between minutes and hours. (These are our tools.) We'll state them in equation form:

$$1 \text{ minute} = 60 \text{ seconds}$$
$$1 \text{ hour} = 60 \text{ minutes}$$

Now that we know how we're going to proceed, we can work out the solution.

Assembling the Tools:

In the analysis section we see that a mathematical concept is at the heart of conversions. That is expressed in the process called dimensional analysis.

Solution:

First, we convert 24 seconds to minutes. Although this is really a simple exercise, let's illustrate how we perform the conversion using the dimensional analysis method. We use the relationship 1 minute = 60 seconds to form a conversion factor that will cancel seconds and give the desired unit, minute.

$$24 \text{ seconds} \times \frac{1 \text{ minute}}{60 \text{ seconds}} = 0.40 \text{ minute}$$

The total time required to go five miles was therefore 4.40 minutes. Now we convert this to hours using 1 hour = 60 minutes. Once again, we form a conversion factor so units cancel correctly.

$$4.40 \text{ minutes} \times \frac{1 \text{ hour}}{60 \text{ minutes}} = 0.0733 \text{ hours}$$

Now we're able to calculate her speed by dividing the distance, 5.00 miles, by the time, 0.0733 hour.

$$\text{speed} = \frac{5.00 \text{ miles}}{0.0733 \text{ hour}} = 68.2 \frac{\text{miles}}{\text{hour}} = 68.2 \text{ mph}$$

The answer, therefore, is that the driver's speed was 68.2 mph.

Is the Answer Reasonable?

Aside from checking the arithmetic, can we feel confident that we've solved the problem correctly? We can if the answer seems reasonable, and here it does. We don't expect that the speedometer will be grossly in error and therefore we expect that the answer should be fairly close to 65 mph, which it is.

Stringing conversion factors together

The dimensional analysis method is especially useful in multi-step calculations where two or more conversion factors are needed to go from the starting units to those of the answer. For example, suppose you wanted to know how many minutes there are in 2.00 weeks.

$$2.00 \text{ weeks} = ? \text{ minutes}$$

We could do this in a one-step calculation if we knew how many minutes are in one week, but most people don't keep such numbers in their heads. However, we do know how many days there are in a week, and we know how many hours there are in a day and how many minutes in an hour.

$$1 \text{ week } = 7 \text{ days}$$

$$1 \text{ day } = 24 \text{ hours}$$

$$1 \text{ hour } = 60 \text{ minutes}$$

We can apply these relationships one after another, being sure the units cancel, until we have the final desired unit, minute (we will drop the plural form of these units for clarity in canceling)

$$\boxed{1 \text{ week } = 7 \text{ day}} \qquad \boxed{1 \text{ day } = 24 \text{ hour}} \qquad \boxed{1 \text{ hour } = 60 \text{ minute}}$$

$$2.00 \text{ week} \times \frac{7 \text{ day}}{1 \text{ week}} \times \frac{24 \text{ hour}}{1 \text{ day}} \times \frac{60 \text{ minute}}{1 \text{ hour}} = 2.02 \times 10^4 \text{ minute}$$

When we do this, we say that we are "stringing together" the conversion factors.

Following the path through the units

In a multi-step calculation, a way to be sure you have all the information necessary to set up the string of conversion factors is to follow the "path" through the units, starting with the units given and going through until you reach the units of the answer. For instance, in the list of relationships we used in the preceding example,

$$1 \text{ week} = 7 \text{ day}$$
$$1 \text{ day} = 24 \text{ hour}$$
$$1 \text{ hour} = 60 \text{ minute}$$

we can follow the path: week → day → hour → minute. Thus, the first equation takes us from *week* to *day*, the second from *day* to *hour*, and the third from *hour* to *minute*. If one of these relationships were missing (for example, the middle one), the path would be interrupted.

We can connect week to day and hour to minute, but we are missing the connection between day and hour

$$1 \text{ week} = 7 \text{ day}$$
$$1 \text{ hour} = 60 \text{ minute}$$

$$\text{week} \rightarrow \text{day} \xrightarrow{\ ?\ } \text{hour} \rightarrow \text{minute}$$

We see that to solve the problem, we need an additional relationship that connects the units day and hour.

Example 1.6   Stringing conversion factors together

The distance between the Earth and the moon is approximately 239,000 miles. Radio waves travel through space at the speed of light, 3.00×10^8 m/s. How many seconds does it take for a radio wave to go to the moon?

Analysis:

We need to convert the distance into time. Speed (of light in this case) has the appropriate units to use as a conversion factor (the units of speed are m/s or meters per second). We need to arrange our equation so that distance cancels and we are left with time. An added problem is that the distance to the moon and our speed do not have the same distance units, one is miles and the other is meters. Therefore we will convert miles to meters and then to time. An important equality is that

$$0.62 \text{ mile} = 1 \text{ km}$$

Assembling the Tools:

As mentioned in Example 1.5 we are using the dimensional analysis concept.

Solution:

We know how to convert miles to km and then to meters. Then we use the speed to convert meters to seconds. We will string the conversions together as

$$239{,}000 \text{ miles} \left(\frac{1 \text{ km}}{0.62 \text{ miles}} \right) \left(\frac{1000 \text{ m}}{1 \text{ km}} \right) \left(\frac{1 \text{ s}}{3.00 \times 10^8 \text{ m}} \right) = 1.3 \text{ s}$$

Is the Answer Reasonable?

We can check the math by rounding 239,000 to 300,000 and rounding the 0.62 to 1.0. The result is 300,000 × 1000 = 3 × 10⁸ in the numerator and 3 × 10⁸ in the denominator. The estimated result is 1 second which is close to the 1.3 second answer we got. The answer has two significant figures because of the two significant figures in 0.62. Notice that abbreviations for units never have the plural s added. For example kilometers = km and miles = mi.

Self-Test

19. The distance between two houses was measured to be 0.27 miles. What is the distance expressed in yards (1 mile = 5280 ft)?

_____

20. What conversion factor could you use to convert

   (a) 62 lb to kg? _____   (b) 45 in.^{2} to cm^2 _____

1.7   Density and Specific Gravity

Learning Objective

To learn to determine the density of substances and to use density as a conversion factor.

Review

Density is the ratio of an object's mass to its volume ($d = m/V$), as expressed in Equation 1.6 in the text. Determining the density therefore involves two measurements and a brief calculation. In performing the calculation, be careful about significant figures and the significant-figure rules that we follow in multiplication and division.

Density is an intensive property, meaning that its value for a given substance is the same regardless of the size of the sample. Increasing the sample size increases both the mass *and* the volume, but their ratio remains the same. Density also serves as a tool for converting between the mass and volume for a sample of a substance. For example, methanol (a substance used in "dry gas" and as the fuel in "canned heat") has a density of 0.791 g/mL, which tells us that each milliliter of the liquid has a mass of 0.791 g. This information provides a relationship between mass and volume.

$$0.791 \text{ g methanol} \Leftrightarrow 1.00 \text{ mL methanol}$$

We can use this equivalence to construct two conversion factors.

$$\frac{0.791 \text{ g methanol}}{1.00 \text{ mL methanol}} \quad \text{and} \quad \frac{1.00 \text{ mL methanol}}{0.791 \text{ g methanol}}$$

Suppose we wanted the mass of 12.0 mL of methanol. We would use the first factor so the units mL will cancel.

$$12.0 \text{ mL methanol} \times \frac{0.791 \text{ g methanol}}{1.00 \text{ mL methanol}} = 9.49 \text{ g methanol}$$

If we wanted the volume occupied by, say 15.0 g of methanol, we would use the second factor.

$$15.0 \text{ g methanol} \times \frac{1.00 \text{ mL methanol}}{0.791 \text{ g methanol}} = 19.0 \text{ mL methanol}$$

The factor we choose simply depends on the units that we want to cancel.

Example 1.7   Using density as a conversion factor

A rectangular piece of steel 12.0 in. long and 18.2 in. wide has a mass of 12.5 lb. Steel has a density of 7.88 g cm^{-1}. What is the thickness of the piece of steel expressed in inches?

Analysis:

We know the mass and density and can calculate the volume after we adjust the units with appropriate conversion factors. Finally, we know that the volume of a rectangular object is determined from its length, width, and thickness. We know the length and width, and the volume, so the thickness can be calculated.

Assembling the Tools:

We need to use the tool that defines density and we will also be using the mathematical formula for the volume of a rectangular object.

Solution:

Let's calculate the volume of the piece of steel by converting pounds to cubic inches. The sequence of conversions is

$$12.5 \text{ lb}\left(\frac{454 \text{ g}}{1 \text{ lb}}\right)\left(\frac{1 \text{ cm}^3}{7.88 \text{ g}}\right)\left(\frac{1 \text{ inch}}{2.54 \text{ cm}}\right)\left(\frac{1 \text{ inch}}{2.54 \text{ cm}}\right)\left(\frac{1 \text{ inch}}{2.54 \text{ cm}}\right) = 43.9 \text{ in}^3$$

We can now write the equation for the volume of the steel sample.

$$\text{volume} = \text{length} \times \text{width} \times \text{thickness}$$

We substitute the numbers we know:

$$43.9 \text{ in}^3 = 12.0 \text{ in} \times 18.2 \text{ in} \times \text{thickness}$$

Solving for the thickness we get

$$\text{thickness} = \frac{43.9 \text{ in}^3}{12 \text{ in} \times 18.2 \text{ in}}$$

$$\text{thickness} = 0.201 \text{ inches}$$

Is the Answer Reasonable?

You should go back to the calculation of the volume and cancel the units to satisfy yourself that the problem was set up correctly. To estimate the volume we'll round 12.5 lb to 10 lb and 454 g to 500 g and the numerator will be $10 \times 500 = 5000$. In the denominator we round 7.88 to 10 and one of the 2.54s to 3. Round the other 2.54s to 2 to get $10 \times 3 \times 2 \times 2 = 120$. We can see that 5000/120 should be a little less than 50 in^3, close to our answer.

Self-Test

21. Why is density more useful for identifying a substance than either the mass or the volume alone?

_____

_____

22. A sample of copper was found to have a mass of 3.42 kg and a volume of 382 cm^3.

    (a) What is the density of copper in g/cm^3? _____

    (b) What is the mass of 175 cm^3 of Cu? _____

    (c) What is the density of copper in units of kg/m^3? _____

23. Aluminum has a density of 2.70 g/cm^3.

    (a) What is the mass of 18.30 cm^3 of aluminum? _____

    (b) What is the volume in mL of 75.0 g of aluminum? _____

    (c) What is the thickness of aluminum foil in a 75 ft roll that is 10 inches wide and weighs 0.852 pound?

Specific gravity

Scientists and engineers often use different sets of units than the SI or SI-derived units we have been discussing. In addition, scientists and engineers often rely on tables of data that show how properties of matter change as temperature or pressure is changed. For density–temperature tables in particular, specific gravity is the preferred method of presenting data. The specific gravity is defined as the ratio of the density of a substance divided by the density of water under identical conditions. Specific gravity is also a *dimensionless* quantity since one density with units of g cm^{-3} is divided by the density of water with the same units.

For example, if wet cement has a density of 138 lb/ft^3 and water has a density of 62.4 lb/ft^3 under the same conditions, what is the specific gravity of the concrete?

$$\text{specific gravity} = \frac{\text{density of substance}}{\text{density of water}} = \frac{138 \text{ lb/ft}^3}{62.4 \text{ lb/ft}^3} = 2.21$$

Specific gravity is useful in that an engineer needing the density of cement in pounds per gallon just needs to multiply the specific gravity of the cement by the density of water having units of pounds per gallon. So, using the cement with a specific gravity of 2.21, we look up the density of water with units of pounds per gallon (8.3436 lb/gal) and then multiply the specific gravity by the needed density of water.

$$\text{density of cement in lb/gal} = 2.21 \times \frac{8.3436 \text{ lb}}{\text{gal}} = 18.44 \text{ lb/gal}$$

Self-Test

24. What is the specific gravity of gold? (*Hint*: Look up the densities.)

_____

_____

25. What is the density of silver in pounds per cubic foot? (sp. gr. Ag = 9.32)

_____

Answers to Self-Test Questions

1. (a) conclusion, (b) observation
2. (a) law, (b) theory
3. Based on observation (experiment) or experience.
4. A theory is a rigorously tested explanation of one or more scientific observations while a hypothesis is a tentative explanation.
5. (a) a mixture, (b) two, (c) Add water, which dissolves the salt but not the sand. This involves a physical change.
6. (a) C, (b) Br, (c) Cu, (d) Ni, (e) Si, (f) Ar
7. (a) aluminum, (b) magnesium, (c) lithium, (d) iodine
8. (a) chemical property, (b) physical property
9. (a) soft, silvery, melts at 97.8 °C, (b) Burns in presence of chlorine to give sodium chloride.
10. (a) mass, volume, length, (b) color, melting point, boiling point, odor
11. (a) kilogram, kg; (b) meter, m; (c) second, s; (d) kelvin, K; (e) mole, mol
12. newton × meter or N m
13. (a) centi, (b) nano, (c) milli, (d) kilo, (e) 10^{-12}, (f) 10^{-6}, (g) 10^{-1}, (h) 10^{-9}
14. (a) 6.3, (b) 23,000, (c) 0.000002450, (d) 24.87
15. (a) 135, (b) 224, (c) 3.26, (d) 1250, (e) 246, (f) 0.525, (g) 22, (h) 0.346, (i) 0.00241, (j) 298, (k) –8, (l) 27
16. (a) –5 °C, (b) 50 °F
17. Precision
18. (a) 15, (b) 40.10, (c) 78.68, (d) 104.6
19. 475.2 yd rounds to 480 yd (2 sig. fig.)
20. (a) $\dfrac{1\ \text{kg}}{2.205\ \text{lb}}$, (b) $\left(\dfrac{2.54\ \text{cm}}{1\ \text{in.}}\right)^2 = \dfrac{6.45\ \text{cm}^2}{1\ \text{in.}^2}$
21. Density doesn't depend on sample size; mass and volume do depend on the size of the sample.
22. (a) 8.95 g/cm^3, (b) 1.57 kg, (c) 8.95 × 10^3 kg/m^3
23. (a) 49.4 g, (b) 27.8 mL, (c) 0.0025 cm (or 25 μm)
24. 19.3
25. 582 lb ft^{-3}

Tools for problem solving

In this chapter you learned to apply the following concepts as tools in solving problems. Study each one carefully so that you know what each is used for. When faced with solving a problem, recall what each tool does and consider whether it will be helpful in finding a solution.

You might want to tear these pages out to use along with solving problems in this chapter.

Base SI Units (Table 1.2; Section 1.4)

The seven base units of the SI system are used to derive the units for all scientific measurements.

SI Prefixes (Table 1.5, Section 1.4)

These prefixes are used to create larger and smaller units.

Units in laboratory measurements (Section 1.4)

Often we must convert among units commonly used for laboratory measurements.

Length:	1 m = 100 cm = 1000 mm
Volume:	1 L = 1000 mL = 1000 cm^3

Temperature conversions (Equations 1.2 and 1.3, Section 1.4)

$$t_F = \left(\frac{9\ {}^\circ F}{5\ {}^\circ C}\right) t_C + 32\ {}^\circ F \qquad\qquad T_K = (t_C + 273.15)\ {}^\circ C \left(\frac{1\ K}{1\ {}^\circ C}\right)$$

Counting significant figures (Section 1.5)

- All nonzero digits are significant.
- Zeros to the *left* of the first nonzero digit are not significant.
- Zeros between significant digits, imbedded zeros, are significant.
- Zeros on the end of a number (a) with a decimal point are significant; (b) those without a decimal point are assumed not to be significant (to avoid confusion, scientific notation should be used.)

Significant figures: multiplication and division (Section 1.5)

Round the answer to the same number of significant figures as the factor with the fewest significant figures.

Significant figures: addition and subtraction (Section 1.5)

Round the answer to the same number of decimal places as the quantity with the fewest decimal places.

Significant figures: exact numbers (Section 1.5)

Exact numbers such as those that arise from definitions, do not affect the number of significant figures in the result of a calculation.

Density (Section 1.7)

$$d = \frac{m}{V}$$

Specific gravity (Section 1.7)

$$\text{Specific gravity} = \frac{d_{\text{substance}}}{d_{\text{water}}}$$

Notes:

Chapter 2

Elements, Compounds, and the Periodic Table

In Chapter 0, you learned about atoms and we expanded that briefly to elements and compounds in Chapter 1. Now we start with organizing the elements in the periodic table. The periodic table then informs us that metals, nonmetals and metalloids occupy distinct parts of the table. We now pay more attention to compounds, which are substances formed when two or more elements combine. Elements and compounds participate in chemical reactions and the basics of chemical reactions are presented in this next section. Finally, compounds can be roughly divided into two general types, molecular and ionic. We will discuss how the two types differ, the kinds of elements that form each type, and the characteristic properties that help us identify each. We also describe the systematic method used to name molecular and ionic substances.

2.1   The Periodic Table

Learning Objective

To describe the information in and the organization of the periodic table.

Review

The number of facts in chemistry is enormous. Besides the many similarities among various elements, many differences also exist. Progress in understanding and explaining these similarities and differences required a search for order and organization. The product of this search—the modern periodic table—has become our primary tool for organizing chemical facts.

Mendeleev discovered that similar properties are repeated at regular intervals when the elements are arranged in order of increasing atomic mass. By breaking this sequence at the right places, Mendeleev arranged elements with similar properties in vertical columns (groups) within his periodic table. Mendeleev's genius was his insistence on having elements with similar properties in the same column, even though it sometimes meant leaving empty spaces for (presumably) undiscovered elements.

The modern periodic table

Atomic number forms the basis for the sequence of the elements in the modern periodic table shown in Section 2.1 and on the inside front cover of the textbook. The rows are called periods and the columns are called groups, just as in Mendeleev's original table. Sometimes the term *family* is used when speaking of a group of elements. The periods are numbered using a variety of systems. The IUPAC recommends that the groups be numbered with Arabic numerals (1, 2, etc.) from 1 through 18. In the North American form of the table, which we shall use throughout the remainder of the Study Guide, the groups are given Arabic numerals (sometimes Roman numerals) and a letter (e.g., Group 2A). We will use the North American system that has eight groups numbered from 1A to 8A and ten groups numbered 1B to 8B as shown on the inside cover of the text..

In the table that we will use, the representative elements are the A-group elements. The B-group elements are the transition elements. The two long rows of elements below the main body of the table are the inner transition elements (the lanthanides and actinides). Examine Figure 2.1 in Section 2.1 of the text to be sure you understand where they properly fit into the periodic table. Also, remember the names of the following families:

| Group IA | Alkali metals | Group VIA | Halogens |
| Group IIA | Alkaline earth metals | Group VIII | Noble gases |

Self-Test

1. What problem did Mendeleev face with the elements iodine and tellurium? How did he solve it?

    _____

2. Argon and potassium do not fit in atomic mass order in the periodic table. What does this suggest about Mendeleev's basis for constructing the periodic table?

    _____

3. According to the IUPAC system, in which groups are the following elements found?

    (a) lithium   _____

    (b) iron   _____

    (c) silicon   _____

    (d) sulfur   _____

4. Among the elements Mg, Al, Cr, U, Kr, K, Br, and Ce,

    (a) which are representative elements?   _____

    (b) which is a transition element?   _____

    (c) which is a halogen?   _____

    (d) which is a noble gas?   _____

    (e) which is an alkaline earth metal?   _____

    (f) which is an alkali metal?   _____

    (g) which is an actinide element?   _____

    (h) which is a lanthanide element?   _____

5. Fill in the blanks.

    (a) The ability of copper to be drawn into wire depends on its

    _____.

    (b) Gold can be hammered into very thin sheets because of its

    _____.

    (c) The property that makes mercury useful in thermometers is

    _____.

    (d) The reason that tungsten is used as filaments in electric light bulbs is

    _____.

    (e) A nonmetal having a yellow color is _____.

(f)   Two elements that are liquids at room temperature are

_____ and _____ .

(g)   The color of bromine is _____ .

(h)   The reason that helium is used to inflate the Goodyear blimp rather than hydrogen is

_____ .

(i)   Sodium is rarely seen as a free metal because _____

_____ .

(j)   Two common semiconductors used in electronic devices are

_____ and _____ .

2.2 Metals, Nonmetals, and Metalloids

Learning Objective

Understand the distribution of metals, nonmetals, and metalloids within the periodic table.

Review

Metals are *shiny*, are *good conductors of heat and electricity*, and many are *malleable* and *ductile*. They are found in the lower left portion of the periodic table and make up most of the known elements. (Study Figure 2.3 in the text.)

Nonmetals lack the luster of metals, are nonconductors of electricity, and are poor conductors of heat. Many nonmetals are gases (H_2, O_2, N_2, F_2, Cl_2, and the elements of Group VIII) and those that are solids tend to be brittle. Nonmetals are found in the upper right portion of the periodic table.

Metalloids are semiconductors, but resemble nonmetals in many of their properties. They are located on opposite sides of the step-like line running from boron (B) to astatine (At) in the periodic table.

We can use the periodic table as a tool to help us compare the metallic and nonmetallic properties of elements. Within the periodic table, elements become more metallic (less nonmetallic) going from right to left in a period and more metallic going from top to bottom in a group.

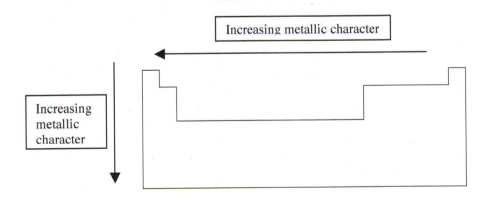

Increasing metallic character

Increasing metallic character

Example 2.1 Using the periodic table to compare physical properties

Use the periodic table to decide which is more metallic in character, antimony or bismuth? Which has the most nonmetallic character, indium or tellurium?

Analysis:

For both parts of this problem we are going to use the concept that elements become less metallic moving horizontally across the periodic table from left to right. Similarly, elements are less metallic if they are closer to the top of the table.

Assembling the Tools:

The periodic table, and trends in physical properties within the table, are our tools.

Solution:

Bismuth and antimony are in the same column and bismuth is closer to the bottom of the table and must be more metallic in character. For the second part, indium is to the left of tellurium and is predicted to have greater metallic character.

Is the Answer Reasonable?

We need to check that the more metallic elements are to the left and lower in the periodic table, and they are.

Self-Test

6. Which element is more metallic, Ga or Ge? _____

7. Which element is more nonmetallic, As or P? _____

2.3 Molecules and Chemical Formulas

Learning Objective

We will learn how to explain the information embodied in a chemical formula.

Review

There are two important topics to learn in this section. First, it's important to understand that when a chemical reaction takes place, the observable properties of the chemicals usually change significantly. As a reaction progresses, new substances are formed (with their unique properties) as the original substances (with their own unique properties) disappear.

The second topic is a skill that's critical to much of the rest of the course. You must be able to interpret a chemical formula in terms of the number of atoms of each element present. An example is given below.

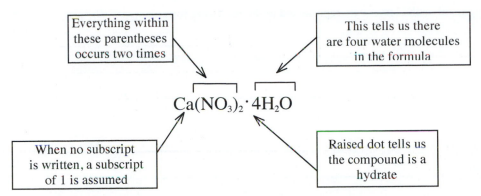

This formula shows 1 calcium atom, 2 nitrogen atoms, 8 hydrogen atoms, and 10 oxygen atoms (4 from the four H_2O molecules plus 6 from the two NO_3 units within parentheses).

A number of the nonmetals occur as diatomic molecules in the *free*, uncombined state (i.e., not combined with any other element). Be sure to study Table 2.1.

The *reactants* in a chemical equation appear on the left side of the arrow and are the substances present before the reaction begins. The *products* appear on the right side of the arrow and are the substances formed by the reaction. An equation is balanced by placing numbers called coefficients (often called *stoichiometric coefficients*) in front of the chemical formulas so that for each element there are the same number of atoms on both sides of the arrow. You will learn more about balancing equations later; for now you just need to recognize when an equation is balanced. This means you have to be able to count atoms in the formulas. Remember that in a chemical equation, a coefficient multiplies everything in the formula that follows. Consider, for example, the following:

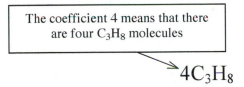

Therefore, in the expression $4C_3H_8$ we have $4 \times 3 = 12$ atoms of carbon and $4 \times 8 = 32$ atoms of hydrogen.

Sometimes in a chemical equation we specify the physical states of the chemicals involved or whether they are dissolved in a solution. The following symbols are used and are placed after the chemical formula:

solid	(*s*)	liquid	(*l*)
gas	(*g*)	aqueous solution	(*aq*)

Self-Test

8. A simple experiment you can perform in your kitchen at home or in an apartment is to add a small amount of milk of magnesia to some vinegar in a glass. Stir the mixture and observe what happens. Then add some milk of magnesia to the same amount of water and stir. What evidence did *you* observe that suggests that there is a chemical reaction between the milk of magnesia and the vinegar?

    _____

    _____

9. Drop an Alka Seltzer tablet into a glass of water. Observe what happens. What evidence is there that a chemical reaction is taking place?

    _____

10. How many atoms of each element are represented in the following formulas

    (a) Al_2Cl_6 _____

    (b) $CaSiO_3$ _____

    (c) $K_2Cr_2O_7$ _____

    (d) $C_{12}H_{22}O_{11}$ _____

    (e) $Na_2CO_3 \cdot 10H_2O$ _____

    (f) $(NH_4)_2SO_4$ _____

    (g) Al_2O_3 _____

    (h) $CaSO_4 \cdot 2H_2O$ _____

    (i) $Ca_3(PO_4)_2$ _____

11. Write the formulas of the diatomic nonmetals.

    _____

12. How many atoms of each kind are represented by the following expressions

    (a) $3KMnO_4$ _____

    (b) $5Al_2(SO_4)_3$ _____

13. How many atoms of each element are found on each side of the following equations? Are the equations balanced?

    (a) $Al_2O_3 + 6HCl \longrightarrow 2AlCl_3 + 3H_2O$

    _____

    (b) $2(NH_4)_3PO_4 + 3CaO \longrightarrow 3NH_3 + Ca_3PO_4 + 3H_2O$

    _____

14. Rewrite the equation in part (a) of the preceding question to show that the Al_2O_3 is a solid, the HCl and $AlCl_3$ are dissolved in water, and that H_2O is a liquid.

    _____

2.4   Chemical Reactions and Chemical Equations

Learning Objective

To understand the nature of a balanced chemical equation and how it relates to atomic theory..

Review

Chemists are noted for "making things," specifically new compounds. This is done by combining raw materials called reactants in a manner that allows the atoms to interact and rearrange into new compounds, or products. We can describe the process in words such as "exposing molten sodium to chlorine gas produces sodium chloride" or "mixing hydrochloric acid with potassium hydroxide produces potassium chloride and water." These descriptions are a bit cumbersome so chemists use the symbols from the periodic table to represent the elements and compounds in a reaction. By convention chemists place all the reactants to the left of an arrow and all the products to the right of that arrow. The arrow is interpreted to mean "react to make," "yield," "produce," or some similar expression. The two reactions described above can be written as:

$$Na + Cl_2 \longrightarrow NaCl$$

$$HCl + KOH \longrightarrow KCl + H_2O$$

While these chemical equations represent the written statements of the reactions we described in words, there is one fatal flaw, the first reaction does not obey the law of conservation of mass. In fact, all chemical equations must be balanced so that we end up (on the product side) with the same number of each element that we started with (on the reactant side).

Our two foci in this section are the relationship between a chemical equation and a written (or oral) description of a chemical reaction and developing the skill to recognize if an equation is balanced or unbalanced.

Self-Test

In the questions below, all of the names of the compounds are mentioned in the previous section or the review above. Those that are not mentioned are given in the question.

15. $$SO_2 + O_2 \longrightarrow SO_3$$

    Describe this reaction in words. Is this reaction balanced? Which elements, if any, are balanced?

    _____

    _____

16. Write out in words (carbon dioxide is one of the products) what the following equation represents. Is this equation balanced? Which elements, if any, are balanced?

    $$CH_4 + O_2 \longrightarrow CO_2 + 18H_2O$$

    _____

    _____

17. $$CCl_4 + 2H_2 \longrightarrow CH_4 + Cl_2$$

    Describe this reaction in words. Is this reaction balanced? Which elements, if any, are balanced?

    _____

    _____

18. $$C + H_2O \longrightarrow CO_2 + H_2$$

    Describe this reaction in words (carbon dioxide is one of the products). Is this reaction balanced? Which elements, if any, are balanced?

    _____

    _____

2.5 Ionic Compounds

Learning Objective

To use the periodic table and ion charges to write chemical formulas of ionic compounds.

Review

Metals combine with nonmetals to form ionic compounds by the transfer of electrons from one atom to another. This produces electrically charged particles that we call ions. Study the diagram that illustrates the transfer of an electron from a sodium atom to a chlorine atom.

In an ionic compound we cannot identify individual groups of ions as belonging to each other. We simply specify in the formula the smallest whole-number ratio of the ions. The group of ions specified in such a formula is called a formula unit.

Review

Cations (positive ions) are formed by metals when their atoms lose electrons. Anions (negative ions) are formed by nonmetals when their atoms acquire electrons.

Table 2.2 in this section illustrates how the periodic table can help you remember the ions formed by the representative metals and nonmetals. The number of positive charges on a cation equals the group number; the number of negative charges on an anion equals the number of spaces to the right we have to go to get to a noble gas in the periodic table. Be sure you understand how the formulas of the ions in Table 2.2 correlate with the locations of the elements in the periodic table.

Example 2.2 Using the periodic table to predict ionic formulas

If the symbol M stands for a representative metal and X for a representative nonmetal, determine what groups the metal and nonmetal come from in the following compounds: (a) MX_2 and (b) M_3X_2.

Analysis:

We can assign the groups to the metals and nonmetals if we know their charges. For instance, an ion with a 1+ charge must come from group IA. We can assume a charge for the metal (between 1 and 3) and see if the charge of the nonmetal is reasonable. If so we can then assign them to groups.

Assembling the Tools:

The periodic table, and trends in charges of the representative element's ions based on the group number of the element.

Solution:

(a) Let's start by trying some possible charges for M. If $M = 1+$, then X must be 0.5–. Half charges are not allowed and therefore M cannot be 1+. Similarly, if M is 3+ then X must be 1.5–, another impossible number. If M is 2+ then X can be 1– and M would come from Group IIA and X would come from Group VIIA. We have no representative metals with a 4+ charge and we can stop here.

(b) In the same manner, if M is 1+ then X must be 1.5– which is impossible. If M is 2+ then X must be 3– which is reasonable and M would come from Group IIA and X would come from Group VA. If M is 3+ then X must be 4.5–, another impossible answer.

Is the Answer Reasonable?

All we can do is verify our math to show that only representative elements from Groups IIA and VIIA form the MX_2 compound and elements from Groups IIA and VA form the M_3X_2 compound.

Ionic compounds always contain positive ions and negative ions in ratios that give electrically neutral substances. In an ionic compound, the ions are arranged around one another in a way that maximizes the attractions between oppositely charged ions and minimizes the repulsions between like-charged ions. Since we can't identify unique molecules in an ionic compound, we always write their formulas with the smallest set of whole-number subscripts. This identifies one formula unit of the substance.

Writing the formulas for ionic compounds correctly is a skill you must master, which is why the Rules for Writing Formulas of Ionic Compounds are identified by the icon in the margin. The rules are simple to apply, so be sure to study Example 2.3 and work Practice Exercises 2.7 and 2.8.

The transition and post-transition metals generally are able to form two or more different ions, and there are no simple rules that enable us to figure them out. Therefore, you must simply memorize the symbols (including charges) for these ions that are given in Table 2.3.

Table 2.4 contains the formulas and names of frequently encountered polyatomic ions—ions composed of more than one atom. These should also be memorized. Practice writing both their names and their formulas. Be sure to learn their charges, too. (It might help to prepare a set of flash cards.) When you feel you have learned the contents of Tables 2.3 and 2.4 and can use the periodic table to write the formulas of the ions in Table 2.2, try the Self-Test below.

Self-Test

19. Without looking at Table 2.2, but using the periodic table on the inside front cover of your textbook, write the formula for the ion formed by each of the following elements:

    (a) K _____   (e) Ba _____

    (b) Al _____   (f) Na _____

    (c) N _____   (g) Br _____

    (d) Mg _____   (h) Se _____

20. What is the general formula for an ion formed by a metal from

    (a) Group IA? _____   (b) Group IIA? _____

21. What is the general formula for an ion formed by a nonmetal from

    (a) Group VA? _____   (b) Group VIIA? _____

22. Write the formula for the ionic compound formed by

    (a) Al and Se _____

    (b) Mg and Cl _____

    (c) Na and S _____

    (d) Sr and F _____

23. Write the formula for the ionic compound formed from

    (a) sodium ion and perchlorate ion _____

    (b) barium ion and hydroxide ion _____

    (c) ammonium ion and dichromate ion _____

    (d) magnesium ion and sulfite ion _____

    (e) nickel ion and phosphate ion _____

    (f) silver ion and sulfate ion _____

24. Write the formulas for *two* different ionic compounds of

    (a) iron and oxygen _____ _____

    (b) copper and chlorine _____ _____

    (c) mercury and chlorine _____ _____

    (d) tin and sulfur _____ _____

    (e) lead and oxygen _____ _____

2.6 Nomenclature of Ionic Compounds

Learning Objective

To name ionic compounds and write chemical formulas from chemical names.

Review

This is the first section in the book that discusses naming compounds (nomenclature). Examples of how to name ionic compounds are reviewed below. However, the flowchart in Figure 2.32 in Section 2.8 summarizes the basics of the decision process needed to name most compounds. Try to remember and use this flowchart style of thinking when naming compounds.

Binary compounds containing a metal and a nonmetal
In writing the name of an ionic compound, the cation is always named first followed by the anion. The rules are designed to permit us to derive the names of cations and anions. Once we have them, we write them in the order just mentioned.

If the cation is of metal that forms only one positive ion, it simply is given the English name of the metal. For example, the cation Mg^{2+} is specified as *magnesium*. When the metal forms two or more cations, the preferred method for naming the cation is called the Stock system. In this case, the number of positive charges on the cation is specified by placing a Roman numeral in parentheses following the English name for the metal. Thus, Fe^{2+} is the called the iron(II) ion and Fe^{3+} is the iron(III) ion. Notice that there is no space between the name of the metal and the parentheses containing the Roman numeral.

Monatomic (one-atom) anions of nonmetals are named by adding the suffix *-ide* to the stem of the nonmetal name. Note that hydroxide, OH^-, and cyanide, CN^-, are the only two polyatomic ions ending in -*ide*. If the name of the anion ends in *-ide* and it isn't hydroxide or cyanide, then you can be sure it's a monatomic anion.

Ionic compounds that contain polyatomic ions

Here we simply use the name of the polyatomic cation or anion instead of the name of a monatomic ion. It's essential that you learn the names, formulas, and charges of the polyatomic ions in Table 2.4. This is especially so when you have to write the formula given the name of a compound.

Self-Test

25. Name these compounds.

   (a) CaI_2 _____

   (b) $FeBr_3$ _____

   (c) $Fe(NO_3)_2$ _____

   (d) KNO_2 _____

   (e) $Cu(BrO_3)_2$ _____

   (f) CaH_2PO_4 _____

   (g) $KHSO_3$ _____

   (h) $Ca(C_2H_3O_2)_2$ _____

26. Write the formulas for these compounds.

   (a) barium arsenide _____

   (b) sodium perbromate _____

   (c) manganese(IV) oxide _____

   (d) sodium permanganate _____

   (e) aluminum nitrate _____

   (f) copper(II) iodate _____

   (g) cobalt(II) acetate _____

   (h) lead(II) chloride _____

2.7   Molecular Compounds

Learning Objective

To understand the difference between ionic and molecular compounds.

Review

Molecules are electrically neutral particles composed of two or more atoms. Evidence for the existence of molecules is Brownian motion, which is believed to be caused by the battering about of tiny particles by collision with molecules of a liquid. Molecules are held together by chemical bonds between their atoms. These bonds arise from the sharing of electrons between atoms, which we will discuss more fully in later chapters. The chemical formulas we write for molecules are called **molecular formulas**; they describe the number of atoms of each kind in one molecule of a substance.

It is useful to remember that nonmetals combine with nonmetals to form molecular compounds. Examples presented in this section include the simple hydrogen compounds of the nonmetals. Also notice that most of the nonmetals exist in their elemental forms as molecules. Be sure you learn the ones that are diatomic: H_2, N_2, O_2, F_2, Cl_2, Br_2, and I_2. (Only the noble gases occur in nature as single atoms.)

Compounds of nonmetals with hydrogen

The formulas of the simple hydrogen compounds of the nonmetals are given in Table 2.6. Notice that we can use the periodic table to figure out their formulas: For a nonmetal in a particular group in the periodic table, the number of hydrogens in the formula equals the number of steps to the right we have to go to get to Group VIII (the noble gases). Thus, for any element in Group VIIA, we have to move *one* step to the right to get to Group VIII; the formula of the hydrogen compound contains *one* hydrogen (e.g., HF). Similarly, for an element in Group IVA, we have to move *four* elements to the right to get to Group VIII, and each element in Group IVA forms a hydrogen compound with *four* hydrogens (e.g., CH_4).

Compounds of carbon

Carbon and hydrogen form compounds called hydrocarbons in which atoms of carbon are attached to each other, often in chains of varying lengths. Hydrocarbons make up the chief constituents of petroleum, and they serve as the foundation of a class of substances called organic compounds. Organic chemistry is the study of hydrocarbons and compounds formed from them by substituting various groups of atoms in place of hydrogen atoms.

Remember the general formula for an *alkane* hydrocarbon, C_nH_{2n+2}, where n is the number of carbon atoms. Butane (the fuel in cigarette lighters) is a four-carbon alkane. Its formula is

$$C_4H_{(2(4)\,+\,2)} = C_4H_{10}$$

An important class of organic compounds is the alcohols, obtained by substituting OH in place of a hydrogen atom. Be sure you know the formulas for methanol (methyl alcohol), CH_3OH, and ethanol (ethyl alcohol), C_2H_5OH.

Self-Test

27.   In terms of their composition, what is the major difference between a molecular compound and an ionic compound?

_____

_____

28.   (a)   What name is used for the tiny individual particles that

occur in molecular compounds?   _____

(b)   What is the name for the individual particles that make up

an ionic compound?   _____

29.   What kind of compound (ionic or molecular) would be expected if the following pairs of atoms combine?

(a)   magnesium and chlorine   _____

(b)   sulfur and oxygen   _____

(c)   carbon, hydrogen, and nitrogen_____

30. Without referring to anything but the periodic table, write the formula for the simplest compound formed from hydrogen and

    (a) sulfur _____

    (b) iodine _____

    (c) phosphorus _____

    (d) silicon _____

31. Write the general formula for all alkanes. _____

32. Write the formula for the alkane hydrocarbon that has

    (a) three carbon atoms _____

    (b) nine carbon atoms _____

2.8   Nomenclature of Molecular Compounds

Learning Objective

To learn to name molecular compounds.

Review

We begin with simple inorganic molecular compounds. To name them, you must learn the Greek prefixes given in Section 2.6. For binary compounds of two elements, the first is specified by giving its English name, preceded by a prefix if necessary. The second element is specified by appending the suffix -ide to the stem of the element's name; study Example 2.9.

For binary nonmetal hydrides, it is not necessary to use Greek prefixes to specify the number of hydrogen atoms in the formula. Once we know the other element, we can figure out how many hydrogen atoms go with it.

Common names

Some well-known compounds, such as H_2O (water) and NH_3 (ammonia) were given names before a systematic system was developed. In most instances, we do not attempt to name them according to the IUPAC rules and simply use their common names. Once again, our goal is to have a system that allows us to derive a unique name for each compound and that enables us to derive the formula given the name.

Self-Test

33. Name the following compounds.

    (a) N_2O_5 _____

    (b) H_2Se _____

    (c) P_2S_3 _____

    (d) $SbCl_5$ _____

    (e) IF_7 _____

34. Write formulas for the following.

    (a) dinitrogen trioxide _____

    (b) disulfurdichloride _____

    (c) selenium tetrachloride _____

    (d) nitrogen monoxide _____

    (e) oxygen difluoride _____

35. Use the concepts in the flowchart to determine the formulas of the following compounds.

    (a) cobalt(II) nitrate _____

    (b) iron(III) chloride _____

    (c) ammonium sulfate _____

    (d) silicon dioxide _____

Answers to Self-Test Questions

1. If placed in the table in order of atomic weight, their properties do not match those of other elements in the same columns. He put them in the table according to their properties rather than according to their atomic weights.

2. Atomic weight is not really the basis for the periodic law.

3. (a) Group 1, (b) Group 8, (c) Group 14, (d) Group 16

4. (a) Mg, Al, Kr, K, Br   (b) Cr   (c) Br
   (d) Kr   (e) Mg   (f) K   (g) U   (h) Ce

5. (a) ductility, (b) malleability, (c) low melting point and fairly high boiling point, (d) it conducts electricity and has a very high melting point.
   (e) sulfur, (f) bromine, mercury, (g) red-brown, (h) helium is very unreactive, (i) it is very reactive toward oxygen, (j) silicon and germanium

6. Ga

7. P

8. You should have observed that the milk of magnesia dissolves in the vinegar, but not in plain water.

9. The fizzing is evidence that a reaction is taking place.

10. (a) 2Al, 6Cl (b) 1Ca, 1Si, 3O (c) 2K, 2Cr, 7O (d) 12C, 22H, 11O (e) 2Na, 1C, 13O, 20H
    (f) 2N, 8H, 1S, 4O (g) 2Al, 3O (h) 1Ca, 1S, 6O, 4H (i) 3Ca, 2P, 8O

11. $H_2, N_2, O_2, F_2, Cl_2, Br_2, I_2$

12. (a) 3K, 3Mn, 12O (b) 10Al, 15S, 60O

13. (a) On the left: 2Al, 3O, 6H, 6Cl; on the right: 2Al, 3O, 6H, 6Cl (Balanced) (b) On the left: 6N, 24H, 2P, 11O, 3Ca; on the right: 3N, 15H, 1P, 7O, 3Ca (Not balanced)

14. $2Al_2O_3(s) + 6HCl(aq) \rightarrow 2AlCl_3(aq) + 3H_2O(l)$

15. Sulfur dioxide reacts with molecular oxygen forming sulfur trioxide. The equation is not balanced. Only sulfur is balanced.

16. Methane reacts with oxygen forming carbon dioxide and water. Only carbon is balanced.

17. Carbon tetrachloride reacts with hydrogen to form methane and chlorine. The reaction is balanced, meaning that all atoms are balanced.

18. Carbon reacts with water to form carbon dioxide and hydrogen. Reaction is not balanced. Carbon and hydrogen are balanced.

19. (a) K^+, (b) Al^{3+}, (c) N^{3-}, (d) Mg^{2+}, (e) Ba^{2+}, (f) Na^+, (g) Br^-, (h) Se^{2-}

20. (a) M^+, (b) M^{2+}

21. (a) X^{3-}, (b) X^-

22. (a) Al_2Se_3, (b) $MgCl_2$, (c) Na_2S, (d) SrF_2

23. (a) $NaClO_4$, (b) $Ba(OH)_2$, (c) $(NH_4)_2Cr_2O_7$,
    (d) $MgSO_3$, (e) $Ni_3(PO_4)_2$, (f) Ag_2SO_4

24. (a) FeO and Fe_2O_3, (b) $CuCl$ and $CuCl_2$,
    (c) Hg_2Cl_2 and $HgCl_2$, (d) SnS and SnS_2, (e) PbO and PbO_2

25. (a) calcium iodide, (b) iron(III) bromide, (c) iron(II) nitrate, (d) potassium nitrite,
    (e) copper(II) bromate, (f) calcium dihydrogen phosphate, (g) potassium hydrogen sulfite,
    (h) calcium acetate

26. (a) Ba_3As_2, (b) $NaBrO_4$, (c) MnO_2, (d) $NaMnO_4$, (e) $Al(NO_3)_3$, (f) $Cu(IO_3)_2$,
    (g) $Co(C_2H_3O_2)_2$, (h) $PbCl_2$

27. A molecular compound is made up of neutral particles (molecules); an ionic compound is composed of charged particles (ions).

28. (a) molecules, (b) ions

29. (a) ionic, (b) molecular

30. (a) H_2S, (b) HI, (c) PH_3 (or H_3P), (d) SiH_4 (or H_4Si)

31. C_nH_{2n+2}

32. (a) C_3H_8, (b) C_9H_{20}

33. (a) dinitrogen pentaoxide, (b) hydrogen selenide, (c) diphosphorus trisulfide,
    (d) antimony pentachloride, (e) iodine heptafluoride

34. (a) N_2O_3, (b) S_2Cl_2, (c) $SeCl_4$, (d) NO, (e) OF_2

35. (a) $Co(NO_3)_2$, (b) $FeCl_3$, (c) $(NH_4)_2SO_4$, (d) SiO_2

Tools for problem solving

We have learned the following concepts that can be applied as tools in solving problems. Study each one carefully so that you know what each is used for. When faced with solving a problem, recall what each tool does and consider whether it will be helpful in finding a solution. This will aid you in selecting the tools you need. If necessary, refer to this table when working on the exercises in the chapter. Remember that tools from Chapter 1 may be needed at times to solve problems in this chapter.

You may wish to tear out these pages to use while solving problems.

Periodic Table (Section 2.1)

The periodic table lists the atoms by atomic number and organizes them by their properties in periods and groups. We can obtain atomic numbers and average atomic masses from the periodic table.

Periodic Table: Metals, nonmetals, or metalloids (Section 2.2)

From an element's position in the periodic table, we can tell whether it's a metal, nonmetal, or metalloid.

Chemical symbols and subscripts in a chemical formula (Section 2.3)

In a chemical formula, the chemical symbol stands for an atom of an element. Subscripts show the number of atoms of each kind, and when a subscript follows a parentheses, it multiplies everything within the parentheses.

Coefficients in an equation (Section 2.4)

Coefficients are only used to balance an equation and indicate the number of chemical units of each type that are present in reactants and products.

Ionic compounds (Section 2.5)

Ionic compounds are formed when metals react with nonmetals and the ions formed combine into a compound.

Predicting cation charge (Section 2.5)

The metals in Group 1A have a +1 charge and metals in Group 2A have a +2 charge.

Predicting anion charge (Section 2.5)

From a nonmetal's position in the periodic table, we can determine the charge of the monatomic anions.

Formulas for ionic compounds (Section 2.5)

Following the rules gives us the correct formulas for ionic compounds with electrically neutral formula units and with subscripts in the smallest set of whole numbers.

Polyatomic ions (Section 2.5)

The names, formulas and charges of these ions are listed in Table 2.4.

Monatomic anion names (Section 2.6)

The list in Table 2.5 gives the common names of anions that must be remembered so you can use them to name ionic compounds.

Naming ionic compounds (Section 2.6)

These rules give us a systematic method for naming ionic compounds.

Using the Stock system (Section 2.6)

For metals that can form ions with more than one possible charge, the Stock system specifies the charge of a cation by placing a Roman numeral in parentheses just after the name of the cation,

Naming with polyatomic ions (Section 2.6)

Naming compounds that contain polyatomic anions is done by specifying the cation name, using the Stock system if needed, and then specifying the polyatomic anion name as given in Table 2.4.

Greek prefixes (Section 2.6)

This is a list of the Greek prefixes from one to ten. They are used in naming hydrates and molecular compounds.

Predicting formulas of nonmetal hydrogen compounds (Section 2.7)

From a nonmetal's position in the periodic table we can write the formula of its simple hydrogen compound. These are given in Table 2.6.

Naming binary molecular compounds (Section 2.8)

These rules give us a logical system for naming binary molecular compounds by specifying the number of each type of atom using Greek prefixes.

Flowchart for naming compounds (Section 2.8)

This flowchart in Figure 2.32 can be followed for naming both ionic and molecular compounds.

Notes:

Chapter 3

The Mole and Stoichiometry

Atoms and molecules react and combine in whole-number ratios. Therefore, to deal with formulas and reactions quantitatively, we must have a way to measure and count numbers of atoms and molecules. However, these particles are so tiny that we can't count them in the same way that we count dimes and quarters. Instead, we count them using a balance, taking advantage of the fact that each element has atoms with a characteristic atomic mass. The mole concept allows us to relate numbers of atoms and molecules to mass measurements we can perform in the laboratory.

In this chapter you will learn how to measure moles of atoms and molecules, how to determine chemical formulas from experimental data, and how to use balanced chemical equations to relate amounts of substances involved in chemical reactions. All of these are aspects of *stoichiometry*. The most important skill to develop is the ability to use the mole concept to perform quantitative reasoning.

3.1 The Mole and Avogadro's Number

Learning Objective

Explain how the mole and Avogadro's number serve as conversion factors between the molecular and laboratory scales of matter.

Review

The main point of this section is that we can use the known mass of an object to count the number of such objects in a large collection of them. We simply measure the total mass of all the objects and then divide by the mass of one of them. In counting atoms of an element in a lab-sized sample, we need the atomic mass of the element expressed in atomic mass units, u, as well as the mass in grams of the atomic mass unit.

Using the atomic mass unit, u, you can relate numbers of individual atoms or molecules to the lab-sized mass unit gram. When we work with realistic samples of substances, they contain enormous numbers of atoms or molecules, and we are usually not concerned with the actual number of particles. Instead, we're interested in the *ratios* of their numbers. For example, in a molecule of CO_2, the ratio of C to O atoms is 1 to 2. This same atom ratio will exist regardless of the sample size. If the sample had 100 molecules of CO_2, it would be composed of 100 atoms of C and 200 atoms of O; we still have a 1 to 2 atom ratio of C to O.

The mole is a quantity that allows us to conveniently measure, using a balance, large amounts of atoms (or molecules) in any desired ratio. A mole contains a fixed number of particles and has a mass, in grams, that is numerically the same as the atomic mass (for atoms), formula mass (for ionic compounds) or molecular mass (for molecules). Thus, one mole of carbon atoms (1 mol C) has a mass of 12.01 g, and one mole of magnesium atoms (1 mol Mg) has a mass of 24.30 g. Similarly, the molecular mass of CO_2 is 44.02, so one mole of CO_2 molecules (1 mol CO_2) has a mass of 44.02 g. Notice that atomic, formula, and molecular masses are all determined the same way, by adding the masses of the atoms. We simplify these three names to just *molar mass*.

The point of the preceding discussion is that a table of atomic masses (or a periodic table that incorporates atomic masses) provides us with a tool for relating moles of elements to masses measured in grams. All we need to do is look up the atomic mass of the element and we can write down how much one mole of that element weighs. For example, the periodic table gives the atomic mass of sodium (Na) as 22.98977, so we can write

$$1 \text{ mol Na} = 22.98977 \text{ g Na}$$

Usually, these are more significant figures than we require, so we round it as needed. If four significant figures are sufficient, then we write

$$1 \text{ mol Na} = 22.99 \text{ g Na}$$

Here are two more examples, also to four significant figures.

$$1 \text{ mol Cl} = 35.45 \text{ g Cl}$$

$$1 \text{ mol Fe} = 55.85 \text{ g Fe}$$

A similar situation applies when we're working with compounds. For example, if we wished to know the molar mass of hydrogen sulfide, H_2S, we would add the atomic masses of the elements to obtain the molecular mass. For H_2S, we obtain 34.082. Therefore, we can write

$$1 \text{ mol } H_2S = 34.082 \text{ g } H_2S$$

Let's look at a sample problem that requires us to relate grams to moles.

Example 3.1 Converting between grams and moles for an element

Suppose we needed to measure 2.50 mol of sulfur for a particular experiment. How many grams of sulfur would we need?

Analysis:

We'll begin by stating the problem in equation form.

$$2.50 \text{ mol S} = ? \text{ g S}$$

To solve the problem, we need to convert grams of sulfur to moles of sulfur. The tool to accomplish this is the atomic mass of sulfur written as an equality.

$$1 \text{ mol S} = 32.066 \text{ g S}$$

To be sure to have enough precision in the calculation, we'll round this to one more significant figure than in the given data or 32.07 g S.

Our general pathway for this calculation is

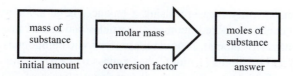

Assembling the Tools:

Now we see that molar mass will be the tool needed to convert mass to moles.

Solution:

We form the conversion factor from the equality so mol S will cancel.

$$2.50 \text{ mol S} \times \frac{32.07 \text{ g S}}{1 \text{ mol S}} = 80.2 \text{ g S}$$

Is the Answer Reasonable?

Some simple arithmetic will answer the question. One mole of sulfur is about 32 g. We have between two and three moles of S, so the answer should lie between 2×32 g = 64 g and 3×32 g = 96 g. Our answer, 80.2 g, is between these two values, so the answer does seem reasonable.

The mole and Avogadro's number

As mentioned above, the mole is a unit that stands for a fixed number of things. This number is called Avogadro's number and is equal to 6.02×10^{23}.

$$1 \text{ mol} = 6.02 \times 10^{23} \text{ things}$$

For most purposes, it isn't necessary to know the size of this number. Usually, it is sufficient to understand that if we have equal numbers of moles of two elements, they contain equal numbers of atoms. However, there are some instances when we do need to know the number of things in a mole, and we will study those now.

To understand when we need to use Avogadro's number, we need to examine how we view chemical substances from two perspectives—the large-scale *macroscopic* world of the laboratory (where we deal with physical samples of substances and describe amounts in grams and moles) and the very tiny *submicroscopic* world of atoms and molecules (where we count individual atoms and molecules). As long as we stay entirely within one of these two views of matter, we don't need Avogadro's number. Thus, to calculate the number of *moles* of carbon in 5 *moles* of glucose, $C_6H_{12}O_6$, Avogadro's number isn't necessary. As we discuss in the next section, we simply use the chemical formula to tell us how many moles of carbon are in one mole of $C_6H_{12}O_6$. Similarly, if we wished to know how many *atoms* of carbon are in 5 *molecules* of glucose, $C_6H_{12}O_6$, we use the chemical formula to tell us how many carbon atoms are in one molecule of $C_6H_{12}O_6$, and then multiply by five. Once again, we don't need Avogadro's number.

Avogadro's number becomes necessary when we wish to relate an amount of something in the macroscopic world to an amount in the submicroscopic world. An example would be finding the mass in *grams* (a unit in the macroscopic world) of some number of carbon *atoms* (a unit in the submicroscopic world).

$$25 \text{ atoms Ni} \left(\frac{1 \text{ mol Ni}}{6.022 \times 10^{23} \text{ atoms Ni}} \right) \left(\frac{58.7 \text{ g Ni}}{1 \text{ mol Ni}} \right) = 2.44 \times 10^{-21} \text{ g Ni}$$

Besides understanding *when* you need to use Avogadro's number, you also need to know *how* to use it. As a tool in calculations, Avogadro's number provides a conversion between moles and individual units of a substance. For example, if we are dealing with an element such as sodium, we can write

$$1 \text{ mol Na} = 6.02 \times 10^{23} \text{ atom Na}$$

For the compound C_3H_8, we can write

$$1 \text{ mol } C_3H_8, = 6.02 \times 10^{23} \text{ molecule } C_3H_8,$$

For an ionic compound such as NaCl, which doesn't contain discrete molecules, we write

$$1 \text{ mol NaCl} = 6.02 \times 10^{23} \text{ formula units NaCl}$$

There are two important things to observe in these equations. First, notice that we are relating *moles* in the macroscopic world to atoms, molecules, or formula units in the submicroscopic world. Second, notice that when we write these equations, we are very careful to specify, using chemical formulas, the exact nature of the substances involved. Now let's look at a sample problem.

Example 3.2   Working with Avogadro's number

What is the mass in grams of 25 atoms of nickel?

Analysis:

We'll begin by expressing the problem in the form of an equation:

$$25 \text{ atoms Ni} = ? \text{ g Ni}$$

The main link in solving this problem is realizing that we are attempting to convert a unit in the submicroscopic world (atoms) to a unit in the macroscopic world (grams). Because we are connecting these two realms, we will need to use Avogadro's number as the tool. We are dealing with an element, so we write

$$1 \text{ mol Ni} = 6.02 \times 10^{23} \text{ atoms Ni}$$

We can now go from the unit "mol Ni" to "atoms Ni." Next, we need a tool to take us from "mol Ni" to "g Ni." This is provided by the atomic mass of nickel.

$$1 \text{ mol Ni} = 58.7 \text{ g Ni}$$

We now have a path from "atoms Ni" to "g Ni," so we can proceed to the solution step.

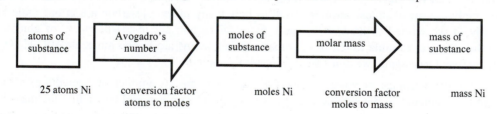

Assembling the Tools:

The new tool we are using is Avogadro's number, also needed is the molar mass.

Solution:

As usual, we assemble conversion factors from the relationships we've established in a way that will allow us to cancel units correctly. (Add the cancel marks yourself.)

$$25 \text{ atoms Ni} \left(\frac{1 \text{ mol Ni}}{6.022 \times 10^{23} \text{ atoms Ni}} \right) \left(\frac{58.7 \text{ g Ni}}{1 \text{ mol Ni}} \right) = 2.44 \times 10^{-21} \text{ g Ni}$$

Thus, 25 atoms of nickel have a mass of 2.44×10^{-21} g.

Is the Answer Reasonable?

In working these kinds of problems, the most common mistake is to use Avogadro's number incorrectly, or to not use it at all. Let's see how you can avoid these pitfalls.

We know atoms are very tiny and that individual atoms have very small masses. The answer we obtained is a very small number, so in that sense, it seems that our answer is reasonable.

If we had used Avogadro's number incorrectly, the answer would have been a huge number. Atoms do not have huge masses, so we should recognize our mistake. Similarly, if we had not used Avogadro's number at all, our calculation would give an answer that is not an extremely tiny number. That should sound an alarm and suggest that we'd better examine our method more closely.

Self-Test

1. Calculate the formula mass of $(NH_4)_3PO_4 \cdot 3H_2O$, and write the two conversion factors relating grams to moles for this compound.

   _____

2. What is the molecular mass of acetone, $(CH_3)_2CO$, a solvent in nail polish?

   _____

3. How many grams of calcium are in 1.00 mol Ca? _____

4. How many moles of phosphorus atoms are in 1.00 g of phosphorus ?_____

5. How many moles of $Ca(NO_3)_2$ are there in 99.9 grams of calcium nitrate?

   _____

6. How many molecules of CO_2 are present in a sample that has a mass of 264 u?

   _____

7. Acetone has a density of 0.788 g cm^{-3}. How many molecules of acetone are in 15.0 cm^3 of acetone?

   _____

8. For ammonia, 17.0 g NH_3 ⇔ 6.02×10^{23} molecules NH_3. Write the two conversion factors that this equivalency makes possible.

   _____

9. If an experiment calls for 0.500 mol of H_2SO_4 (sulfuric acid), how many grams of H_2SO_4 do we have to weigh out?

   _____

10. If 110 g of CO_2 are formed in an experiment, how many moles of CO_2 are formed?

   _____

11. What is the mass, in grams, of 0.250 mol of phenobarbital ($C_{12}H_{12}N_2O_3$)? _____

12. How many moles of pentachlorophenol (C_6HCl_5O) are in 1.00 g of C_6HCl_5O?

   _____

13. How many atoms of silver are in a bracelet that weighs 34.0 g?

   _____

14. How many atoms of iron, Fe, are in 0.400 mol of iron?

   _____

15. How many atoms are there in 0.520 grams of iron?   _____

16. 25.0 moles of phosphoric acid (H_3PO_4) contain how many hydrogen atoms?

   _____

17. What is the mass in grams of 245 atoms of sodium? _____

18. How many atoms of uranium are there in 1.00 g of uranium metal?

   _____

19. How many H_2O molecules are there in a raindrop weighing 0.050 g?

   _____

3.2   The Mole, Formula Mass, and Stoichiometry

Learning Objective

To perform calculations involving moles and amount of substance.

Review

The subscripts in a formula give us the *atom ratio* in which the elements are combined. Those same subscripts also give us the ratio by moles in which the elements are combined. For example, consider the substance propane, the gas used for cooking in rural areas and in nearly all gas barbecue grills. Propane consists of molecules that have the formula C_3H_8.

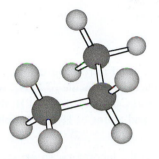

A molecule of propane, C_3H_8. It consists of three carbon atoms and eight hydrogen atoms joined together in a single particle that we call a molecule.

If we had one molecule of propane, it would contain 3 atoms of C and 8 atoms of H. If we had 10 propane molecules, all together they would contain 30 atoms of C and 80 atoms of H. Notice, however, that even though we have more atoms of C and H in the larger sample, they are in the same numerical C-to-H ratio, namely 3-to-8. In fact, in any sample of this compound, the atom *ratio* of C to H will be 3 to 8.

Now suppose we had one dozen propane molecules. In this sample we would find 3 dozen carbon atoms and 8 dozen hydrogen atoms.[1]

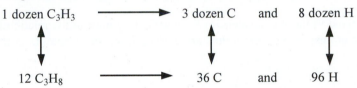

1 dozen C_3H_3	3 dozen C	and 8 dozen H
12 C_3H_8	36 C	and 96 H

Notice that in a dozen molecules, the ratio of atoms *by the dozen* (3 to 8) is numerically the same as the atom ratio in one molecule of the compound.

As we discussed earlier, the mole is a quantity similar to the dozen, because it stands for a fixed number of things. The reasoning with moles is therefore the same as the reasoning with dozens. If we had one mole of C_3H_8, the ratio of atoms *by the mole* would be numerically the same as the atom ratio in one molecule.

$$1 \text{ mol } C_3H_8 \longrightarrow 3 \text{ mol C} + 8 \text{ mol H}$$
$$1 \text{ dozen } C_3H_8 \longrightarrow 3 \text{ dozen C} + 8 \text{ dozen H}$$
$$1 \text{ molecule } C_3H_8 \longrightarrow 3 \text{ molecules C} + 8 \text{ molecules H}$$

Thus, whether we're dealing with individual particles (molecules and atoms), particles by the dozen, or particles by the mole, the ratios are the same.

Interpreting a chemical formula for problems in stoichiometry

In chemical problem solving, the chemical formula serves as a tool that establishes the ratios among atoms in a single molecule of the compound. It is also a tool that establishes the ratios among moles of atoms in a mole of the compound. We can use these relationships to construct equivalencies that we can use in chemical calculations. For propane, we can write the following, which apply when we are dealing with numbers of individual molecules and atoms.

$$1 \text{ molecule } C_3H_8 \Leftrightarrow 3 \text{ atoms C}$$

---

[1] When we write "1 dozen C_3H_8" we mean "1 dozen C_3H_8 *molecules*," and when we write "3 dozen C" and "8 dozen H," we mean "3 dozen C *atoms*" and "8 dozen H *atoms*." We generally omit the word molecule or atom because the formula C_3H_8 stands for a molecule of C_3H_8, and the symbols C and H stand for atoms of C and H.

$$1 \text{ molecule } C_3H_8 \iff 8 \text{ atoms H}$$

$$3 \text{ atoms C} \iff 8 \text{ atoms H}$$

For example, the first equivalence would be used when we are interested in relating the number of carbon atoms to a certain number of C_3H_8 molecules. It tells us that for every 1 molecule of C_3H_8, we will find 3 atoms of C. If we were interested in the relation between molecules of C_3H_8 and atoms of H, then we would use the second equivalence, and if we wanted to relate atoms of C and atoms of H in a sample of propane, we would use the third equivalence.

Individual atoms and molecules are too small to work with in the laboratory, so we scale everything up to mole-sized quantities and write the following equivalencies.

$$1 \text{ mol } C_3H_8 \iff 3 \text{ mol C}$$

$$1 \text{ mol } C_3H_8 \iff 8 \text{ mol H}$$

$$3 \text{ mol C} \iff 8 \text{ mol H}$$

Notice that the numbers involved in these equivalencies are identical to those in the relationships given for the individual molecules and atoms; in both cases they are derived from the subscripts in the formula for propane.

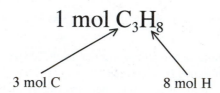

Let's look at an example that illustrates how we use the concepts described above to solve problems.

Example 3.3　Using the mole concept in stoichiometry

When butane burns in air, the carbon in the compound becomes incorporated entirely in molecules of carbon dioxide, CO_2. How many moles of CO_2 would be formed if 0.200 mol of C_4H_{10} is burned?

Analysis:

Let's begin by expressing the problem in an equation format:

$$0.200 \text{ mol } C_4H_{10} \iff ? \text{ mol } CO_2$$

The key to solving this problem is finding the amount of carbon (in moles) in the sample of C_4H_{10}. Because all the carbon in the C_4H_{10} goes into the CO_2, we can then use this amount of carbon to figure out how much CO_2 is formed. The tools we will use are the subscripts in the formulas of C_4H_{10} and CO_2. Here's an outline of the path we will use to solve the problem.

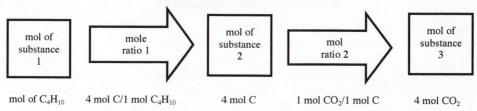

Assembling the Tools:

The tools we will use are the subscripts in the formulas of C_4H_{10} and CO_2 to determine the needed mole ratios.

Solution:

We use the tools to form conversion factors. As usual, we are careful to select the factors that allow the units to cancel.

$$0.200 \text{ mol } C_4H_{10} \times \frac{4 \text{ mol C}}{1 \text{ mol } C_4H_{10}} \times \frac{1 \text{ mol } CO_2}{1 \text{ mol C}} = 0.800 \text{ mol } CO_2$$

The calculation tells us that 0.800 mol of CO_2 would be formed.

Is the Answer Reasonable?

First check the setup by canceling terms in the *Solution* above. Now let's do some whole-number reasoning to see how the relative amounts of C_4H_{10} and CO_2 compare. If we had 1 mol C_4H_{10}, it would contain 4 mol C. But it takes only 1 mol of C to make 1 mol of CO_2, so 4 mol of C is enough to make 4 mol of CO_2. The conclusion, then, is that 1 mol of C_4H_{10} has enough carbon in it to make 4 mol of CO_2. Another way of looking at this is that the number of moles of CO_2 formed is four times the number of moles of C_4H_{10} that react. If we start with 0.200 mol C_4H_{10}, then the amount of CO_2 formed would be $4 \times 0.200 = 0.800$ mol. That's the answer we obtained.

In the preceding example, we worked entirely in moles. Translating between grams and moles is a task you learned how to do in Section 3.1. Study Examples 3.5 and 3.6 in the textbook to see how we are able to combine these two skills to solve more complex problems.

Relationships between masses of elements in a compound

There are times when we need to be able to calculate the mass of an element in a sample of a compound. For example, suppose we wished to know how many grams of sodium are in 3.58 g of Na_2CO_3. One way to do the calculation is the following:

1. Use the molar mass of Na_2CO_3 to calculate the moles of Na_2CO_3 in the sample.

2. Next, use the subscripts in Na_2CO_3 to calculate the moles of Na in the sample.

3. Finally, use the atomic mass of Na to calculate the grams of Na in the sample.

Now all we have to do is string these conversion factors together to calculate the answer. We start with

$$3.58 \text{ g } Na_2CO_3 \Leftrightarrow \text{? g Na}$$

We can draw the sequence of steps as before.

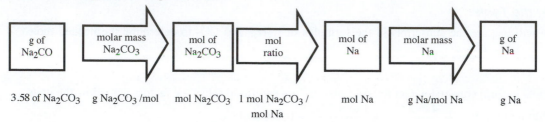

$$3.58 \text{ of Na}_2\text{CO}_3 \quad \text{g Na}_2\text{CO}_3 / \text{mol} \quad \text{mol Na}_2\text{CO}_3 \quad \frac{1 \text{ mol Na}_2\text{CO}_3 /}{\text{mol Na}} \quad \text{mol Na} \quad \text{g Na/mol Na} \quad \text{g Na}$$

Inserting the appropriate conversion factors we get

$$3.58 \text{ g Na}_2\text{CO}_3 \left(\frac{1 \text{ mol Na}_2\text{CO}_3}{106.0 \text{ g Na}_2\text{CO}_3} \right) \left(\frac{2 \text{ mol Na}}{1 \text{ mol Na}_2\text{CO}_3} \right) \left(\frac{22.99 \text{ g Na}}{1 \text{ mol Na}} \right) = 1.55 \text{ g Na}$$

Example 3.4 Using moles to deduce the atomic scale formula

Compound *X* was known to be either ammonia (NH_3), hydrazine (N_2H_4), or hydrazoic acid (HN_3). In a sample of 0.012 mol of compound *X* there was found to be 0.024 mol of N atoms and 0.048 mol of H atoms. What was compound *X*?

Analysis:

The chemical formula can be viewed on the atomic scale where we say that three atoms of hydrogen and one atom of nitrogen make up one ammonia molecule. Similarly two atoms of nitrogen and four atoms of hydrogen make up the hydrazine molecule. The ratio of hydrogen atoms divided by nitrogen atoms is three (3) for ammonia and two (2) for hydrazine. If we counted atoms by dozens, the three to one ratio of hydrogen to nitrogen in ammonia would be three dozen hydrogen atoms to one dozen nitrogen atoms. If we count in groups of 6.02×10^{23} (one mole) we would have three moles of hydrogen atoms for each mole of nitrogen atoms. For hydrazine the ratio is four moles of hydrogen to each two moles of nitrogen, which can be simplified to two moles of hydrogen for each mole of nitrogen. In any case, the ratio of moles of hydrogen to moles of nitrogen will reveal which of the two compounds we have.

Assembling the Tools:

The tools we will use are the subscripts in the formulas of NH_3 and N_2H_4 to determine the needed mole ratios.

Solution:

We need to calculate the ratio of moles of the hydrogen divided by moles of nitrogen. Since we are given the moles of each in the problem we just need to do a simple division.

$$\frac{0.048 \text{ mol H}}{0.024 \text{ mol N}} = 2.00 \text{ mol H/mol N}$$

This ratio tells us we have N_2H_4.

Is the Answer Reasonable?

The data given in the problem also allows us to calculate the moles of each element in one mole of substance. We are given 0.012 moles of compound. Dividing that into 0.024 mol N tells us there are 2 moles of N in the molecule. Dividing 0.012 moles of compound into 0.048 moles of N tells us there are 4 moles of N in the compound. These correspond to the formula for hydrazine.

Self-Test

20. How many mol of O and how many mol of H are present in 0.800 mol of H_2O?

_____

21. How many mol of Ca, O, and H are present in 0.600 mol of $Ca(OH)_2$?

_____

22. How many g of O are present in 0.800 mol of $CaSO_4 \cdot 5H_2O$?

_____

23. How many g of H are present in 0.600 mol of $Ca(OH)_2$?

_____

24. How many grams of phosphorus are needed to make 24.00 g of tetraphosphorus hexaoxide, P_4O_6? How many grams of O are needed?

_____

25. If a sample of $(NH_4)_3PO_4 \cdot 3H_2O$ contains 0.30 mol N, how many mol O does it contain?

_____

3.3   Chemical Formula and Percentage Composition

Learning Objective

To be able to calculate the percentage composition of a compound.

Review

Qualitative analysis is used in the lab to find out which elements are present in a substance. Then *quantitative analysis* is used to obtain their relative amounts, often expressed as the percentage by mass of each element. If a compound consists, say, of 60.26% carbon, then to make 100.00 g of the compound from its elements requires 60.26 *grams* of carbon. Another way of looking at a percentage like 60.26% carbon is that it is the number of grams of carbon that could be obtained by changing 100.00 g of the compound back into its elements.

We can find the *percentage by mass* of an element in a compound in two ways that involve two different kinds of data. One way is experimental; it uses the actual masses of the individual elements found in a weighed sample of the compound. Equation 3.1 in the text is used to perform the calculation.

The other method is theoretical; it uses the formula to calculate the percentages. The procedure is straightforward. Consider, for example, the compound Na_2CO_3. One mole of this substance contains two moles of sodium, one mole of carbon, and three moles of oxygen. We can use these amounts and the atomic masses of the elements to calculate the masses of each in one mole of the compound.

Element	Moles	Mass
Sodium	2 mol	2×22.99 g $= 45.98$ g
Carbon	1 mol	1×12.01 g $= 12.01$ g
Oxygen	3 mol	3×16.00 g $= 48.00$ g
	Total mass	105.99 g

Once we have the masses of each element and the formula mass, we can calculate the percentages by mass of each. Let's start with the percent sodium.

$$\boxed{\text{mass of sodium in 1 mol } Na_2CO_3}$$

$$\text{percent sodium} = \frac{45.88 \text{ g}}{105.99 \text{ g}} \times 100\% = 43.38\% \text{ Na}$$

$$\boxed{\text{mass of 1 mol } Na_2CO_3}$$

Similarly, for carbon and oxygen,

$$\text{percent carbon} = \frac{12.01 \text{ g}}{105.99 \text{ g}} \times 100\% = 11.33\% \text{ C}$$

$$\text{percent oxygen} = \frac{48.00 \text{ g}}{105.99 \text{ g}} \times 100\% = 45.29\% \text{ O}$$

$$\text{Sum of percentages} = 100.00\%$$

Calculating a percentage composition doesn't involve a great deal of reasoning. It's just a matter of learning how to do the calculation. It is helpful to remember that the sum of the percentages should add up to 100 % (although sometimes the calculated sum will be slightly more or slightly less than 100 % because of rounding during the calculations).

A list of the percentages by mass of a compound's elements makes up the compound's *percentage composition*. As Example 3.8 in the text demonstrates, a calculated percentage composition can be compared with one obtained from experimental analysis to check the identity of an unknown substance.

Self-Test

26. A 0.9278 g sample of a bright orange compound consists of 0.3683 g of chromium, 0.1628 g of sodium, and 0.3967 g of oxygen. Calculate the percentage composition.

27. An analysis of a sample believed to be photographers hypo, $Na_2S_2O_3$, gave the following percentage composition: Na, 29.10%; S, 40.50%; and O, 30.38%. Do these data correspond to the formula given? Use a separate sheet of paper to solve the problem.

3.4   Determining Empirical and Molecular Formulas

Learning Objective

To describe how to determine the empirical and molecular formulas of a given compound.

Review

In the previous section we saw how to determine the percentage composition of a compound. Only the chemical formula and the atomic masses of the elements were needed to obtain the answer. It seems reasonable then that the formula can be determined if the percentage composition and the atomic masses are known. That is true except that we can only determine the empirical formula (described below), not the complete molecular formula. The procedure takes advantage of the fact that the empirical formula is a simple ratio of the moles of each element in a compound.

Empirical and molecular formulas

When the smallest whole-number subscripts are used to describe the ratio of the atoms of different elements in a compound the resulting formula is an *empirical formula*. The empirical formula of caffeine, for example, is $C_4H_5N_2O$ but its *molecular formula*, the actual composition of one molecule, is $C_8H_{10}N_4O_2$.

To calculate an empirical formula, *the critical data needed are the masses of the elements in a sample of the compound*. These masses can be converted to moles, from which the ratios by moles can then be obtained. The ratios by moles have to be identical to the ratios by atoms, so the mole ratios give us the subscripts in the empirical formula.

Data to calculate an empirical formula can be obtained in several ways. A chemical analysis might give the masses directly, as illustrated in Example 3.9 in the text. Immediately after Example 3.9 is a discussion of how we handle situations in which whole-number subscripts are not obtained initially.

Sometimes the results of an analysis are provided as the percentages by mass of each of the elements. The percentages by mass of the elements in a compound are numerically equal to the *grams* of each element in 100 grams of the compound. We can therefore take the grams of each element in a 100 g sample (as given by the percentages), multiply each mass by a conversion factor obtained from the atomic masses to find the corresponding moles of each element in the sample. The moles are converted to mole ratios, from which we obtain the subscripts, as illustrated in Example 3.10.

In an indirect analysis, chemical reactions are carried out that separate the elements in a sample and capture them quantitatively (i.e., without loss) in compounds of known composition. An example is a combustion analysis, in which a compound containing carbon and hydrogen is burned to give CO_2 and H_2O. The experimental data obtained in this way are masses not of elements but of compounds, so the masses of C and H in the original sample have to be calculated from the masses of CO_2 and H_2O formed in the combustion reaction. This is easy because we know that 1 mol of CO_2 contains 1 mol C; and 1 mol H_2O

contains 2 mol of H.   In other words, the calculations are nothing more than grams-to-moles-to-grams calculations.

Example 3.5 Calculating an empirical formula from mass data

A compound was known to contain only potassium and oxygen. When a sample with a mass of 0.2564 g was analyzed, there was found to be 0.2128 g of potassium. How many grams of oxygen were also present? What is the empirical formula of the compound?

Analysis:

We are given the mass of sample, the mass of potassium in that sample, and the information that oxygen is the only other element that may be present. We can use the law of conservation of mass to calculate the mass of oxygen. Once the mass of oxygen is determined it can be used to determine the empirical formula of this compound as described above.

Assembling the Tools:

The tools we will use are the law of conservation of mass, molar mass, use of subscripts and the method for determining empirical formulas.

Solution:

The mass of oxygen can be calculated since

$$\text{mass of sample} = \text{mass of potassium} + \text{mass of oxygen}$$

and we know two of the three items in this equation.

$$\text{mass of oxygen} = 0.2564 \text{ g sample} - 0.2128 \text{ g potassium} = 0.0436 \text{ g O}$$

The empirical formula is the simplest whole-number ratio of the moles of each element in the compound. We calculate the moles of potassium and oxygen as

$$0.0436 \text{ g O} \times \frac{1 \text{ mol O}}{16.00 \text{ g O}} = 0.002725 \text{ mol O}$$

and

$$0.2128 \text{ g K} \times \frac{1 \text{ mol K}}{39.098 \text{ g K}} = 0.005443 \text{ mol K}$$

To obtain whole numbers we divide *both answers* by the smallest, 0.002725. The result is 1 mol O and 2 mol K and using these as subscripts a formula of K_2O is found.

Is the Answer Reasonable?

The whole numbers obtained for each element suggest we are right. In addition the formula is consistent with what we learned in Chapter 2 about ionic formulas. We expect potassium to form 1+ ions, K^+, and oxygen to form the 2− ion, O^{2-}. The formula of the compound made with these two ions is also K_2O.

Finding molecular formulas from empirical formulas

Often the experimentally measured molecular mass is not the same as the one we calculate from an empirical formula. It is some simple multiple of it, such as 2× or 3× or 4× and so forth. To find out which multiple, we divide the experimental molecular mass by the calculated empirical formula mass. That will give the *whole* number multiple we need (or one so close to a whole number we can round to it). Finally, multiply each subscript in the empirical formula by the whole number and the result is the molecular formula. For example, the true molecular mass of benzene is 78 and its empirical formula is CH (with a formula mass of 13). We simply divide 78 by 13 to get 6. This tells us that there are really 6 units of CH in each benzene molecule, and so the molecular formula is $C_{(1 \times 6)}H_{(1 \times 6)} = C_6H_6$.

Self-Test

28. Carotene, the pigment responsible for the color of carrots, has a percentage composition of 89.49% C and 10.51% H. Its molecular mass was found to be 546.9. Calculate its empirical formula and its molecular formula.

_____

29. When 0.8788 g of a liquid isolated from oil of balsam was burned completely, 2.839 g of CO_2 and 0.9272 g of H_2O were obtained. The molecular mass of the compound was found to be 136.2. Calculate the empirical and molecular formulas of this compound.

_____

3.5   Stoichiometry and Chemical Equations

Learning Objective

To perform calculations involving moles of reactants and products in a chemical reaction.

Review

In Section 2.4, we introduced you to the concept of a balanced equation and coefficients preceding chemical formulas. We're now ready to begin to write equations and balance them. As you proceed, keep in mind that writing and balancing equations should be viewed as two separate steps; first write the formulas for all the reactants and products, then adjust the coefficients to achieve balance. To obtain a *balanced equation*:

1. Be sure all the chemical formulas are correct—both the atomic symbols and the *subscripts*.

2. Remember that all atoms among the reactants must end up somewhere among the products and that a *coefficient* is a multiplier for all of the atoms in a formula.

3. Balance elements other than H and O first.

4. Balance as a group those polyatomic ions that appear unchanged on both sides of the arrow.

5. Be sure you have the smallest set of whole-number coefficients.

Example 3.6 Writing and balancing chemical equations

Pentanol ($C_5H_{11}OH$) is burned in oxygen to produce carbon dioxide and water. Write the chemical equation and balance it.

Analysis:

The statement of the problem indicates that pentanol and oxygen are reactants and that carbon dioxide and water are products of this reaction. We write the correct formulas for the reactants and products and then balance the equation. Recall that oxygen is a diatomic element.

Assembling the Tools:

The tool for balancing equations is appropriate here.

Solution:

We first construct the basic form of the chemical equation as

$$C_5H_{11}OH + O_2 \longrightarrow CO_2 + H_2O$$

$$C_5H_{11}OH + O_2 \longrightarrow 5CO_2 + H_2O$$

Next the hydrogen atoms are balanced.

$$C_5H_{11}OH + O_2 \longrightarrow 5CO_2 + 6H_2O$$

$$C_5H_{11}OH + 7.5O_2 \longrightarrow 5CO_2 + 6H_2O$$

We have a fractional coefficient, so we multiply all coefficients by 2 to eliminate the fraction.

$$2C_5H_{11}OH + 15O_2 \longrightarrow 10CO_2 + 12H_2O$$

Is the Answer Reasonable?

The best check is to be sure that we have the same number of each atom on each side of the arrow. Count the atoms to verify that the equation is balanced.

You've already learned that mole relationships among the elements in a compound are given by the subscripts in the formula. In this section we see that quantitative relationships among substances that react and form during chemical reactions are provided by the coefficients in balanced chemical equations.

For stoichiometry problems dealing with chemical reactions, *balanced* chemical equations are essential. This is because the coefficients in an equation provide the *only way* to relate the relative amounts of reactants and products. This relationship is on both a molecular and mole basis. Consider, for example, the balanced equation for the combustion of methanol (CH_3OH), the fuel in "canned heat" products such as Sterno®.

$$2CH_3OH + 3O_2 \longrightarrow 2CO_2 + 4H_2O$$

On a molecular basis, we see that 2 molecules of CH_3OH combine with 3 molecules of O_2 to form 2 molecules of CO_2 and 4 molecules of H_2O. Scaling up to laboratory amounts, we find that when 2 moles of CH_3OH

burn, they consume 3 moles of O_2 and form 2 moles of CO_2 and 4 moles of H_2O. Notice that the *mole ratios* among reactants and products are numerically equal to the ratios expressed in terms of the individual molecules of the reactants and products.

When we face a stoichiometry problem involving substances in a chemical reaction, the critical links are the coefficients of the balanced equation. These coefficients *alone* provide the relationships among moles of reactants and moles of products. For example, if we wished to relate the amounts of CO_2 and H_2O formed in the combustion of CH_3OH, we would use the coefficients of CO_2 and H_2O to establish the mole relationship.

$$H_2O \Leftrightarrow 2 \, mol \, CO_2$$

Similarly, if we wished to relate the amount of O_2 consumed when a certain amount of CO_2 is formed in the reaction, we would use the coefficients of O_2 and CO_2 in the equation to write

$$3 \, mol \, O_2 \Leftrightarrow 2 \, mol \, CO_2$$

Examples 3.14 through 3.16 in the text illustrate the principles above. Let's look closely at two more examples.

Example 3.7 Calculations using balanced chemical equations

In a gas barbecue, the combustion of propane, C_3H_8, follows the equation

$$C_3H_8(g) + 5O_2(g) \longrightarrow 3CO_2(g) + 4H_2O(g)$$

How many moles of water vapor are formed if 0.250 mol of C_3H_8 are burned?

Analysis:
Let's begin by stating the problem in equation form.

$$0.250 \, mol \, C_3H_8 \Leftrightarrow \, ? \, mol \, H_2O$$

To solve the problem we need to find a relationship between the amounts of these substances. Because the C_3H_8 and H_2O are involved in the combustion reaction, the tools we need are their coefficients in the balanced equation. The coefficients of C_3H_8 and H_2O give us the following:

$$1 \, mol \, C_3H_8 \Leftrightarrow \, 4 \, mol \, H_2O$$

This provides the connection we need to convert moles of C_3H_8 to moles of H_2O.

Assembling the Tools:

Our tools are the sequence of conversions needed for a stoichiometry question as well as the mol ratio tool and the equivalences obtained from a balanced chemical equation.

Solution:
To solve the problem we apply the mole relationship between C_3H_8 and H_2O as a conversion factor.

$$0.250 \, mol \, C_3H_8 \times \frac{4 \, mol \, H_2O}{1 \, mol \, C_3H_8} = 1.00 \, mol \, H_2O$$

We conclude that combustion of 0.250 mol C_3H_8 will produce 1.00 mol H_2O.

Is the Answer Reasonable?

This is a pretty simple problem, and you could probably see the answer without having to work it out in such great detail. The coefficients tell us that the number of moles of water formed is four times the number of moles of propane consumed. Therefore, when 0.250 mol C_3H_8 is burned, we will get 4×0.250 mol = 1.00 mol of water as one of the products.

Example 3.8   Calculations using balanced chemical equations

Magnesium hydroxide, $Mg(OH)_2$ (the creamy white substance in milk of magnesia), reacts with hydrochloric acid, HCl (found in stomach acid), to give magnesium chloride and water. The reaction is

$$Mg(OH)_2(s) + 2HCl(aq) \longrightarrow MgCl_2(aq) + 2H_2O$$

How many grams of $Mg(OH)_2$ are required to react completely with 12.8 g of HCl?

Analysis:

We'll begin by expressing the problem in the form of an equation.

$$12.8 \text{ g HCl} \Leftrightarrow ? \text{ g Mg(OH)}_2$$

We see that we are attempting to relate amounts of substances involved in a chemical reaction. This immediately suggests that we will need to use their coefficients in the balanced equation to relate them on a mole basis. Let's write that relationship.

$$1 \text{ mol Mg(OH)}_2 \Leftrightarrow 2 \text{ mol HCl}$$

We have part of the solution, but we are still missing some connections.

Assembling the Tools:

We will need conversion mole ratios implicit in the balanced equation and the molar masses of the compounds involved.

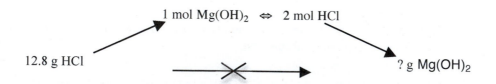

There is no direct path between grams of HCl and grams of $Mg(OH)_2$, but we do have the path between moles. To complete the connections between units, we need to relate grams to moles for both substances. By now you should realize that the tool to accomplish this is the molar mass. For HCl, the molar mass is 36.46 and for $Mg(OH)_2$ it is 58.32. Therefore, we can write

$$1 \text{ mol HCl} = 36.46 \text{ g HCl}$$
$$1 \text{ mol Mg(OH)}_2 = 58.32 \text{ g Mg(OH)}_2$$

Now we have a complete path from the units given (g HCl) to the units desired (g Mg(OH)$_2$), so we can proceed to the solution step.

Solution:

We have all the relationships we need, so we apply them as conversion factors, being sure the units cancel.

$$12.8 \text{ g HCl} \times \frac{1 \text{ mol HCl}}{36.46 \text{ g HCl}} \times \frac{1 \text{ mol Mg(OH)}_2}{2 \text{ mol HCl}} \times \frac{58.32 \text{ g Mg(OH)}_2}{1 \text{ mol Mg(OH)}_2} = 10.2 \text{ g Mg(OH)}_2$$

The calculations tell us that 12.8 g of HCl will require 10.2 g of Mg(OH)$_2$ for complete reaction.

Is the Answer Reasonable?

We can do some proportional reasoning and approximate arithmetic to get an idea whether our answer is "in the right ballpark." The equation tells us that 2 mol HCl are needed to react with 1 mol Mg(OH)$_2$. The mass of 2 mol HCl is approximately $2 \times 36 = 72$ g, and the mass of 1 mol of Mg(OH)$_2$ is approximately 60 g. Therefore, we expect that 72 g of HCl will react with approximately 60 g of Mg(OH)$_2$. In other words, the mass of Mg(OH)$_2$ consumed will be a little less than the mass of HCl that reacts. Our answer, 10.2 g Mg(OH)$_2$, is a little less than 12.8 g HCl, so the answer seems to be reasonable.

The preceding example demonstrates an important lesson about problems that deal with chemical reactions. The critical connections among the substances involved in a reaction are mole relationships provided by the coefficients of the balanced equation. Notice that this was the first relationship we established. Then we proceeded to find connections that take us from: (1) the given data to moles of the first substance, and (2) from moles of the second substance to the units of the desired answer. The following "flow diagram" shows the tools (arrows) we used to convert units (boxes) in solving the problem.

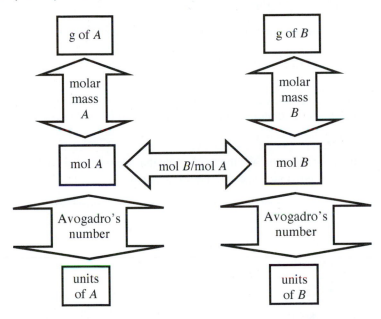

With practice, you will find that all stoichiometry problems involving chemical reactions are pretty much the same. The first step is to establish the balanced chemical equation. Then, using the coefficients, write the mole relationship between the substances of interest. After this, look for any additional connections you may need between moles and the units of the given data and/or moles and the units of the answer. Once you've reached this point, the hard work is done. All that's left is to assemble the relationships into conversion factors and calculate the answer.

Example 3.9   Using chemical equations to convert moles to moles

Sulfur dioxide will react with oxygen (O_2) under special conditions to form sulfur trioxide. How many moles of SO_2 are needed to react with 13.5 mol of O_2, and how many moles of SO_3 will form?

Analysis:

This problem asks us to convert moles of one substance to moles of another. This requires a balanced chemical equation so that we can construct valid conversion factors. Once we have the balanced chemical equation we need to apply the factor-label method to the given data to find the answers.

Assembling the Tools:

We need the mole ratios implicit in a balanced chemical equation.

Solution:

It is important to recall that no problem in reaction stoichiometry can be worked without a balanced equation, so the first step is to identify the products and reactants to set up the equation. Sulfur dioxide and oxygen are said to react with each other, they must be reactants and sulfur trioxide must be the product, then we write

$$SO_2 + O_2 \longrightarrow SO_3$$

Now we add coefficients to balance the equation.

$$2SO_2 + O_2 \longrightarrow 2SO_3$$

Now we know (from the coefficients) that the ratio of moles of SO_2 to moles of O_2 is 2:1. So,

$$13.5 \text{ mol } O_2 \times \frac{2 \text{ mol } SO_2}{1 \text{ mol } O_2} = 27 \text{ mol } SO_2$$

The coefficients tell us that for each 2 moles of SO_2 used, two moles of SO_3 will form; the mole ratio is 2:2 and can be simplified to 1:1. So 27.0 mol of SO_3 can form from 27.0 mol SO_2 calculated in the first part.

Is the Answer Reasonable?

We can check our balancing of the equation and the math (note that we start with 13.5 moles and multiply by 2 so the answer should be 27, and it is). Our logic and math seem okay.

Self-Test

30. Balance each of these equations.

(a) $Li + H_2 \longrightarrow LiH$ _____

(b) $Na + O_2 \longrightarrow Na_2O$ _____

(c) $Sr + O_2 \longrightarrow SrO$ _____

(d) $HCl + SrO \longrightarrow SrCl_2 + H_2O$ _____

(e) $HBr + Na_2O \longrightarrow NaBr + H_2O$ _____

(f) $Al + S \longrightarrow Al_2S_3$ _____

(g) $CH_4 + F_2 \longrightarrow CF_4 + HF$ _____

(h) $CO_2 + H_2 \longrightarrow CH_4 + H_2O$ _____

31.   Phosphorus burns according to the following equation.

$$4P + 5O_2 \longrightarrow P_4O_{10}$$

(a)   How many moles of O_2 react with 10.0 mol of P?

_____

(b)   How many moles of P_4O_{10} can be made from 2.00 mol of P?

_____

(c)   To make 4.00 mol of P_4O_{10}, how many moles of O_2 are needed?

_____

(d)   To make 3.60 mol of P_4O_{10}, how many moles of P are needed?

_____

How many moles of O_2 are needed?

_____

32.   Arsenic combines with oxygen to form As_2O_3 according to the following equation.

$$4As + 3O_2 \longrightarrow 2As_2O_3$$

(a)   To make 9.68 g As_2O_3, how many grams of As are needed?

_____

(b)   How many grams of O_2 are needed to make 8.92 g As_2O_3?

_____

(c)   If 5.85 g As are used, how many grams of As_2O_3 form?

_____

3.6   Limiting Reactants

Learning Objective

To explain how to determine the limiting reactant in a given reaction and the amount of substance remaining unreacted.

Review

When substances are combined for a chemical reaction, they are not always mixed together in just the right proportions to get complete reaction. As a result, sometimes one of the reactants will be completely used up before the rest, and once this happens, no further products can form. The reaction will simply come to a halt. At this point, the reaction mixture will contain the products of the reaction plus some of the reactant that had not been able to react. The following diagram illustrates this for the reaction of carbon with oxygen to form carbon dioxide.

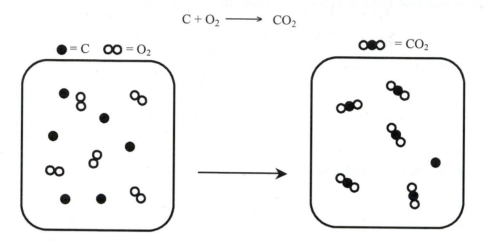

Notice that in both the before and after views, there are the same numbers of atoms of C and O. However, there are not enough oxygen atoms to combine with all the carbon, so after the oxygen is all used up, no more CO_2 can form. At the end of the reaction, there is still some unreacted carbon along with the CO_2 that has formed.

In this reaction mixture, we call the oxygen the **limiting reactant** because we don't have enough of it to consume all the carbon; its amount is what *limits* the amount of product (CO_2) that forms. We might call the carbon the **excess reactant**, because it is present in *excess* amount — that is, in an amount greater than is needed to react with all the oxygen.

If we wish to predict the amount of product in a reaction, we need to know which of the reactants is the limiting reactant. For example, consider the following mixture of C and O_2. How many molecules of CO_2 can form from this mixture?

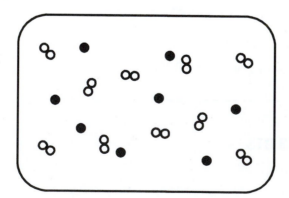

In this mixture, we have 10 molecules of O_2 and 8 atoms of C. It's pretty easy to see which is the limiting reactant. We can figure it out two ways (although we only need to do it one way if all we need to know is which is limiting). The reasoning we do here will be used again later, so be sure you understand it.

One way to find the limiting reactant is to examine the O_2. Each O_2 requires one C to form O_2, so 10 O_2 molecules would require 10 carbon atoms. We have only 8 C atoms, so C must be the limiting reactant. It will be the one used up first and its amount limits the amount of CO_2 that can form.

The second way is to look at the carbon. Each carbon requires one O_2 molecule. Therefore, 8 C atoms require 8 O_2 molecules. We have more than 8 O_2 molecules, so there is an excess of O_2. If O_2 is present in excess, then carbon must be the limiting reactant.

Notice that both approaches gave us the same answer, carbon is the limiting reactant. That means we only have to use one of the two approaches, not both.

Once we've identified the limiting reactant, we can figure out how much product will form and how much of which reactant is left over. *We always use the limiting reactant to determine how much product is formed.* When all 8 atoms of C react, they will yield 8 molecules of CO_2, so this is the amount of product that forms. There will be O_2 left over, and the amount left over is the difference between the amount of O_2 originally available (10 molecules) and the amount that reacts with the carbon (8 molecules). This difference, 2 molecules, is the amount of O_2 left over.

$$10 \text{ molecules } O_2 \quad - \quad 8 \text{ molecules } O_2 \quad = \quad 2 \text{ molecules } O_2$$

| Amount of O_2 available initially | | Amount of O_2 that reacts with C | | Amount of O_2 left over |

How to recognize a limiting reactant problem

There is a particular strategy we will employ in solving limiting reactant problems, but before we can use it, we must be able to recognize when a problem falls into this class. The key is in the way the problem is worded. The following are two stoichiometry problems that deal with the same chemical reaction, the reaction of $Mg(OH)_2$ with HCl.

$$Mg(OH)_2 + 2HCl \longrightarrow MgCl_2 + 2H_2O$$

Problem 1

How many grams of $MgCl_2$ can be formed when 2.48 g of $Mg(OH)_2$ reacts with HCl?

Problem 2

How many grams of $MgCl_2$ can be formed when 2.48 g of $Mg(OH)_2$ are combined with 2.88 g of HCl?

In Problem 1, we are not told about the amount of HCl available, so the only way we can solve the problem is to assume that there is more than enough HCl available. We would therefore base the calculation on the amount of $Mg(OH)_2$ given in the problem. Problem 1 is *not* a limiting reactant problem.

Problem 2 *is* a limiting reactant problem. *Notice that we are given amounts of both reactants;* this is the clue that identifies it as a limiting reactant problem. We can't tell, reading the problem, which one of the reactants will be completely used up, and therefore we don't know at this point which reactant to use to calculate the amount of product. To solve the problem, we must first identify the limiting reactant and then use the amount of that reactant to calculate the amount of product that can form.

Strategy for Solving Limiting Reactant Problems

The approach we will take in solving limiting reactant problems is as follows:

1. We will first determine the number of moles of both reactants.

2. We will select one reactant and calculate the amount of the other that is required for complete reaction. This will require the coefficients of the balanced equation.

3. We will compare the amount needed for complete reaction with the amount available and determine which of the reactants is limiting. We described this reasoning above in our example of the reaction of C with O_2 to form CO_2. Review it if necessary.

4. We will use the amount of limiting reactant to calculate the amount of product formed. We can also calculate the amount of the excess reactant that is actually consumed and, by difference, determine the amount of excess reactant left over.

With this as background, let's work on Problem 2, which we'll restate.

Example 3.10 Solving limiting reactant problems

How many grams of $MgCl_2$ can be formed when 2.48 g of $Mg(OH)_2$ are combined with 2.88 g of HCl?

$$Mg(OH)_2 + 2HCl \longrightarrow MgCl_2 + 2H_2O$$

Analysis:

You already know this is a limiting reactant problem, but in other situations, you should look for the clue that identifies it as a problem of this type — namely, that we are given amounts of both reactants. Then we proceed as described in the strategy described above. Recall that the most important points about limiting reactant problems are that the limiting reactant must be identified first and second, ALL calculations are based on the initial amount of the limiting reactant given in the problem.

Assembling the Tools:

We need all the tools for solving stoichiometry problems, mass to mole conversions, mole to mole conversions, and balancing equations as needed. We also need the tools for determining the limiting reactant:

Solution:

The first step is to determine the number of moles of the reactants that are available. We are given grams, so we convert to moles using the formula masses. The relationships are

$$1 \text{ mol HCl} = 36.46 \text{ g HCl}$$

$$1 \text{ mol Mg(OH)}_2 = 58.32 \text{ g Mg(OH)}_2$$

Applying them gives

$$2.48 \text{ g Mg(OH)}_2 \times \frac{1 \text{ mol Mg(OH)}_2}{58.32 \text{ g Mg(OH)}_2} = 0.0425 \text{ mol Mg(OH)}_2$$

$$2.88 \text{ g HCl} \times \frac{1 \text{ mol HCl}}{36.46 \text{ g HCl}} = 0.0790 \text{ mol HCl}$$

Next, we select one reactant and calculate the number of moles of the other required for complete reaction. For this step we need the coefficients of the balanced equation, from which we obtain

$$1 \text{ mol Mg(OH)}_2 \Leftrightarrow 2 \text{ mol HCl}$$

It doesn't matter which reactant we choose, so let's select the HCl and calculate how many moles of $Mg(OH)_2$ are needed to use it up.

Amount of HCl available		Amount of $Mg(OH)_2$ needed to use up all the HCl

$$0.0790 \text{ mol HCl} \times \frac{1 \text{ mol Mg(OH)}_2}{2 \text{ mol HCl}} = 0.0395 \text{ mol Mg(OH)}_2$$

Now we compare the amount of $Mg(OH)_2$ that's available in the reaction mixture (0.0425 mol) with the amount of $Mg(OH)_2$ that's required to react with all the HCl.

moles of $Mg(OH)_2$ available		moles of $Mg(OH)_2$ needed to react with all the HCl

$$0.0425 \text{ mol Mg(OH)}_2 \longleftrightarrow 0.0395 \text{ mol Mg(OH)}_2$$

Notice that there is more $Mg(OH)_2$ available than is needed to use up all the HCl. This means that $Mg(OH)_2$ is the reactant in excess and that some $Mg(OH)_2$ will be left over. It also means that *HCl must be the limiting reactant.*

Now that we've identified the limiting reactant, we can calculate the amount of $MgCl_2$ that can form. We use the moles of the limiting reactant, HCl, to calculate the moles of $MgCl_2$ that can form, and then convert that to grams.

Coefficients of HCl and $MgCl_2$ in the equation		Formula mass of $MgCl_2$

$$\text{moles HCl} - \boxed{\text{tool}} \longrightarrow \text{moles MgCl}_2 - \boxed{\text{tool}} \longrightarrow \text{grams MgCl}_2$$

The chemical equation gives us

$$2 \text{ mol HCl} \Leftrightarrow 1 \text{ mol MgCl}_2$$

and the formula mass of $MgCl_2$ gives us

$$1 \text{ mol MgCl}_2 = 95.2 \text{ g MgCl}_2$$

We use these and the amount of the limiting reactant to calculate the mass of $MgCl_2$ that could be formed in the reaction.

$$0.0790 \text{ mol HCl} \times \frac{1 \text{ mol MgCl}_2}{2 \text{ mol HCl}} \times \frac{95.2 \text{ g MgCl}_2}{1 \text{ mol MgCl}_2} = 3.76 \text{ g MgCl}_2$$

The results tell us that the maximum amount of $MgCl_2$ that could form in the reaction using these amounts of reactants is 3.76 g.

Is the Answer Reasonable?

There are a few things we can check without too much trouble. Let's start with the number of moles of $Mg(OH)_2$ and HCl in the reaction mixture. For $Mg(OH)_2$, 0.1 mol would weigh 5.8 g, so 2.5 g is about half of 0.1 mol, or 0.05 mol; the value we obtained (0.0425 mol) seems about right. Similarly, 0.1 mol HCl weighs about 3.6 g, so 2.88 g HCl is a little less than 0.1 mol. The value we obtained (0.0790 mol) seems okay.

We can check the limiting reactant conclusion by working with the $Mg(OH)_2$. If 0.0425 mol $Mg(OH)_2$ were to react, it would require twice as many moles of HCl, or 0.0850 mol HCl. We have only 0.0790 mol

HCl, so there isn't enough HCl to react with all the $Mg(OH)_2$. The reactant we "don't have enough of" is the limiting reactant, so once again, we conclude that HCl is the limiting reactant.

According to the coefficients in the equation, the number of moles of $MgCl_2$ that could form should equal half the number of moles of HCl that reacts. Half of 0.079 is about 0.04, so we expect about 0.04 moles of $MgCl_2$ could form. The formula mass of $MgCl_2$ is 95.2, so a mole of $MgCl_2$ weighs about 100 g. Therefore, 0.04 mol of $MgCl_2$ would weigh about 4 g. Our answer of 3.76 is not far from this, so the answer does seem to be reasonable.

In the preceding problem, we could also have been asked to calculate the amount of the excess reactant that would remain after the reaction is complete. This is a fairly simple calculation. We determined that the HCl would consume 0.0395 mol of $Mg(OH)_2$. We also know that we began with 0.0425 mol $Mg(OH)_2$. The difference between these two values is the amount of $Mg(OH)_2$ that will be left over.

Amount available in the reaction mixture	Amount that would react with the HCl	Amount left over

$$0.0425 \text{ mol } Mg(OH)_2 - 0.0395 \text{ mol } Mg(OH)_2 = 0.0030 \text{ mol } Mg(OH)_2$$

The amount of $Mg(OH)_2$ left over would be 0.0030 mol. If we wished to know its mass, we would multiply by the mass of a mole of $Mg(OH)_2$.

$$0.0030 \text{ mol } Mg(OH)_2 \times \frac{58.32 \text{ g } Mg(OH)_2}{1 \text{ mol } Mg(OH)_2} = 0.17 \text{ g } Mg(OH)_2$$

Example 3.11 Limiting reactant problem without a chemical reaction given

Aluminum and sulfur, when strongly heated, react to give aluminum sulfide, Al_2S_3. How many grams of Al_2S_3 can form from a mixture of 5.65 g of Al and 12.4 g of S?

Analysis:

We recognize this as a limiting reactant problem because the masses of *two* reactants are given. We are not given a balanced chemical reaction so that is our first step. We've discussed the general strategy on the preceding pages.

Assembling the Tools:

The balanced equation can be written from the information given. Once we have the balanced chemical equation we will need masses of Al, S and Al_2S_3 that are obtained from the periodic table. We then apply the strategies just discussed to find the limiting reactant and then to solve the problem.

Solution:

Our first step was to write the chemical equation and balance it. The result should be

$$2Al + 3S \longrightarrow Al_2S_3$$

We must convert the masses given into moles, using the grams-to-moles tool, so that we can compare the *mole* ratios of the reactants as they were taken.

$$5.65 \text{ g Al} = ? \text{ mol Al}$$

We need the atomic mass of Al, 26.98 g/mol

$$5.65 \text{ g Al} \times \frac{1 \text{ mol Al}}{26.98 \text{ g Al}} = 0.209 \text{ mol Al}$$

Doing the same kind of calculation on 12.4 g S (atomic mass 32.07) tells us that 12.4 g S = 0.387 mol S. Now we have to do the comparison. The mole ratio of Al to S is supposed to be 2 Al : 3 S. Arbitrarily beginning with Al, we can see how many moles of S are needed for complete reaction with 0.209 mol of Al.

$$0.209 \text{ mol Al} \times \frac{3 \text{ mol S}}{2 \text{ mol Al}} = 0.314 \text{ mol S}$$

More than 0.314 mol of S was taken, so all of the Al will be used up. Al limits the yield, so we calculate the mass of Al_2S_3 based on Al.

$$0.209 \text{ mol Al} \times \frac{1 \text{ mol Al}_2\text{S}_3}{2 \text{ Al}} = 0.104 \text{ mol Al}_2\text{S}_3$$

Finally, using a mole-to-mass tool—from the formula mass of Al_2S_3 (150.17) on 0.104 mol of Al_2S_3 would show that a mass of 15.6 g of Al_2S_3 can be obtained.

Is the Answer Reasonable?

In most mass-to-mass calculations the result is not smaller than 1/5 of the starting mass nor greater than 5 times the starting mass. Our result falls in this range and gives us some confidence. To be more sure, we can estimate the answers. In the equations above, round the data to one significant figure and we see that all estimated answers are close to the ones we calculated.

Self-Test

33. To prepare some $AlBr_3$, a chemist mixed 3.13 g of Al with 28.8 g of Br_2. The equation is

$$2Al + 3Br_2 \longrightarrow 2AlBr_3$$

(a) Identify the limiting reactant. _____

(b) How many grams of $AlBr_3$ can be made? _____

(c) Of the reactant that was taken in excess, how many grams of it are left over?

_____

3.7 Theoretical Yield and Percentage Yield

Learning Objective

To calculate the percentage yield of a given reaction.

Review

Experiments designed to prepare a substance do not always go "perfectly." The *main reaction* often has one or more *competing reactions,* sometimes called *side reactions,* that channel some of the reactants into the formation of one or more *by-products.* By products do not appear in the balanced equation for the main reaction. To see how well the experiment has gone, the chemist uses the balanced equation for the main reaction and the mass of the limiting reactant to calculate the "perfect" result, the *theoretical yield,* and then assess the success of forming the desired product in terms of a calculated *percentage yield.*

We can calculate a *percentage yield* of a product by the following equation:

$$\frac{\text{mass of product obtained in lab}}{\text{calculated(theoretical) mass of product}} \times 100 = \text{percentage yield}$$

The quantity in the numerator is called the *actual yield* and the quantity in the denominator is the theoretical yield. "Amount" may be in moles or in grams, but it must be in the same unit in both the numerator and denominator. The calculation of the theoretical yield *must* be based on the limiting reactant (Section 3.6).

Example 3.12 Percentage yield combined with a limiting reactant problem

Aluminum and bromine combine to give aluminum bromide. When 10.00 g of Al was mixed with 80.00 g of Br_2 there was obtained 80.00 g of $AlBr_3$. Calculate the percentage yield of $AlBr_3$.

Analysis:

If you have a problem getting started, note that the desired quantity is the percentage yield. That is one of our tools. From this equation we will see that the only thing that is not given is the calculated mass of product. Since we see the masses of two reactants given we'll get ready to solve a limiting reactant problem too.

$$\frac{\text{mass of product obtained in lab}}{\text{calculated(theoretical) mass of product}} \times 100 = \text{percentage yield}$$

From this we see that we know the mass obtained in the lab (80.00 g) and the only thing we need to calculate is the theoretical yield.

To determine the theoretical yield we need a balanced chemical equation to do the stoichiometric calculation. We also see, because a mass for both reactants is given, that this is a limiting reactant problem. We will use the techniques for a limiting reactant problem here.

Assembling the Tools:

The tool for the percentage yield is needed. Then we'll use the tools for writing and balancing equations, determining moles from masses, and the strategy for limiting reactant problems.

Solution:

First we write the percentage yield equation, which shows that we need to determine the theoretical yield in mass units.

$$\frac{\text{mass of product obtained in lab}}{\text{calculated(theoretical) mass of product}} \times 100 = \text{percentage yield}$$

We next need the balanced equation.

$$2Al + 3Br_2 \rightarrow 2AlBr_3$$

Because we're given the masses of *two* reactants, we have to determine which reactant is the limiting reactant using the following process.

We find the following moles of reactants using their formula masses.

$$10.00 \text{ g Al} = 0.3706 \text{ mol Al}$$

$$80.00 \text{ g Br}_2 = 0.5006 \text{ mol Br}_2$$

Since the equation's coefficients show that every two moles of Al react with three moles of Br_2; in other words, we need to multiply the given moles of Al by 1.5 to find out how many moles of Br_2 are needed. If we round 0.3706 mol Al up to 0.38 and multiply by 1.5 to get 0.57 mol Br_2. Rounding 0.3706 mol Al down to 0.36 mol gives us 0.54 mol. So 0.57 mol Br_2 is our upper estimate and 0.54 mol Br_2 is our lower estimate; in both cases we need more than the 0.5006 moles of Br_2 we were given. Our conclusion is that the Br_2 is used up before the aluminum and that Br_2 is the limiting reactant. Now all of our calculations from here on out will be based on the 80.00 g (0.5006 mol) of Br_2 that was given to us.

Now a stoichiometric calculation is done to see how much $AlBr_3$ we could obtain in theory from 0.5006 mol of Br_2.

$$0.5006 \text{ mol Br}_2 \times \frac{2 \text{ mol AlBr}_3}{3 \text{ mol Br}_2} = 0.3337 \text{ mol AlBr}_3$$

Because the formula mass of $AlBr_3$ calculates to be 166.68, we find the mass of $AlBr_3$ as follows.

$$0.3337 \text{ mol AlBr}_3 \times \frac{166.68 \text{ g AlBr}_3}{1 \text{ mol AlBr}_3} = 88.99 \text{ g AlBr}_3$$

Only 80.00 g $AlBr_3$ was produced by experiment, so the percentage yield is

$$\frac{80.00 \text{ g}}{88.99 \text{ g}} \times 100\% = 89.90\%$$

Is the Answer Reasonable?

All percentage yields must be less than 100% (remember, we cannot create matter). We can also reason that the minimum theoretical yield must be 80 g (the mass of product obtained in the lab). The maximum theoretical yield must be 90 g, the total mass of all reactants. From those numbers we see that the percentage yield must be between 100% (80/80 x 100) and 89% (80/90 x 100). Our result falls in that range and is reasonable.

Self-Test

34. Ethyl alcohol can be converted to diethyl ether (an anesthetic) by the following reaction.

$$2C_2H_5OH \longrightarrow C_4H_{10}O + H_2O$$
$$\text{ethyl alcohol} \qquad \text{diethyl ether}$$

Some of the diethyl ether is lost during its synthesis because it evaporates very easily. If 65.3 g of ethyl alcohol gives 48.2 g of diethyl ether, what is the percentage yield?

35. Solid sodium carbonate, Na_2CO_3, was sprinkled onto a spill of hydrochloric acid, HCl, to destroy the acid by the following reaction.

$$Na_2CO_3 + 2HCl \longrightarrow 2NaCl + H_2O + CO_2$$

If the spill contained 50.0 g of HCl (dissolved in water), was the HCl entirely destroyed by 50.0 g of Na_2CO_3? If not, how many grams of HCl were destroyed? If more Na_2CO_3 was used than necessary, then how many grams of it were in excess?

Answers to Self-Test Questions

1. Formula mass = 203.14 $\dfrac{1 \text{ mol } (NH_4)_3PO_4 \cdot 3H_2O}{203.14 \text{ g } (NH_4)_3PO_4 \cdot 3H_2O}$ and $\dfrac{203.14 \text{ g } (NH_4)_3PO_4 \cdot 3H_2O}{1 \text{ mol } (NH_4)_3PO_4 \cdot 3H_2O}$

2. 58.1 u
3. 40.078 g Ca
4. 0.0323 mol P
5. 0.609 moles
6. 6 molecules CO_2
7. 1.22×10^{23} molecules

8. $\dfrac{6.02 \times 10^{23}\,\text{molecules NH}_3}{17.0\,\text{g NH}_3}$ and $\dfrac{17.0\,\text{g NH}_3}{6.02 \times 10^{23}\,\text{molecules NH}_3}$

9. 49.04 g H_2SO_4

10. 2.50 mol CO_2

11. 58.1 g $C_{12}H_{12}N_2O_3$

12. 3.75×10^{-3} mol C_6HCl_5O

13. 1.90×10^{23} atoms Ag

14. 2.41×10^{23} atoms Fe

15. 5.60×10^{21} atoms Fe

16. 1.51×10^{25} atoms H

17. 9.36×10^{-21} g Na

18. 2.53×10^{21} atoms U

19. 1.7×10^{21} molecules H_2O

20. 0.800 mol O; 1.60 mol H

21. 0.600 mol Ca; 1.20 mol O; 1.20 mol H

22. 115 g O

23. 1.21 g H

24. 13.52 g P; 10.48 g O. (Note the sum of these; it's 24.00 g.)

25. 0.70 mol O

26. 39.70% Cr; 17.55% Na; 42.76% O

27. The formula mass of $Na_2S_2O_3$ is 158.12. The calculated percentages are Na, 29.08; S, 40.56; and O, 30.36, which are very close to the experimental percentages, so the formula is correct.

28. empirical formula C_5H_7, molecular formula $C_{40}H_{56}$

29. empirical formula C_5H_8, molecular formula $C_{10}H_{16}$

30. (a) $2Li + H_2 \longrightarrow 2LiH$

    (b) $4Na + O_2 \longrightarrow 2Na_2O$

    (c) $2Sr + O_2 \longrightarrow 2SrO$

    (d) $2HCl + SrO \longrightarrow SrCl_2 + H_2O$

    (e) $2HBr + Na_2O \longrightarrow 2NaBr + H_2O$

    (f) $2Al + 3S \longrightarrow Al_2S_3$

    (g) $CH_4 + 4F_2 \longrightarrow CF_4 + 4HF$

    (h) $CO_2 + 4H_2 \longrightarrow CH_4 + 2H_2O$

31. (a) 12.5 mol O_2, (b) 0.500 mol P_4O_{10}, (c) 20.0 mol O_2
    (d) 14.4 mol P, 18.0 mol O_2

32. (a) 7.33 g As, (b) 2.16 g O_2, (c) 7.72 g As_2O_3

33. (a) Al limits, (b) 30.9 g $AlBr_3$
    (c) 1 g Br_2 left over (rounded from 0.959 g)

34. 91.8%

35. 34.5 g HCl were destroyed

Tools for problem solving

In this chapter you learned to apply the following concepts as tools in solving problems. Study each one carefully so that you know what each is used for. When faced with solving a problem, recall what each tool does and consider whether it will be helpful in finding a solution. This will aid you in selecting the tools you need. In this chapter we see that many of the tools from Chapters 1 and 2 must be used with the new tools from this chapter.

You may wish to tear out these pages to use while solving problems.

Atomic mass (Section 3.1)

$$\text{gram atomic mass of } X = \text{molar mass } X = 1 \text{ mole } X$$

Molecular mass and formula mass (Section 3.1)

Used to form a conversion factor to calculate mass from moles of a compound, or moles from the mass of a compound.

$$\text{gram molecular mass of } X \; = \; \text{molar mass } X \; = \; 1 \text{ mole } X$$
$$\text{gram formula mass of } X \; = \; \text{molar mass } X \; = \; 1 \text{ mole } X$$

Molar mass (Section 3.1)

This is a general term encompassing atomic, molecular and formula masses. All are the sum of the masses of the elements in the chemical formula.

$$\text{molar mass of } X \; = \; 1 \text{ mole of } X$$

Avogadro's number (Section 3.1)

$$1 \text{ mole } X \; = \; 6.02 \times 10^{23} \text{ particles of } X$$

Mole ratios (Section 3.2)

Subscripts in a formula establish atom ratios and mole ratios between the elements in the substance.

Mass-to-mass conversions using formulas (Section 3.2)

These steps are required for a mass-to-mass conversion problem using a chemical formula; also see Figure 3.6.

Percentage composition (Section 3.3)
This describes the composition of a compound and can be the basis for computing the empirical formula.

$$\text{percent of } X = \frac{\text{mass of } X \text{ in the sample}}{\text{mass of the entire sample}} \times 100$$

Empirical formula (Section 3.4) The empirical formula expresses the simplest ratio of the atoms of each element in a compound.

Determining integer subscripts (Section 3.4)
When determining an empirical formula, dividing all molar amounts by the lowest value often normalizes subscripts to integers. If decimals remain, multiply the subscript by a small whole number until integer subscripts result.

Empirical formulas from percentage compositions (Section 3.4)
Percentage composition correlates information from different experiments, to determine empirical formulas.

Balancing chemical equations (Section 3.5)
Balancing equations involves writing the unbalanced equation and then adjusting the coefficients to get equal numbers of each kind of atom on both sides of the arrow.

Equivalencies obtained from balanced equations (Section 3.5)

The coefficients in balanced chemical equations give us relationships between all reactants and products that can be used in factor-label calculations.

Mass-to-mass calculations using balanced chemical equations (Section 3.5)

A logical sequence of conversions allows calculation of all components of a chemical reaction. See Figure 3.6.

Limiting reactant calculations (Section 3.6)
When the amounts of at least two reactants are given, one will be used up before the other and dictate the amount of products formed and the amount of excess reactant left over.

Theoretical, actual, and percentage yields (Section 3.7)

The theoretical yield is based on the limiting reactant whether stated, implied, or calculated. The actual yield must be determined by experiment and the percentage yield relates the magnitude of the actual yield to the percentage yield.

$$\text{percentage yield } = \frac{\text{actual mass by experiment}}{\text{theoretical mass by calculation}} \times 100\%$$

Multi-step percentage yield (Section 3.7)

Modern chemical synthesis often involves more than one distinct reaction step. The overall percentage yield of a multi-step synthesis is

$$\text{Overall \% yield } = \left(\frac{\% \text{ yield}_1}{100} \times \frac{\% \text{ yield}_2}{100} \times ... \right) \times 100\%$$

Notes:

Chapter 4

Molecular View of Reactions in Aqueous Solutions

Solutions of compounds are often used for carrying out chemical reactions, and we begin this chapter with the basic terms used to describe solutions. The principal focus of the chapter is on reactions that involve ionic substances as well as substances that form ions when dissolved in water. The latter are usually acids or bases, and we introduce you here to these important kinds of chemicals and the reactions they undergo with each other. You will learn about several classes of reactions for which it is possible to predict the products. These are common reactions often encountered in the laboratory, so you should learn how to deal with them.

To deal with the stoichiometry of reactions in solutions, the concentration unit *moles per liter* or *molarity* is studied. If you know a solution's molarity, you can obtain the number of moles of its solute required for an experiment by dispensing a certain volume of the solution instead of weighing the reactant. You will learn how to prepare solutions of a desired molarity, how to dilute solutions quantitatively, and how to use molarity in stoichiometric calculations when substances are in solution.

4.1   Describing Solutions

Learning Objective

To describe solutions qualitatively and quantitatively.

Review

The chief reason that reactions are carried out using solutions of the reactants is to let their molecules or ions have the freedom to move around and find each other. Several new terms are given in this section for describing solutions. Be sure to learn their meanings, because they will be used often in the classroom and the lab.

Self-Test

1.   When sugar dissolves in water, which is the *solvent* and which is the *solute*?

Solvent _____   Solute _____

2. When a crystal of sodium acetate was added to a solution of sodium acetate in water, the crystal dissolved. What term would be used to describe this solution (saturated, unsaturated, or supersaturated)?

_____

3. When a crystal of sodium acetate was added to a different solution of sodium acetate in water, considerable crystalline sodium acetate separated from the solution. What term should be used to describe the sodium acetate solution before the crystal was introduced?

_____

4.2 Electrolytes and Nonelectrolytes

Learning Objective

To understand that electrolytes are soluble ionic compounds and nonelectrolytes do not produce ions in solutions.

Review

Ionic compounds *dissociate* (break apart) into their individual ions when they dissolve in water. These ions become surrounded by water molecules and are said to be *hydrated*. Because the solute yields charged particles in the solution, the solution conducts electricity, which is why such solutes are called *electrolytes*. Ionic compounds are essentially 100% dissociated in water, so their solutions conduct electricity well and ionic compounds are called strong electrolytes. Even ionic compounds that have very low solubilities in water are called strong electrolytes because they are effectively completely dissociated.

Compounds such as sugar that are not able to undergo dissociation and do not yield ions in solution are called *nonelectrolytes*; their solutions do not conduct electricity. You should be sure that you can write equations for the *dissociation* of an ionic compound in water. To do this, it is necessary that you know the formulas and the charges of the ions. If the formulas of polyatomic ions are still difficult to remember, review them in Table 2.4 in the text.

In studying this section, notice how the formula of the salt determines the number of each kind of ion that is found in the solution. For example, the dissociation of chromium(III) sulfate, $Cr_2(SO_4)_3$, in water produces two Cr^{3+} ions and three SO_4^{2-} ions for each formula unit of the salt.

$$Cr_2(SO_4)_3(s) \longrightarrow 2Cr^{3+}(aq) + 3SO_4^{2-}(aq)$$

Self-Test

4. What are the formulas of the ions that would be found in aqueous solutions of the following salts?

(a) $AgC_2H_3O_2$ _____

(b) $(NH_4)_2Cr_2O_7$ _____

(c) $Ba(OH)_2$ _____

5. Write chemical equations for the dissociation of the compounds in Question 4 when they are dissolved in water.

(a) _____

(b) _____

(c) _____

4.3   Equations for Ionic Reactions

Learning Objective

To write balanced molecular, ionic, and net ionic equations.

Review

Reactions between ionic compounds in aqueous solutions are really reactions between their ions because ionic compounds exist in solution in the form of ions. There are three ways that we normally represent these reactions by chemical equations. A *molecular equation* shows complete formulas for all reactants and products. Generally we indicate whether the substances are soluble or insoluble by writing (*aq*) or (*s*) following their formulas. In an *ionic equation*, we write the formulas for all soluble strong electrolytes in dissociated form. In a *net ionic equation*, we show only those ions that are actually involved in the chemical reaction. We obtain the net ionic equation by dropping *spectator ions,* those that do not react, from the ionic equation.

As an example, consider the reaction of solutions of silver nitrate ($AgNO_3$) and sodium chloride ($NaCl$) to give solid silver chloride ($AgCl$) and a solution that contains sodium nitrate ($NaNO_3$). Here are the molecular, ionic, and net ionic equations for the reaction. Note that we use (*s*) to indicate a solid in chemical equations.

Molecular Equation:

$$AgNO_3(aq) + NaCl(aq) \longrightarrow AgCl(s) + NaNO_3(aq)$$

Ionic Equation:

$$Ag^+(aq) + NO_3^-(aq) + Na^+(aq) + Cl^-(aq) \longrightarrow AgCl(s) + Na^+(aq) + NO_3^-(aq)$$

Net Ionic Equation:

$$Ag^+(aq) + Cl^-(aq) \longrightarrow AgCl(s)$$

Notice that to obtain the net ionic equation, we have dropped the Na^+ and NO_3^- ions from the ionic equation. Because these ions do not change during the reaction, they are called spectator ions.

When you write ionic and net ionic equations, be sure you follow the two criteria for having them balanced: (1) There must be the same number of atoms of each kind on both sides of the equation, and (2) the net electrical charge must be the same on both sides.

Example 4.1 Writing ionic and net ionic equations

Write the balanced molecular, ionic, and net ionic equations based on the unbalanced equation

$$FeCl_3(aq) \ + \ Pb(NO_3)_2(aq) \longrightarrow Fe(NO_3)_3(aq) \ + \ PbCl_2(s)$$

Analysis:

We see that balancing the equation at the start should also result in balanced equations for the ionic and net ionic equations too. Balancing here will be done by counting the ions and adding coefficients to equalize the numbers of each ion on both sides of the equation, (balancing by inspection). Next we need to write the ions for each substance in solution (those that have the (*aq*) symbol). Each substance will have a positive and a negative ion and subscripts will tell us how many of each ion we have. Once we have all of the ions for the ionic equation we then need to identify and cancel spectator ions. When identifying spectator ions, they must be absolutely identical or they do not cancel.

Assembling the Tools:

First we need to use the criteria for a balanced reaction. Next, the tools for constructing formulas from ions will be needed, but we will reverse the process to deduce the ions present. Additionally we use the principles for writing compounds as ions; that is, solids, nonelectrolytes, weak electrolytes, and gases are always written in their molecular format. Finally, the process just described for writing ionic and net ionic equations will be needed.

Solution:

When we count the atoms we end up with the balanced equation below

$$2FeCl_3(aq) \ + \ 3Pb(NO_3)_2(aq) \longrightarrow 2Fe(NO_3)_3(aq) \ + \ 3PbCl_2(s)$$

We will break all of the substances except $PbCl_2$ into ions. For $FeCl_3$ we have $3Cl^-$ and therefore the iron ion must be Fe^{3+}. Since the coefficient for $FeCl_3$ is two, we have a total of $6Cl^-$ and $2Fe^{3+}$ ions for our ionic equation. Similarly, for $Pb(NO_3)_2$ we obtain two nitrate ions $2NO_3^-$ ions and the lead ion must be Pb^{2+}, and these are multiplied by three for the ionic equation. We get one Fe^{3+} and three NO_3^- for each $Fe(NO_3)_3$ and will have $2Fe^{3+}$ and $6NO_3^-$ on the product side of the reaction. $PbCl_2$ does not ionize since it is a solid. Putting it all together we have the ionic equation

$$\boxed{2Fe^{3+}(aq)} + 6Cl^-(aq) + 3Pb^{2+}(aq) + \boxed{6NO_3^-(aq)} \longrightarrow \boxed{2Fe^{3+}(aq)} + \boxed{6NO_3^-(aq)} + 3PbCl_2(s)$$

Our next task is to identify and cancel identical numbers of identical ions from the reactant and product sides of the reaction. We see the ovals around $2Fe^{3+}(aq)$ and the rectangles around the $6\ NO_3^-$ (*aq*) ions are spectator ions. Canceling them will leave us with

$$3Pb^{2+}(aq) \ + \ 6Cl^-(aq) \longrightarrow 3PbCl_2(s)$$

We now simplify this equation by dividing each coefficient by three to get the net ionic equation.

$$Pb^{2+}(aq) \ + \ 2Cl^-(aq) \longrightarrow PbCl_2(s)$$

Is the Answer Reasonable?

The easiest check of an ionic equation is to see if the charges are balanced, and they are. The second check is to recount each atom to be sure that there are equal numbers as reactants as there are as products.

Self-Test

6. Balance the following molecular equations and then write their ionic and net ionic equations:

    (a) ___ $Cu(NO_3)_2(aq)$ + ___ $KOH(aq)$ $\longrightarrow$ ___ $Cu(OH)_2(s)$ + ___ $KNO_3(aq)$

    Ionic equation:

    _____

    Net ionic equation:

    _____

    (b) ___ $NiCl_2(aq)$ + ___ $AgNO_3(aq)$ $\longrightarrow$ ___ $Ni(NO_3)_2(aq)$ + ___ $AgCl(s)$

    Ionic equation:

    _____

    Net ionic equation:

    _____

    (c) ___ $Cr_2(SO_4)_3(aq)$ + ___ $BaCl_2(aq)$ $\longrightarrow$ ___ $CrCl_3(aq)$ + ___ $BaSO_4(s)$

    Ionic equation:

    _____

    Net ionic equation:

    _____

4.4   Introducing Acids and Bases

Learning Objective

To identify acids and bases based on their names and formulas and write balanced equations showing hydronium and hydroxide ions.

Many substances can be categorized as acids or bases because they possess certain characteristic properties and because they undergo predictable reactions with each other called *neutralization*. Be sure to review the general properties of acids and bases described in this section in the textbook.

According to the description of acids and bases presented in this chapter (a modern version of the Arrhenius definition), an *acid* is a substance that produces H_3O^+ by transferring H^+ ions to water molecules. A base produces OH^- in water. The neutralization reaction of an acid and a base generally produces water plus a *salt*. We use the term salt to mean any ionic compound not containing OH^- or O^{2-}.

In general, acids are molecular substances. Their reaction with water is called an *ionization reaction* because ions are formed from neutral molecules—before the reaction there are no ions and afterwards there are. Study the general equation for the reaction given in the text. Acids are classified according to the number of hydrogens of the parent molecule that are capable of ionizing (e.g., monoprotic, diprotic, triprotic, polyprotic). Polyprotic acids ionize stepwise, losing one hydrogen ion at a time. Some oxides of nonmetals, such as SO_3, N_2O_5, and CO_2, react with water to give acid molecules, which subsequently undergo ionization in water. Such oxides are called *acid anhydrides*, meaning "acids without water."

There are two kinds of Arrhenius bases. One consists of the metal hydroxides—compounds such as NaOH and $Ca(OH)_2$. These are ionic substances and simply dissociate in water to give the metal ion and the hydroxide ion. Because they are ionic compounds, they are completely dissociated. Soluble metal oxides react with water to form hydroxides and are said to be *base anhydrides*. (Notice that nonmetal oxides are acidic anhydrides and metal oxides are base anhydrides; this is a chemical distinction between metals and nonmetals.)

The second kind of base consists of molecules that react with water and release OH^- ions. An example is ammonia. When a molecular base ionizes, an H^+ ion is transferred from a water molecule to a molecule of the base. Study the general equation for the reaction of a molecular base with water illustrated in the text.

Strong electrolytes are substances that are completely divided into ions in aqueous solution. Salts are examples. Some acids are also strong electrolytes and are classified as *strong acids*; they are 100% ionized in water. A list of seven strong acids is tabulated in the text. You should memorize them because if you encounter an acid and it's *not* on the list, you can be fairly confident that it is *not* a strong acid. When strong acids dissolve in water a single arrow is used in the equation.

The *strong bases* are the soluble metal hydroxides. These are the Group IA hydroxides and the hydroxides of the Group IIA metals from calcium on down. (Actually, you don't need to memorize them because they are covered by the solubility rules in Table 4.1 in Section 4.6.) When strong bases are dissolved in water a single arrow is used in the equation. Metal hydroxides with low solubilities, such as $Mg(OH)_2$, are also strong electrolytes, but their solutions contain so little hydroxide ion that they are not strongly basic.

For a *weak acid*, only a small fraction of the molecules are ionized. Most of the weak acid is present in the solution as non-ionized molecules, and there is an equilibrium between the ions of the acid and its molecules. Similarly, in solutions of *weak bases* (which are the molecular bases), only a small fraction of the base exists in ionized form. Most of the base exists in solution as molecules. Double arrows are generally used for equations representing a dynamic equilibrium for instance, the ionization of weak acids or bases.

Dynamic equilibrium in solutions of weak acids and bases

When describing the breakup into ions of weak acids or bases, an equation with oppositely pointing arrows, $\rightleftarrows$, is usually used to describe the *dynamic equilibrium* in the solution. The *forward reaction* (going from left to right) is the ionization and the *reverse reaction* (going right to left) is the reconversion of the ions to the un-ionized or undissociated species. The double arrows symbolize the dynamic nature of an equilibrium—both forward and reverse reactions occur in the solution continuously and at equal rates when there is equilibrium. But there is no net change in the concentrations of either the ions or the un-ionized acid or base. For a weak acid or base, the extent to which the forward reaction proceeds toward completion is small and we say that the *position of equilibrium* lies to the left in the ionization reaction. (For a strong acid, the ionization reaction is essentially complete, so if we were to look at the reaction as an equilibrium, the position of equilibrium would lie far to the right.)

Example 4.2 Acid and base properties

For the following, provide the requested answer and justify your decisions based on acid-base concepts.

1. A white solid readily dissolves in water to form a solution that turns red litmus paper blue. Which of the following compounds could this solid be?

CO_2, Na_2O, HNO_3, KOH, SO_3, $HC_2H_3O_2$

2. Consider the oxide of an element in group IIA of the periodic table. Let Z be the symbol of this element. (a) Write the formula of this oxide using the symbol Z. (b) Will a solution of this oxide in water give a blue or a red color to litmus paper?

3. Is butanoic acid a strong or a weak acid? How can you tell without even knowing its formula and without having access to a table of data?

Analysis:

Each of the three questions relate to the nature of acidic or basic compounds. In the first part the substance changes litmus blue and we learned that bases turn litmus blue. We need to determine which of the compounds given are bases that are also soluble white solids. Next we consider an oxide of any of the Group 2 elements. How do they react with water as a group and are they acids or bases? Finally we can identify butyric acid, not by knowing what acids are weak, but by knowing the small group that are strong acids.

Assembling the Tools:

We will need tools from previous chapters concerning the expected state of common compounds and relationships in the periodic table, particularly expected charges of monatomic ions. Finally we can take advantage of the table of strong acids to determine if an acid is weak or strong.

Solution:

For Question 1, we know that CO_2 and SO_3 are not white solids (they are gases) and would be acids (H_2CO_3 and H_2SO_4) if dissolved in water. HNO_3 and $HC_2H_3O_2$ are acids, leaving only KOH and Na_2O. KOH is an obvious base and Na_2O is the oxide of a metal that results in $NaOH$ when dissolved in water. Both are soluble, white solids that will turn pink litmus paper blue.

For Question 2, we see that Group II elements form ions with a 2+ charge. A formula of ZO (or CaO, BaO, SrO, etc.) is expected. The reaction of a metal oxide with water is

$$ZO + H_2O \longrightarrow Z(OH)_2$$

And we recognize $Z(OH)_2$ as a base that will turn pink litmus paper blue.

The third question is answered by a process of elimination. You should be able to recognize the short list of strong acids. Butanoic acid is not one of the strong acids, therefore butanoic acid must be a weak acid.

Are the Answers Reasonable?

The only real check is to review our logic to be sure there are no flaws in interpreting fundamental concepts.

Self-Test

7. Write chemical equations for the ionization of the following acids in water.

    (a) $HClO_2$ _____

    (b) H_2SeO_3 _____

    _____

    (c) H_3AsO_4 _____

    _____

    _____

8. Tetraphosphorus decaoxide is the acid anhydride of phosphoric acid, H_3PO_4. Write the chemical equation showing the reaction of this oxide with water to form the acid.

    _____

9. Write an equation for the ionization reaction of the base hydrazine, N_2H_4, in water.

    _____

10. The radioactive element radium forms a soluble oxide with the formula RaO. Write an equation for its reaction with water.

    _____

11. Why is Na_2O called a *base anhydride*? Write the equation for its reaction with water.

    _____

    _____

12. Why is SO_3 called an *acid anhydride*? Write the equation for its reaction with water.

    _____

    _____

13. Hydrogen cyanide, HCN, is a weak acid when dissolved in water. Write the chemical equation that illustrates the equilibrium that exists in the aqueous solution.

    _____

    Which reaction, the forward or the reverse, is far from completion?

    _____

14. Chloric acid, $HClO_3$ is a strong acid. Write a chemical equation that shows the reaction to this acid with water.

    _____

15. Aniline, $C_6H_5NH_2$, is a weak base in water. Write a chemical equation that illustrates the equilibrium that exists in aqueous aniline solutions.

   _____

16. From memory, write the names and the formulas of the seven strong acids.

   _____

   _____

4.5　Acid–Base Nomenclature

Learning Objective

Learn how to translate formulas to names and names to formulas for common acids and bases.

Review

In this section, we extend the system of chemical nomenclature introduced in Chapter 2 to cover acids and bases.

Binary acids and their salts

The binary acids are the binary hydrogen compounds of elements in Groups VIA and VIIA. Examples include hydrogen chloride, HCl, and hydrogen sulfide, H_2S.

$$HCl(g) + H_2O \longrightarrow H_3O^+(aq) + Cl^-(aq)$$

$$H_2S(g) + H_2O \rightleftharpoons H_3O^+(aq) + HS^-(aq)$$

These two acids are called binary acids because they contain only one element besides hydrogen, and so are binary compounds in their pure states. In naming the acids, we add the prefix *hydro-* and the suffix *-ic acid* to the stem of the nonmetal name. Thus, when dissolved in water, hydrogen chloride becomes *hydro*chloric *acid* and hydrogen sulfide becomes *hydro*sulfuric *acid*.

As mentioned earlier, acids react with bases in a reaction called neutralization. The products are water and an ionic compound. Salts formed by neutralizing binary acids contain a monatomic anion of the acid and have the *-ide* ending. Thus, hydro*chlor*ic acid gives *chlor*ide salts.

Oxoacids and their salts

An oxoacid contains hydrogen, oxygen, plus a third element. Examples include HNO_3 (nitric acid) and H_2SO_4 (sulfuric acid). Many nonmetals form more than one oxoacid. If only two of them are possible, the name of the one having the larger number of oxygen atoms ends in *-ic* and the name of the acid with the smaller number of oxygens ends in *-ous*. The following examples were given in the text.

H_2SO_4	sulfuric acid	HNO_3	nitric acid
H_2SO_3	sulfurous acid	HNO_2	nitrous acid

When there are more than two oxoacids for a given nonmetal (as there are for the halogens), the following prefixes are used (in the order of increasing numbers of oxygen atoms).

hypo ... ous acid	(For example, hypo*chlor*ous acid — HClO)
... ous acid	(For example, *chlor*ous acid — $HClO_2$)
... ic acid	(For example, *chlor*ic acid — $HClO_3$)
per ... ic acid	(For example, per*chlor*ic acid — $HClO_4$)

Remember that ...*ic* acids give anions that end in -*ate*, and that ...*ous* acids give anions that end in -*ite*.

Acid salts

Monoprotic acids such as HCl have just one hydrogen to be neutralized, so they form only one kind of salt. However, polyprotic acids, such as H_2SO_4 and H_3PO_4, have more than one hydrogen to be neutralized, and the neutralization can be incomplete to give anions that still contain "acidic hydrogens." Compounds formed by these anions are called acid salts. In naming them, the hydrogen is specified either as "hydrogen" or, when only one acid salt is possible, by the prefix *bi-* before the name of the anion (for example, HSO_4^- is named as hydrogen sulfate ion or bisulfate ion).

Bases

Ionic bases contain hydroxide ions and are named just as any other ionic compound by specifying the cation first, followed by the anion (hydroxide). Molecular bases (nitrogen bases related to ammonia) are named just by giving the name of the molecule.

Example 4.3   Names for formulas and formulas from names

Using the principles of nomenclature in this section, name HBr, HIO_3, and H_2S. Similarly give the formulas for bromic acid, hydroselenic acid, and nitrous acid.

Analysis:

The first step is to determine if the formula of the acid classifies it as a binary or oxoacid. For oxoacids of the halogens, we recall the four chlorine-containing oxoacids and then relate those formulas to those that contain bromine or iodine in place of the chlorine. For binary acids we recall that they all start with the prefix hydro- appended to the name of the anion with an –ic ending. Converting names into formulas involves the reverse process.

Assembling the Tools:

We will use the rules for assigning names to acids and we can use those rules to determine the formulas from the names.

Solution:

We see that HBr and H_2S are both binary acids. Bromine becomes bromic for its anion and sulfur takes the name sulfuric. Finally we add acid to the end of the name. The results are hydrobromic acid and hydrosulfuric acid. The acid HIO_3 is related to the $HClO_3$ acid. We know the ClO_3^- is the chlorate ion and by analogy, the IO_3^- must then be the iodate ion. We then change the –ate ending to –ic to get iodic acid.

Converting names to formulas we see that one of the compounds, hydroselenic acid, must be a binary acid. Since selenium is in the same group with oxygen, we expect the selenide ion to have a 2– charge. The correct formula for the acid must be H_2Se. For potassium bromate, we deduce that the formula for the bromate ion must be BrO_3^- from the formula for the chlorate ion ClO_3^-. Now we can add a potassium ion to get $KBrO_3$ for potassium bromate. For nitrous acid we know that the anion must be the nitrite ion. The nitrite ion is most easily recalled as having one fewer oxygen atom than the nitrate ion and it is written as NO_2^-. Adding the hydrogen ion to make an acid gives us HNO_2.

Are the Answers Reasonable?

Once again the best check is to be sure we have followed the rules correctly in identifying the type of acid, binary or oxo-, and the name of the polyatomic ion on the oxoacids.

Self-Test

17. Potassium arsenite has the formula K_3AsO_3. Name the following.

(a) H_3AsO_3 _____

(b) H_3AsO_4 _____

(c) $Ca_3(AsO_4)_2$ _____

18. Name the following.

(a) $H_2Se(g)$ _____

(b) $H_2Se(aq)$ _____

(c) $HF(g)$ _____

(d) $HF(aq)$ _____

(e) $KHSO_3$ _____

(f) HIO_3 _____

(g) $HBrO$ _____

(h) CuH_2AsO_4 _____

19. Write formulas for the following.

(a) sodium perbromate _____

(b) periodic acid _____

(c) sodium bioxalate _____

(d) perbromic acid _____

(e) hydrotelluric acid _____

(f) barium hydroxide _____

4.6   Double Replacement (Metathesis) Reactions

Learning Objective

Use the principles of metathesis to predict reaction products and to plan chemical synthesis.

Review

To start, we learn how to predict whether a *metathesis reaction* will occur between a pair of reactants to form a precipitate. To do this, you will apply the solubility rules in Table 4.1. The first step is to write a molecular equation for the reaction. To do this, it is necessary to determine what the products are. This is done by exchanging the anions between the two cations. However, *we must be very careful to write the correct formulas of the products*. That is, we must be sure that the formulas represent electrically neutral formula units. For example, in the reaction between $Cr_2(SO_4)_3$ and $BaCl_2$, we have to be sure to check the charges on the anions and cations before we write the products. The anions are SO_4^{2-} and Cl^-; the cations are Cr^{3+} and Ba^{2+}. When we exchange the anions, the Cl^- goes with the Cr^{3+} to give $CrCl_3$ and the SO_4^{2-} goes with the Ba^{2+} to give $BaSO_4$. Therefore, the equation before balancing is

$$\text{Contains } Cr^{3+} \text{ and } SO_4^{2-} \qquad Cr^{3+} \text{ requires 3 } Cl^- \text{ for electrical neutrality}$$

$$Cr_2(SO_4)_3 + BaCl_2 \rightarrow CrCl_3 + BaSO_4$$

$$\text{Contains } Ba^{2+} \text{ and } Cl^- \qquad Ba^{2+} \text{ requires one } SO_4^{2-} \text{ for electrical neutrality}$$

One way to balance this type of equation involves balancing ions from one side of the equation to the next until all have been balanced. To try this method, we start with the most complex formula, in this case $Cr_2(SO_4)_3$. It has two chromium atoms that can be balanced with a coefficient of 2 in front of the product $CrCl_3$. That now gives us six chlorine atoms that can be balanced with a coefficient of 3 in front of the reactant $BaCl_2$. That then gives us three barium atoms that are balanced with a coefficient of 3 in front of the product $BaSO_4$. All of the coefficients have been assigned at this point. The fact that the three sulfate ions from the $BaSO_4$ balance with the three sulfates in the $Cr_2(SO_4)_3$ is a check that the balancing was correct. Many metathesis reaction equations can be balanced in this manner and our final result is

$$Cr_2(SO_4)_3 + 3BaCl_2 \longrightarrow 2CrCl_3 + 3BaSO_4$$

The next step is to write the ionic equation for the reaction. To do this we must know whether the reactants and products are soluble in water. This is where we apply the solubility rules. After we've identified any solid that might form, we can then divide the equation into ionic and net ionic equations. For this reaction, the molecular, ionic, and net ionic equations are, respectively,

$$Cr_2(SO_4)_3(aq) \ + \ 3BaCl_2(aq) \ \longrightarrow \ 2CrCl_3(aq) \ + \ 3BaSO_4(s)$$

$$2Cr^{3+}(aq) + 3SO_4{}^{2-}(aq) + 3Ba^{2+}(aq) + 6Cl^-(aq) \ \longrightarrow \ 2Cr^{3+}(aq) + \ 6Cl^-(aq) + 3BaSO_4(s)$$

$$3SO_4{}^{2-}(aq) + 3Ba^{2+}(aq) \ \longrightarrow \ 3BaSO_4(s)$$

Strong acids and strong bases react with each other to form a salt and water in a reaction we call *neutralization*. If both are soluble, the net reaction is

$$H^+ + OH^- \ \longrightarrow \ H_2O$$

The anion of the acid and the cation of the base are spectator ions in the reaction and all of the ions do not cancel because of the formation of the weak electrolyte H_2O.

Polyprotic acids can be completely neutralized or they can be partially neutralized to form acid salts.

When constructing an ionic equation, weak acids and bases are written in molecular form because in a solution most of the weak acid or base is present as molecules, not ions. This changes the nature of the spectator ions and the form of the net ionic equation. Study Examples 4.6 – 4.8 in the text.

Insoluble oxides and hydroxides react with both strong and weak acids. In constructing the ionic equation, remember that insoluble solids and weak acids are written in "molecular" form.

Another type of metathesis reaction is one in which a gas is produced. In this section we describe five kinds of reactions in which gases are formed.

1.   Acids react with soluble sulfides and many insoluble sulfides to give gaseous hydrogen sulfide and a salt.

Because H_2S has a low solubility in water, it leaves a solution when it is formed as a product in a metathesis reaction.

2.   Cyanides react with acids to give gaseous hydrogen cyanide.

HCN has a low solubility in water and will be released when a strong acid reacts with a metal cyanide.

3.   Acids (strong and weak) react with soluble *and* insoluble carbonates and bicarbonates to give carbon dioxide and water.

When you write a metathesis equation and find H_2CO_3 as one of the products, remember that it decomposes to give CO_2 and H_2O.

4.   Acids react with sulfites and bisulfites to give sulfur dioxide and water.

When you write a metathesis equation and find H_2SO_3 as one of the products, remember that it decomposes to give SO_2 and H_2O.

5.   Bases react with ammonium salts to give gaseous ammonia.

When you write a metathesis equation and find NH_4OH as one of the products, you should rewrite it as NH_3 plus H_2O.

Be sure to study Table 4.2. Ask your instructor whether you are expected to learn all the substances released as gases that are described in this table.

In this section you learn that if a net ionic equation exists for a metathesis reaction, then a reaction does occur. However, if all the ions are spectator ions, and therefore cancel, there is no net reaction. A net ionic equation will be expected to exist under the following conditions.

1. A solid (often called a precipitate) forms from a solution of soluble reactants.

2. Water forms in the reaction of an acid and a base.

3. A weak electrolyte forms from a solution of strong electrolytes.

4. A gas forms that escapes from the reaction mixture.

The preceding provides some general guidelines, but the real test is to write the equation as a metathesis using entire formulas for reactants and products (i.e., the molecular equation). Then write the ionic equation taking into account that (1) weak electrolytes are written in molecular form, (2) insoluble substances are written in molecular form, and (3) gases may be produced as described in Section 4.5. After writing the ionic equation, cancel any spectator ions. If an equation remains, there is a net reaction. If all the ions cancel, there is no net reaction. Study Examples 4.6 through 4.8 in the text.

Example 4.4 Predicting metathesis reactions

Write the ionic and net ionic equations for the reaction of chromium(III) nitrate with potassium sulfite.

Analysis:

To write chemical equations we need the formulas of the compounds. Therefore our first task is to translate the names into formulas. Next, in a metathesis reaction we need to predict the products by pairing the positive ion of one compound with the negative ion of the other. Then we check that the formulas of the products are written correctly (the charges of the ions must add up to zero). We then balance the molecular equation and identify whether substances are soluble (aq), solid (s), gaseous (g), or weak electrolytes. The soluble substances in the molecular equation are written as ions while the others are left in molecular form. Finally, for the net ionic equation the spectator ions are identified and canceled. Now that we have the needed steps, let's solve the problem.

Assembling the Tools:

Nomenclature tools will enable you to write the correct formulas. Then the rules for determining the products when solutions of two ionic compounds are mixed will be needed. We may need the solubility rules, and lists of common gases and weak electrolytes.

Solution:

We first translate the names of the reactants into their chemical formulas. Chromium(III) nitrate contains Cr^{3+} and NO_3^- ions and its formula must be $Cr(NO_3)_3$. Potassium sulfite contains K^+ and SO_3^{2-} ions and its formula must be K_2SO_3. Using the ions we have identified, the products must be $Cr_2(SO_3)_3$ and KNO_3. We write the unbalanced equation as

$$Cr(NO_3)_3. + K_2SO_3. \longrightarrow Cr_2(SO_3)_3 + KNO_3$$

The balanced equation is

$$2Cr(NO_3)_3(aq) + 3K_2SO_3(aq) \longrightarrow Cr_2(SO_3)_3(s) + 6KNO_3(aq)$$

We now determine that all of the compounds are soluble except $Cr_2(SO_3)_3$, recalling that sulfites are generally insoluble. We will break apart each compound into its ions, except, of course, the $Cr_2(SO_3)_3$.

$$2Cr^{3+} + 6NO_3^- + 6K^+ + 3SO_3^{2-} \longrightarrow Cr_2(SO_3)_3 + 6K^+ + 6NO_3^-$$

The six nitrate ions and six potassium ions are easy to identify as spectator ions and they are canceled to obtain the net ionic equation

$$2Cr^{3+}(aq) + 3SO_3^{2-}(aq) \longrightarrow Cr_2(SO_3)_3(s)$$

Is the Answer Reasonable?

All we can do is double-check our work. Starting at the end we count charges in the ionic equations to be sure they are balanced. We then verify the solubilities of the substances and finally check the identities of each formula in the original equation.

Self-Test

20. Test your knowledge of the solubility rules by applying them to the salts listed below. *Circle* the formulas of those that are soluble in water. *Box* those compounds that will produce a gas if placed in acidic or basic solutions.

    Na_2SO_4 $CuCO_3$ $Ni(NO_3)_2$ Hg_2Cl_2 $PbBr_2$ Cr_2O_3 $Ca_3(PO_4)_2$

    $(NH_4)_2CO_2$ $Ba(ClO_4)_2$ AgI ZnS $MgSO_3$ $Sr(C_2H_3O_2)_2$ $(NH_4)_2S$

21. Write molecular, ionic, and net ionic equations for the reactions that occur between the following compounds.

    (a) $AgC_2H_3O_2 + (NH_4)_2S \longrightarrow$

    Molecular equation:

    _____

    Ionic equation:

    _____

    Net ionic equation:

    _____

    (b) $Cr(ClO_4)_3 + KOH \longrightarrow$

    Molecular equation:

    _____

    Ionic equation:

    _____

    Net ionic equation:

    _____

(c) $Pb(NO_3)_2 + K_2SO_4 \longrightarrow$

Molecular equation:

_____

Ionic equation:

_____

Net ionic equation:

_____

22. Write the net ionic equation for the reaction of

(a)   KOH with H_2SO_4 (complete neutralization)

_____

(b) $Ca(OH)_2$ with $HCHO_2$ (a soluble weak acid)

_____

(c) $NH_3(aq)$ with $HNO_3(aq)$

_____

(d) $Ni(OH)_2$ with $HCHO_2(aq)$ (the nickel-containing product is soluble)

_____

(e) $ZnO(s)$ with $HCl(aq)$

_____

23. What salts are formed in the neutralization of H_3PO_4 by $Ca(OH)_2$? Consider both complete and partial neutralization of the acid, but assume complete reaction of the $Ca(OH)_2$.

_____

24. Write the molecular, ionic, and net ionic equations for the complete neutralization of a solution of citric acid by aqueous ammonia. Citric acid has the formula $H_3C_6H_5O_7$.

_____

25. What gas is formed in the reaction of HCl with

(a) $K_2CO_3(aq)$_____ (b) $BaSO_3(s)$_____

(c) $(NH_4)_2S(aq)$_____(d) $CaCO_3(s)$_____

(e) $KHSO_3(aq)$_____(f) $NH_4HCO_3(aq)$ _____

26.   Which of the substances in the preceding question would release a gas if treated with concentrated NaOH solution?

_____

27.   Write net ionic equations for the reactions, if any, that occur between the following compounds. If there is no net reaction, state so by writing N.R. Assume any weak electrolytes are soluble in water, unless they are insoluble gases.

(a) $Pb(C_2H_3O_2)_2 + (NH_4)_2S \longrightarrow$

_____

(b) $Cr(ClO_4)_3 + MgSO_4 \longrightarrow$

_____

(c) $Zn(NO_3)_2 + KOH \longrightarrow$

_____

(d) $Cr_2O_3 + HClO_4 \longrightarrow$

_____

(e) $BaSO_3 + HC_2H_3O_2 \longrightarrow$

_____

(f) $(NH_4)_2SO_4 + Ba(OH)_2 \longrightarrow$

_____

(g) $Ca(C_3H_5O_2)_2 + HCl \longrightarrow$

_____

$$\frac{\text{moles of solute(mol)}}{\text{liters of solution(L)}}$$

Thus, a solution that contains 0.25 mole of solute per liter of solution has a concentration that we can express as 0.25 mol L^{-1}, or 0.25 M. The symbol M stands for the units mol L^{-1}. For example, a solution that is labeled 0.10 M HCl has a concentration of HCl equal to 0.10 mol L^{-1}.

To calculate the molarity of a solution, we need only the number of moles of solute (or the mass of solute, which we can convert to moles) and the volume of the solution. For instance, suppose we had 275 mL of a solution in which is dissolved 0.0264 mol of $CaCl_2$. To calculate the molarity, we divide the number of moles by the volume expressed in liters.

$$\text{molarity} = \frac{0.0264 \text{ mol CaCl}_2}{0.275 \text{ L solution}} = 0.0960 \ M \ \text{CaCl}_2$$

Converting from milliliters to liters involves moving the decimal three places to the left.

Notice how we converted milliliters to liters. You should practice converting between these two units (from mL to L and from L to mL) until it becomes effortless. It is a skill you will use often.

Equivalent ways of expressing molar concentration

Because molarity is a ratio, it is possible to express the ratio in units other than moles and liters, and sometimes it's convenient to do so. One alternative is to convert the unit liter in the denominator to milliliters. Because 1 L = 1000 mL, we can express molarity in the units "mol/1000 mL." Thus, the concentration 0.20 *M* HCl can be expressed as follows:

$$0.20 \, M \text{ HCl} = \frac{0.20 \text{ mol HCl}}{1 \text{ L solution}} = \frac{0.20 \text{ mol HCl}}{1000 \text{ mL solution}}$$

Another alternative is to convert the numerator to millimoles and the denominator to milliliters using 1 mmol = 10^{-3} mol and 1 mL = 10^{-3} L. Let's do this for the 0.20 *M* HCl solution.

$$0.20 \, M = \frac{0.20 \text{ mol HCl} \times \left(\dfrac{1 \text{ mmol HCl}}{10^{-3} \text{ mol HCl}} \right)}{1 \text{ L solution} \times \left(\dfrac{1 \text{ mL solution}}{10^{-3} \text{ L solution}} \right)} = \frac{0.20 \times \dfrac{1}{10^{-3}} \text{ mmol HCl}}{1 \times \dfrac{1}{10^{-3}} \text{ mL solution}}$$

Notice that the result gives the quantity $1/10^{-3}$ in both numerator and denominator. When we cancel this, we obtain 0.20 mmol HCl/1 mL solution.

$$0.20 \, M = \frac{0.20 \times \dfrac{1}{\cancel{10^{-3}}} \text{ mmol HCl}}{1 \times \dfrac{1}{\cancel{10^{-3}}} \text{ mL solution}} = \frac{0.20 \text{ mmol HCl}}{1 \text{ mL solution}}$$

Thus, the ratio of millimoles to milliliters is equal to the ratio of moles to liters. This means we can express molarity in either of these two sets of units. For example, if we had a 0.10 *M* solution of NaOH, we could write

$$0.10 \, M \text{ NaOH} = \frac{0.10 \text{ mol NaOH}}{1 \text{ L solution}} \qquad \text{or} \qquad 0.10 \, M \text{ NaOH} = \frac{0.10 \text{ mmol NaOH}}{1 \text{ mL solution}}$$

Molar concentration is the link between moles of solute and volume of solution

The reason molarity is so useful for stoichiometry is because it relates the amount of solute in a solution to the volume of solution. This makes it easy to dispense moles of solute simply by measuring out volumes of solution.

For calculations, molarity is the link that allows us to relate moles of solute and volumes of solution. Molarity provides the conversion factors needed to calculate the number of moles of solute in a given volume of a solution, or the volume of solution that contains a certain number of moles.

Let's see how we can use molarity to form conversion factors. Suppose we have a solution that is labeled 0.250 *M* HCl. The first thing we have to do is express this in units of moles and volume. For this solution we can write

$$0.250 \, M = \frac{0.250 \text{ mol HCl}}{1 \text{ L solution}}$$

For convenience, we'll write the relationship between moles and volume as an equivalence.

$$0.250 \text{ mol HCl} \Leftrightarrow 1.00 \text{ L solution}$$

(Notice that we've expressed the volume to 3 significant figures. For molarity, the precision is indicated by the numerator; the denominator can be written to the same number of significant figures.)   From this relationship we can form two conversion factors useful in calculations.

$$\frac{0.250 \text{ mol HCl}}{1.00 \text{ L solution}} \qquad \text{and} \qquad \frac{1.00 \text{ L solution}}{0.250 \text{ mol HCl}}$$

A useful thing to remember is that when you multiply molar concentration by the volume of solution (expressed in liters), the result is the number of moles of solute. This is given as Equation 4.5 in the text.

$$\text{molarity x volume(L)} = \text{moles of solute}$$

If in a particular problem you are given both the volume and molarity of a solution, you are given information from which you can calculate the number of moles of solute in the solution. This is a critically important concept to remember.

Let's look at two examples.

Example 4.5   Working with molarity

How many grams of KOH are in 150.0 mL of 0.1638 *M* KOH solution?

Analysis:

We can express the problem in equation form as follows:

$$150.0 \text{ mL KOH solution} \Leftrightarrow ? \text{ g KOH}$$

Let's examine the data and see what we can calculate. We have both the molarity and volume of the solution. We know molarity relates moles and volume, so we should be able to use the molarity to calculate the number of moles of solute in the solution. We can write the molarity as

$$0.138 \text{ } M \text{ KOH} = \frac{0.138 \text{ mol KOH}}{1 \text{ L KOH solution}}$$

As written, this fraction is the conversion factor that will convert liters of solution to moles of KOH. However, to use it, we need the volume in liters, so we'll convert 150.0 mL to liters. In liters, the volume is 0.1500 L. Multiplying this volume by the molarity will give the number of moles of solute.

To convert moles of KOH to grams, the tool is the formula mass of KOH, which gives us the relationship

$$1 \text{ mol KOH} = 56.10 \text{ g KOH}$$

Now we can see the path to the answer. We need to convert the given volume to moles and then the moles to mass.

Assembling the Tools:

We will use the rules for stoichiometry calculations. Also needed is the tool that relates molarity to moles and volumes, and the concept that molarity, correctly written, can be a conversion factor.

Solution:

Lets be more specific about the path to use and the conversion factors. (Recall that arrows represent conversion factors (tools) and rectangles represent some measure of our sample such as moles or grams.)

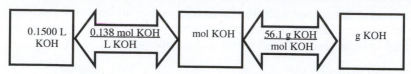

We will calculate the values for mol KOH and for g KOH using this sequence. Now let's apply the conversion factors

$$0.1500 \text{ L KOH} \times \frac{0.138 \text{ mol KOH}}{1 \text{ L KOH solution}} \times \frac{56.10 \text{ g KOH}}{1 \text{ mol KOH}} = 1.16 \text{ g KOH}$$

The 150.0 mL of solution contains 1.16 g of KOH.

Is the Answer Reasonable?

Let's do some proportional reasoning and approximate math. If we had an entire liter of the solution, it would contain 0.138 mol of KOH. 0.15 L is about 1/6 of a L and one sixth of 0.138 is approximately 0.02 mol. One mole of KOH weighs 56 g, so 0.01 mol would weigh 0.56 g. We have about 0.02 mol, so the answer should be about twice 0.56 g or 1.12 g. The value we obtained, 1.16 g, and the estimate, 1.12 g, agree quite well.

It is worth stating again that *whenever you have the molarity and volume of a solution, you can calculate the number of moles of solute*. Just multiply the molarity by the volume (in liters).

$$\text{L solution} \times \frac{\text{mol solute}}{\text{L solution}} = \text{mol solute}$$

volume (L)   molarity

Let's look at another problem.

Example 4.6   Working with molarity

How many milliliters of 0.105 M Na_3PO_4 contain 22.5 g of the solute?

Analysis:

The problem, in equation form, is

$$22.5 \text{ g } Na_3PO_4 \Leftrightarrow \text{ ? mL solution}$$

The molarity (0.105 M or 0.105 mol L^{-1}) will allow us to relate volume of solution and moles of solute. The relationship is

$$0.105 \text{ mol } Na_3PO_4 \Leftrightarrow 1.00 \text{ L solution}$$

Before we can use it, however, we need to have the amount of solute expressed in moles. To convert grams to moles, we use the formula mass.

$$1 \text{ mol Na}_3\text{PO}_4 = 163.94 \text{ g Na}_3\text{PO}_4$$

We now have a way to go from g Na_3PO_4 to liters of solution. After we obtain this, we move the decimal point to change to milliliters.

Assembling the Tools:

We will use the rules for stoichiometry calculations. Also needed is the tool that relates molarity to moles and volumes, and the concept that molarity, correctly written can be a conversion factor.

Solution:

We begin by applying the unit conversions.

$$22.5 \text{ g Na}_3\text{PO}_4 \times \frac{1 \text{ mol Na}_3\text{PO}_4}{163.94 \text{ g Na}_3\text{PO}_4} \times \frac{1.00 \text{ L solution}}{0.105 \text{ mol Na}_3\text{PO}_4} = 1.31 \text{ L solution}$$

Finally, we convert the volume to milliliters to obtain 1.31×10^3 mL.

Is the Answer Reasonable?

One mole of Na_3PO_4 weighs about 160 g, so 0.1 mol would weigh about 16 g. Our amount (approximately 22 g) is about 1.3 times 16 g, or about 0.13 mol. If we round the molarity (0.105 *M*) to 0.10 *M*, then the amount of solution needed to have 0.13 mol would be 1.3 L. Our estimated answer and the calculated answer are very close to each other, giving confidence that the answer is correct.

Preparing solutions of known molarity

If we want to prepare a solution of known molar concentration, we usually have in mind a particular final volume, so the calculation we have to do is to find first the moles of solute we need by using Equation 4.5. Then we convert moles to grams.

Preparing solutions by dilution

Often we prepare solutions by diluting a more concentrated solution with solvent. During this operation, the number of moles of solute remains constant. This leads to Equation 4.6 in the text.

$$V_{dil}M_{dil} = V_{conc}M_{conc}$$

where the subscript "dil" refers to the more dilute solution and "conc" refers to the more concentrated one.

Example 4.7 Diluting solutions to a desired concentration

A solution is prepared by dissolving 65.2 grams of acetic acid in enough water to make 250.0 mL of solution. What volume of this solution is needed to prepare 100.0 mL of a 0.200 *M* solution of acetic acid?

Analysis:

We are asked to dilute the first solution to make exactly 100.0 mL of a 0.200 *M* solution. This requires the use of the dilution concept above. We know the volume and molarity of the dilute solution needed and we are asked what volume of the concentrated solution should be used. We need the molarity of the concentrated solution to solve the problem.

Assembling the Tools:

We need the equation for calculating molarity of the original solution. Next the dilution equation is needed.

Solution:

From the analysis, we need to determine the molarity of the acetic acid solution made from 65.2 grams of acetic acid with a volume of 250.0 mL. First we convert 65.2 g of acetic acid to moles.

$$65.2 \text{ g HC}_2\text{H}_3\text{O}_2 \times \frac{1 \text{ mol HC}_2\text{H}_3\text{O}_2}{60.05 \text{ g HC}_2\text{H}_3\text{O}_2} = 1.09 \text{ mol HC}_2\text{H}_3\text{O}_2$$

The molarity is then calculated as

$$\text{Molarity} = \frac{\text{moles solute}}{\text{liters solution}} = \frac{1.09 \text{ mol HC}_2\text{H}_3\text{O}_2}{0.250 \text{ L solution}} = 4.34 \text{ } M$$

Now, we have all of the information needed to solve the dilution equation, we make the substitutions.

$$(100 \text{ mL})(0.200 \text{ } M) = (x \text{ mL})(4.34 \text{ } M)$$

Solving for *x* mL we obtain 4.61 mL. We must take 4.61 mL of the first solution and dilute it to 100 mL to prepare our 0.200 *M* solution of acetic acid.

Is the Answer Reasonable?

First we check the answer for the first molarity. We use a little more than the mass of one mole in one quarter of a liter and therefore the molarity should be near 4 *M*. If we dilute 4 *M* by 10, we get 0.4 *M* and we can reason that we should dilute by a factor of 20 to get a 0.2 *M* solution. One twentieth of the final volume of 100 mL is 5 mL, which is close to the volume we calculated. This close agreement gives us confidence in our results.

Self-Test

28. A solution is labeled 0.150 M HNO_3. Write the two conversion factors we can create from this information that would allow us to relate moles of HNO_3 and volume of this solution.

_____

29. For the solution in the preceding question, how many millimoles of HNO_3 would be found in 1 mL of the solution?

_____

How many micromoles of HNO_3 would be found in 1 μL of the solution?

_____

30. How would 500 mL of a 0.125 M $CaCl_2$ solution be prepared?

_____

31. How many moles and how many grams of NaCl are in 250 mL of 1.38 M NaCl?

_____

32. How many milliliters of 0.124 M HCl are needed to obtain 0.00244 mol HCl?

_____

33. How would you prepare 100 mL of 0.500 M H_2SO_4 from 0.800 M H_2SO_4?

_____

_____

34. How would you prepare 250 mL of 0.100 M HCl from 2.00 M HCl?

_____

_____

35. How many milliliters of water would have to be added to 50.0 mL of 0.250 M HCl to give a solution with a concentration of 0.100 M?

_____

4.8   Solution Stoichiometry

Learning Objective

Learn to use molarity in stoichiometric calculations.

Review

In previous problems dealing with reaction stoichiometry, you saw that the amounts of reactants and products could be expressed in grams or moles. Molarity and volume provide a third alternative, as illustrated in the following problem.

Example 4.8 Using molarity in a stoichiometry problem

How many grams of calcium carbonate will react with 75.0 mL of 0.250 *M* HCl? The equation for the reaction is

$$CaCO_3(s) + 2HCl(aq) \longrightarrow CaCl_2(aq) + CO_2(g) + H_2O$$

Analysis:

The problem can be expressed as

$$75.0 \text{ mL HCl solution} \Leftrightarrow ? \text{ g } CaCO_3$$

We're dealing with a chemical reaction, so the critical link between HCl and $CaCO_3$ is provided by the coefficients of the equation. These give us the mole relationship between $CaCO_3$ and HCl.

$$1 \text{ mol } CaCO_3 \Leftrightarrow 2 \text{ mol HCl}$$

We have the volume and molarity of the HCl solution, which will allow us to calculate the number of moles of HCl. We just need to change 75.0 mL to liters.

We now know how we are going to go from the volume of the solution to moles of $CaCO_3$. To find grams of $CaCO_3$, the tool will be the formula mass (100.1 g mol^{-1}).

We can diagram the flow of the problem as follows:

Assembling the Tools:

We need our standard tools for stoichiometry problems. We'll also need the molarity to use as a conversion factor between volume and moles.

Solution:

We'll start by changing the volume of the solution to liters: 75.0 mL = 0.0750 L. Then we apply the conversion factors following the path through the problem.

$$0.0750 \text{ L solution} \times \frac{0.250 \text{ mol HCl}}{1.00 \text{ L solution}} \times \frac{1 \text{ mol } CaCO_3}{2 \text{ mol HCl}} \times \frac{100.1 \text{ g } CaCO_3}{1 \text{ mol } CaCO_3} = 0.938 \text{ g } CaCO_3$$

Is the Answer Reasonable?

Let's suppose we had 100 mL (0.1 L) of the HCl solution. One liter of the HCl solution contains 0.25 mol HCl, so in 0.1 L, there would be 0.025 mol HCl. This would consume 0.0125 mol $CaCO_3$, which would weigh about 1.25 g. We have somewhat less than 100 mL, so we expect an answer somewhat less than 1.25 g. The answer we obtained, 0.938 g, seems to be reasonable.

Concentrations of ions in solutions of electrolytes

Multiply the molar concentration given for the salt by the number of ions per formula unit to obtain the concentration of the ion in the solution. For example, if the salt is $Al_2(SO_4)_3$ and its concentration is 0.10 M, the concentration of Al^{3+} is 2 x 0.10 M = 0.20 M, and the concentration of the SO_4^{2-} is 3 x 0.10 M = 0.30 M.

$$0.10\ M\ Al_2(SO_4)_3$$

$$\boxed{0.10 \times 2} \qquad \boxed{0.10 \times 3}$$

$$0.20\ M\ Al^{3+} \qquad 0.30\ M\ SO_4^{2-}$$

Study Examples 4.14, 4.15, and 4.16 in the text, then work Practice Exercises 4.33 to 4.38. Additional exercises are found in the Self-Test below.

Using net ionic equations in stoichiometry calculations

Once you have learned to calculate the concentration of an ion using the formula of a salt and the salt concentration, you are ready to use a net ionic equation in a stoichiometry calculation. Follow Example 4.16 in the text carefully.

Example 4.9 Using net ionic equations in stoichiometry calculations

Two bottles on the lab desk are labeled 0.500 M Al^{3+} and 0.250 M OH^-. How many mL of each must be used to produce 0.685 g of $Al(OH)_3$?

Analysis:

We always need a balanced chemical equation in order to solve stoichiometry problems. This problem suggests that the net ionic equation for aluminum ions reacting with hydroxide ions will produce the insoluble aluminum hydroxide. Next, we need to solve two problems. First we need to calculate the mL of the aluminum solution needed, and second we need to calculate the mL of the hydroxide ion solution. Since we are given the molarities of these solutions, they can be used as conversion factors in our

calculations. Our sequence of steps will be

Assembling the Tools:

We need our standard tools for stoichiometry problems. We'll also need the molarity to use as a conversion factor between volume and moles. This time our calculations will use the concentrations of ions.

Solution:

In our plan, the first step is to write the net ionic equation as

$$Al^{3+}(aq) \ + \ 3OH^-(aq) \longrightarrow Al(OH)_3(s)$$

Now we can set up the conversion to calculate the mol $Al(OH)_3$.

$$0.685 \text{ g } Al(OH)_3 \times \frac{1 \text{ mol } Al(OH)_3}{78.00 \text{ g } Al(OH)_3} = 0.008782 \text{ mol } Al(OH)_3$$

Notice that we keep one extra significant figure at this point. Now we calculate the two answers.

$$0.008782 \text{ mol } Al(OH)_3 \times \frac{1 \text{ mol } Al^{3+}}{1 \text{ mol } Al(OH)_3} \times \frac{1 \text{ L } Al^{3+}}{0.500 \text{ mol } Al^{3+}} = 0.0176 \text{ L } Al^{3+}$$

$$0.008782 \text{ mol } Al(OH)_3 \times \frac{3 \text{ mol } OH^-}{1 \text{ mol } Al(OH)_3} \times \frac{1 \text{ L } OH^-}{0.250 \text{ mol } OH^-} = 0.105 \text{ L } OH^-$$

We can readily convert liters to 17.6 mL Al^{3+}, and 105 mL OH^- solution will be needed for this reaction.

Are the Answers Reasonable?

First we check our balanced ionic equation and the molar mass of $Al(OH)_3$. Finding them correct we estimate the answers by rounding our numbers to one significant figure. For the first calculation we round 0.685 to 0.7 and we round the 78.00 to 70. Now, 0.7/70 is 0.01, which is close to our answer 0.008782. Next we multiply the numerator values and then the denominator values of the next two equations, using 0.01 for the moles of $Al(OH)_3$ to get 0.01/0.5 = 0.02, which is close to the 0.0176 calculated. For the second calculation we get 0.03/0.3 = 0.1. This is close to the 0.105 we calculated. We have confidence that our values are correct.

Self-Test

36. What are the molar concentrations of the ions in

    (a)   0.15 M $Ba(OH)_2$?_____

    (b)   0.30 M Na_3PO_4?_____

37. In a certain solution of $(NH_4)_2SO_4$ the ammonium ion concentration was 0.42 *M*.

    (a) What was the sulfate ion concentration in the solution?_____

    (b) What was the molar concentration of ammonium sulfate in the solution?

                                                  _____

38. How many moles of each ion are in

    (a)   50.0 mL of 0.250 *M* $(NH_4)_2SO_4$? _____

    (b)   80.0 mL of 0.60 *M* K_3PO_4? _____

39. How many milliliters of 0.120 *M* $AgNO_3$ are needed to react with 18.0 mL of 0.0460 *M* $FeCl_3$ solution? The reaction follows the net ionic equation

$$Ag^+(aq) + Cl^-(aq) \longrightarrow AgCl(s)$$

    Answer _____

40. A student mixed 30.0 mL of 0.25 *M* NaI solution with 45.0 mL of 0.10 *M* $Pb(NO_3)_2$ solution.

    (a) Write the molecular equation for the reaction that occurred.

                                                  _____

    (b) How many moles of each of the ions were in the mixture before any reaction took place?

       Na^+ _____ I^- _____

       Pb^{2+} _____ NO_3^- _____

    (c) How many moles of which ions reacted? _____

    (d) How many moles of what compound were formed? _____

    (e) How many moles of each of the ions were present in the solution after the reaction was over?

       Na^+ _____ I^- _____

       Pb^{2+} _____ NO_3^- _____

    (f) What were the concentrations of each of the ions in the solution after the reaction was over?

       Na^+ _____ I^- _____

       Pb^{2+} _____ NO_3^- _____

4.9 Titrations and Chemical Analysis

Learning Objective

Understand the methods and calculations used in titrations and chemical analyses.

Review

Example 4.17 illustrates an important principle often used in chemical analyses. We begin with a mixture with an unknown composition (i.e., the relative amounts of the various components are unknown). A reaction is then carried out that transfers one component of the mixture quantitatively into a different compound, one with a known composition. By measuring the amount of this compound collected, the amount of the component in the original mixture can be calculated. By means of procedures of this sort, repeated for as many components as necessary, the composition of the entire mixture can be determined.

Titrations

This is a useful procedure for chemical analyses involving reactions in solution. In this section we describe acid-base titrations using appropriate indicators. Study Example 4.18 and the example below and be sure to work the Practice Exercises to test your knowledge. Then try the titration calculation in the Self-Test.

Example 4.10 Solving titration problems

You are hired to take charge of an analytical laboratory and one task is to analyze samples of a commercial product used to remove iron stains. It consists of $NaHSO_4$, an acid salt that reacts with sodium hydroxide as follows:

$$NaHSO_4(aq) + NaOH(aq) \longrightarrow Na_2SO_4(aq) + H_2O$$

The intent of the analysis is to ensure that the bottle labels show the actual percentage of $NaHSO_4$. The legal standard is pure $NaHSO_4$. You propose to analyze the samples by titrating 0.300 g portions of the product (dissolved in water) with 0.100 M NaOH. How many milliliters of the NaOH solution will be needed if the sample is pure $NaHSO_4$?

Analysis:

The question that needs to be answered is whether or not the $NaHSO_4$ is pure. If it is pure, the percentage of $NaHSO_4$ in the sample should be 100%. For this we recall the tool for calculating percentage composition.

$$\text{Percentage of } X = \frac{\text{mass of } X}{\text{mass of sample}} \times 100\%$$

If the percentage of **X** is 100%, the mass of **X** must equal the mass of the sample, which is 0.300 g. To answer the question we need to convert 0.300 g $NaHSO_4$ to moles of $NaHSO_4$ and then to the equivalent volume of NaOH.

Assembling the Tools:

We need our standard tools for stoichiometry problems. We see that we need three tools, the gram-to-mol tool, the mol-to-mol tool, and the mol-to-volume tool. The first comes from the molar masses, the second needs the balanced chemical equation that is given above, and the third uses the molarity given in the problem.

Solution:

The brief sequence described in the analysis can be diagramed as

$$\text{g } NaHSO_4 \xrightarrow[\text{tool}]{\text{g to mol}} \text{mol } NaHSO_4 \xrightarrow[\text{tool}]{\text{mol to mol}} \text{mol } NaOH \xrightarrow[\text{tool}]{\text{mol to V}} \text{mL } NaOH$$

After analyzing the problem, we see that all we need to calculate are the molar mass of $NaHSO_4$ and the volume of NaOH needed to react with 0.300 g of pure $NaHSO_4$. We determine the molar mass to be 120.1 g mol^{-1}. Then we set up the conversion equation as (Notice that for convenience we write 1000 mL in place of 1 L in the last conversion factor.)

$$0.300 \text{ g } NaHSO_4 \times \frac{1 \text{ mol } NaHSO_4}{120.1 \text{ g } NaHSO_4} \times \frac{1 \text{ mol } NaOH}{1 \text{ mol } NaHSO_4} \times \frac{1000 \text{ mL } NaOH}{0.100 \text{ mol } NaOH} = 25.0 \text{ mL } NaOH$$

$$\text{(initial mass)} \qquad \text{(molar mass)} \qquad \text{(mole ratio)} \qquad \text{(molarity)}$$

The sources of the conversion factors are shown below the equation.

Is the Answer Reasonable?

We check the cancellations used in the calculation. Next we estimate the answer by rounding all numbers to one or two significant figures. Multiplying all the numerators gives us 300 and multiplying all of the denominators gives us 12. The answer is 300/12 or 100/4 = 25. This agrees with our answer and we conclude the answer is most likely correct.

Self-Test

41. A sample of a weak acid weighing 0.256 g was dissolved in water and titrated with 0.200 M NaOH solution. The titration required 17.80 mL of the base.

    (a) How many moles of H^+ were neutralized in the titration?_____

    (b) If the acid is monoprotic, what is its molecular mass? (*Hint*: How many grams are there per mole of the acid?)

    _____

Answers to Self-Test Questions

1. Solvent is water; solute is sugar.
2. Unsaturated
3. supersaturated
4. (a) Ag^+ and $C_2H_3O_2^-$
   (b) NH_4^+ and $Cr_2O_7^{2-}$
   (c) Ba^{2+} and OH^-
5. (a) $AgC_2H_3O_2\,(aq) \longrightarrow Ag^+(aq) + C_2H_3O_2^-(aq)$
   (b) $(NH_4)_2Cr_2O_7\,(aq) \longrightarrow 2NH_4^+(aq) + Cr_2O_7^{2-}(aq)$
   (c) $Ba(OH)_2(aq) \longrightarrow Ba^{2+}(aq) + 2OH^-(aq)$
6. (a) Coefficients are $1, 2, 1, 2$
   $$Cu^{2+}(aq) + 2NO_3^-(aq) + 2K^+(aq) + 2OH^-(aq) \longrightarrow$$
   $$Cu(OH)_2(s) + 2K^+(aq) + 2NO_3^-(aq)$$
   $$Cu^{2+}(aq) + 2OH^-(aq) \longrightarrow Cu(OH)_2(s)$$
   (b) Coefficients are $1, 2, 1, 2$
   $$Ni^{2+}(aq) + 2Cl^-(aq) + 2Ag^+(aq) + 2NO_3^-(aq) \longrightarrow$$
   $$2AgCl(s) + Ni^{2+}(aq) + 2NO_3^-(aq)$$
   $$Cl^-(aq) + Ag^+(aq) \longrightarrow AgCl(s)$$
   (c) Coefficients are $1, 3, 1, 3$
   $$2Cr^{3+}(aq) + 3SO_4^{2-}(aq) + 3Ba^{2+}(aq) + 6Cl^-(aq) \longrightarrow$$
   $$2Cr^{3+}(aq) + 6Cl^-(aq) + 3BaSO_4(s)$$
7. (a) $HClO_2 \rightleftharpoons H^+ + ClO_2^-$
   (b) $H_2SeO_3 \rightleftharpoons H^+ + HSeO_3^-$
   $HSeO_3^- \rightleftharpoons H^+ + SeO_3^{2-}$
   (c) $H_3AsO_4 \rightleftharpoons H^+ + H_2AsO_4^-$
   $H_2AsO_4^- \rightleftharpoons H^+ + HAsO_4^{2-}$
   $HAsO_4^{2-} \rightleftharpoons H^+ + AsO_4^{3-}$
8. $P_4O_{10} + 6H_2O \longrightarrow 4H_3PO_4$
9. $N_2H_4 + H_2O \rightleftharpoons N_2H_5^+ + OH^-$
10. $RaO + H_2O \longrightarrow Ra(OH)_2$
11. Na_2O reacts with water to give sodium hydroxide. $Na_2O(s) + H_2O \longrightarrow 2NaOH(aq)$
12. SO_3 reacts with water to give sulfuric acid. $SO_3(g) + H_2O \longrightarrow H_2SO_4(aq)$
13. $HCN(aq) + H_2O \rightleftharpoons H_3O^+(aq) + CN^-(aq)$; the forward reaction is far from completion.
14. $HClO_3(aq) + H_2O \longrightarrow H_3O^+(aq) + ClO_3^-(aq)$
15. $C_6H_5NH_2(aq) + H_2O \rightleftharpoons C_6H_5NH_3^+(aq) + OH^-(aq)$
16. perchloric acid, $HClO_4$; chloric acid, $HClO_3$; hydrochloric acid, HCl; hydrobromic acid, HBr;

hydriodic acid, HI; nitric acid, HNO_3; sulfuric acid, H_2SO_4

17. (a) arsenous acid, (b) arsenic acid, (c) calcium arsenate

18. (a) hydrogen selenide, (b) hydroselenic acid, (c) hydrogen fluoride, (d) hydrofluoric acid
    (e) potassium hydrogen sulfite (potassium bisulfite), (f) iodic acid, (g) bromous acid
    (h) copper(II) dihydrogen arsenate

19. (a) $NaBrO_4$, (b) HIO_4, (c) $NaHC_2O_4$, (d) $HBrO_4$, (e) $H_2Te(aq)$, (f) $Ba(OH)_2$

20. Soluble: Na_2SO_4, $Ni(NO_3)_2$, $(NH_4)_2CO_3$, $Ba(ClO_4)_2$, $Sr(C_2H_3O_2)_2$, $(NH_4)_2S$
    Gas forming: ZnS, $(NH_4)_2CO_3$, $(NH_4)_2S$, $CuCO_3$, $MgSO_3$

21. (a) $2AgC_2H_3O_2(aq) + (NH_4)_2S(aq) \longrightarrow Ag_2S(s) + 2NH_4C_2H_3O_2(aq)$
          $2Ag^+(aq) + 2C_2H_3O_2^-(aq) + 2NH_4^+(aq) + S^{2-}(aq) \longrightarrow Ag_2S(s) + 2NH_4^+(aq) + 2C_2H_3O_2^-$
    (aq)

          $2Ag^+(aq) + S^{2-}(aq) \longrightarrow Ag_2S(s)$

    (b) $Cr(ClO_4)_3(aq) + 3KOH(aq) \longrightarrow Cr(OH)_3(s) + 3KClO_4(aq)$
          $Cr^{3+}(aq) + 2ClO_4^-(aq) + 3K^+(aq) + 3OH^-(aq) \longrightarrow Cr(OH)_3(s) + 3K^+(aq) + 3OH^-(aq)$
          $Cr^{3+}(aq) + 3OH^-(aq) \longrightarrow Cr(OH)_3(s)$

    (c) $Pb(NO_3)_2(aq) + K_2SO_4(aq) \longrightarrow PbSO_4(s) + 2KNO_3(aq)$
          $Pb^{2+}(aq) + 2NO_3^-(aq) + 2K^+(aq) + SO_4^{2-}(aq) \longrightarrow PbSO_4(s) + 2K^+(aq) + 2NO_3^-(aq)$
          $Pb^{2+}(aq) + SO_4^{2-}(aq) \longrightarrow PbSO_4(s)$

22. (a) $OH^-(aq) + H^+(aq) \longrightarrow H_2O$

    (b) $OH^-(aq) + HCHO_2(aq) \longrightarrow H_2O + CHO_2^-(aq)$

    (c) $NH_3(aq) + H^+(aq) \longrightarrow NH_4^+(aq)$

    (d) $Ni(OH)_2(s) + 2HCHO_2(aq) \longrightarrow Ni^{2+}(aq) + CHO_2^-(aq) + 2H_2O$

    (e) $ZnO(s) + 2H^+(aq) \longrightarrow Zn^{2+}(aq) + H_2O$

23. $Ca(H_2PO_4)_2$, $CaHPO_4$, $Ca_3(PO_4)_2$

24. Molecular equation: $H_3C_6H_5O_7(aq) + 3NH_3(aq) \longrightarrow (NH_4)_3C_6H_5O_7(aq)$
    Ionic and net ionic equations: $H_3C_6H_5O_7(aq) + 3NH_3(aq) \longrightarrow 3NH_4^+(aq) + C_6H_5O_7^{3-}(aq)$

25. (a) CO_2 (b) SO_2 (c) H_2S (d) CO_2 (e) SO_2 (f) CO_2

26. $(NH_4)_2S$ and NH_4HCO_3

27. (a) $Pb^{2+}(aq) + S^{2-}(aq) \longrightarrow PbS(s)$ (b)   N.R.

    (c) $Zn^{2+}(aq) + 2OH^-(aq) \longrightarrow Zn(OH)_2(s)$

    (d) $Cr_2O_3(s) + 6H^+(aq) \longrightarrow 2Cr^{3+}(aq) + 3H_2O$

    (e) $BaSO_3(s) + 2HC_2H_3O_2(aq) \longrightarrow Ba^{2+}(aq) + 2C_2H_3O_2^- + SO_2(g) + H_2O$

    (f) $NH_4^+(aq) + OH^-(aq) \longrightarrow NH_3(g) + H_2O$

    (g) $C_3H_5O_2^-(aq) + H^+(aq) \longrightarrow HC_3H_5O_2(aq)$

28. $\dfrac{0.150 \text{ mol } HNO_3}{1 \text{ L solution}}$ and $\dfrac{1 \text{ L solution}}{0.150 \text{ mol } HNO_3}$

29. 0.150 mmol HNO_3 and 1.0 μmol HNO_3

30. Dissolve 6.94 g $CaCl_2$ in water and make the final volume 500 mL.

31. 0.345 mol NaCl; 20.2 g NaCl

32. 19.7 mL HCl solution

33. Add water to 62.5 mL of 0.800 M H_2SO_4 to a final volume of 100 mL.

34. Add water to 12.5 mL of 2.00 M HCl to a final volume of 250 mL.

35. Add 75.0 mL of water to give a total volume of 125 mL.

36. (a) 0.15 M Ba^{2+}, 0.30 M OH^- (b) 0.90 M Na^+, 0.30 M PO_4^{3-}

37. (a) 0.21 M SO_4^{2-}, (b) 0.21 M $(NH_4)_2SO_4$

38. (a) 0.0250 mol NH_4^+, 0.0125 mol SO_4^{2-} (b) 0.144 mol K^+, 0.048 mol PO_4^{3-}
39. 20.7 mL $AgNO_3$ solution
40. (a) $Pb(NO_3)_2(aq) + 2NaI(aq) \longrightarrow PbI_2(s) + 2NaNO_3(aq)$
    (b) 0.0075 mol Na^+, 0.0075 mol I^-, 0.0045 mol Pb^{2+}, 0.0090 mol NO_3^-
    (c) 0.0075 mol I^- and 0.0038 mol Pb^{2+}
    (d) 0.0038 mol PbI_2
    (e) 0.0075 mol Na^+, 0.0 mol I^-, 0.0007 mol Pb^{2+}, 0.0090 mol NO_3^-
    (f) 0.10 M Na^+, 0.0 M I^-, 0.009 M Pb^{2+}, 0.12 M NO_3^-
41. (a) 3.56×10^{-3} mol H^+, (b) 71.9 g mol^{-1}

Tools for problem solving

In this chapter you learned to apply the following concepts as tools in solving problems. Study each one carefully so that you know what each is used for. When faced with solving a problem, recall what each tool does and consider whether it will be helpful in finding a solution.

You might want to tear these pages out to use along with solving problems in this chapter.

Criteria for a balanced ionic equation (Section 4.3)
To be balanced, an equation that includes the formulas of ions must satisfy two criteria: (1) the number of atoms of each kind must be the same on both sides of the equation, and (2) the net electrical charge shown on each side of the equation must be the same.

List of strong acids (Section 4.4)
The common strong acids are percholoric acid, $HClO_4$, chloric acid, $HClO_3$, hydrochloric acid, HCl, hydrobromic acid, HBr, hydroiodic acid, HI, nitric acid, HNO_3, and sulfuric acid, H_2SO_4.

Ionization of acids and bases in water (Section 4.4)
The anion of a strong acid is often given the symbol X^- while the symbol for the anion of a weak acid is A^-.

$$HX + H_2O \longrightarrow H_3O^+ + X^- \quad \text{strong acids}$$
$$HA + H_2O \rightleftharpoons H_3O^+ + A^- \quad \text{weak acids}$$

The cation of an ionic strong base is usually given the symbol M^{n+} and a molecular weak base is given the symbol B. Their reactions are

$$M(OH)_n \longrightarrow M^{n+} + nOH^- \quad \text{strong bases}$$
$$B + H_2O \rightleftharpoons BH^+ + OH^- \quad \text{weak bases}$$

Acid and polyatomic anion names (Section 4.5)
We use the names of the polyatomic anions to derive the names of acids. When the anion ends in -ate the acid ends in -ic. When the anion ends in -ite the acid ends in -ous.

Predicting net ionic equations (Section 4.6)
A net ionic equation will exist and a reaction will occur when:
- A precipitate is formed from a mixture of soluble reactants.
- An acid reacts with a base. This includes strong or weak acids reacting with strong or weak bases or insoluble metal hydroxides or oxides.
- A weak electrolyte is formed from a mixture of strong electrolytes.
- A gas is formed from a mixture of reactants.

Solubility rules (Section 4.6)
The rules in Table 4.1 serve as the tool we use to determine whether a particular salt is soluble in water.

Neutralization of an acid and a base (Section 4.6)
The reaction of an acid and a base that forms water and a salt is a neutralization reaction.

Neutralization of an acid and a base (Section 4.6)
The reaction of an acid and a base that forms water and a salt is a neutralization reaction.

Gases formed in metathesis reactions (Section 4.6)
Hydrogen sulfide, hydrogen cyanide, carbon dioxide, sulfur dioxide, and ammonia can be formed in metathesis reactions.

Synthesis by metathesis (Section 4.6)
To synthesize a salt by a metathesis reaction, there are three approaches: form an insoluble salt, conduct a neutralization reaction to form the salt and water, and add an acid to a metal carbonate, sulfide or sulfite to form a gas and the salt.

Molarity (Section 4.7)

$$\text{molarity } (M) = \frac{\text{moles of solute (mol)}}{\text{liters of solution (L)}}$$

Dilution equation (Section 4.7)

$$V_{dil} \times M_{dil} = V_{conc} \times M_{conc}$$

Stoichiometry pathways (Section 4.8)
Figure 4.24 gives a summary of the various paths and conversion factors used in stoichiometry calculations to convert from one set of chemical units to another. This diagram adds molarity as a conversion factor to Figure 3.6.

Molarity of ions in a salt solution (Section 4.8)
When using a net ionic equation to work stoichiometry problems, we need the concentrations of the ions in the solutions. The concentration of a particular ion equals the concentration of the salt multiplied by the number of ions of that kind in one formula unit of the salt.

Solubility Rules

Soluble Compounds

1. All compounds of the alkali metals (Group IA) are soluble.

2. All salts containing NH_4^+, NO_3^-, ClO_4^-, ClO_3^-, and $C_2H_3O_2^-$ are soluble.

3. All chlorides, bromides, and iodides (salts containing Cl^-, Br^-, and I^-) are soluble *except* when combined with Ag^+, Pb^{2+}, and Hg_2^{2+} (note the subscript "2").

4. All sulfates (salts containing SO_4^{2-}) are soluble *except* those of Pb^{2+}, Ca^{2+}, Sr^{2+}, Hg_2^{2+}, and Ba^{2+}.

Insoluble Compounds

5. All metal hydroxides (ionic compounds containing OH^-) and all metal oxides (ionic compounds containing O^{2-}) are insoluble *except* those of Group IA and of Ca^{2+}, Sr^{2+}, and Ba^{2+}.

    When metal oxides do dissolve, they react with water to form hydroxides. The oxide ion, O^{2-}, does not exist in water. For example: $Na_2O(s) + H_2O \rightarrow 2NaOH(aq)$

6. All salts that contain PO_4^{3-}, CO_3^{2-}, SO_3^{2-}, and S^{2-} are insoluble, *except* those of Group IA and NH_4^+.

Notes:

Chapter **5**

Oxidation–Reduction Reactions

In Chapter 4 we discussed various kinds of chemical reactions, including metathesis (double replacement) and acid/base reactions. These reactions have certain features that make discussing them together convenient. Now we turn our attention to another class of chemical reactions. These are often characterized by the transfer of electrons from one species to another, and they include many of our most important chemical changes, as you will learn in this chapter.

5.1   Oxidation–Reduction Reactions

Learning Objective

We will be able to define oxidation, reduction, oxidizing agents, reducing agents, and oxidation numbers.

Review

Oxidation–reduction reactions (often called *redox* reactions) are very common, so it is important that you learn how to identify them and know the terminology used in discussing them. These reactions can be thought of as involving the transfer of electrons from one substance to another. (Sometimes an electron transfer actually takes place, as when Na and Cl_2 react to form ions, but for many reactions it is just a convenience to think of them as involving electron transfer.)

Remember the following definitions:

> *Oxidation* can be viewed as a loss of electrons.

> *Reduction* can be viewed as a gain of electrons.

Oxidation and reduction always occur simultaneously during a chemical reaction. If one substance loses electrons and is oxidized, then another substance must gain electrons and be reduced.

The terms *oxidizing agent* and *reducing agent* often cause confusion. The basis for assigning these terms is given in the text, but the simplest way to remember how to apply the terms properly is to find which substance is oxidized and which is reduced. Then, switch words—if a certain substance is oxidized, then it's the reducing *agent*; if the substance is reduced, then it's the oxidizing *agent*.

When you write separate equations showing electron gain or loss, as in Example 5.1, keep in mind that electrons are negatively charged particles. That way, you will be sure to write the electrons on the correct side. For instance, in Example 5.1 we find that magnesium atoms become magnesium ions.

Charge is becoming more positive, so Mg
must be losing *negative* electrons.

$$Mg \rightarrow Mg^{2+}$$

The magnesium is becoming more positive, so it must be losing electrons. Electrons are written on the product side of the equation to show that they have been separated from the magnesium atom, which has become a magnesium ion.

$$Mg \longrightarrow Mg^{2+} + 2e^-$$

Notice that the net charge on both sides of the arrow is the same. Recall that this is a requirement for balanced equations involving ions, so the balanced charge assureS us that we've placed the electrons on the correct side.

Oxidation numbers

Oxidation numbers are a bookkeeping device. We assign them following the rules given in the text, which you should study carefully. The best way to learn how to apply the rules is to practice assigning oxidation numbers, so study Examples 5.2 through 5.5, work the Practice Exercises, and then work the questions in the Self–Test here in the Study Guide. As you apply these rules, remember that if you encounter a conflict between two rules, the one with the lower number (the one higher up on the list) is the one that applies and we ignore the rule with the higher number. This is illustrated in Examples 5.2 and 5.3.

For binary compounds of a metal and a nonmetal (such as $NaCl$, $CaCl_2$, or Al_2O_3), the ions are simple monatomic ions and their oxidation numbers are equal to their charges. The formulas of the ions formed by the representative elements were given in Table 2.2. If necessary, review them and learn how to use the periodic table to obtain their correct charges.

If you recognize a polyatomic ion in a formula, for example, SO_4^{2-} in $Cr_2(SO_4)_3$, you can use its charge as the *net* oxidation number of the ion. For the "SO_4" in this example, we can assign it a net oxidation number of -2, which means that the chromium must have an oxidation number of $+3$ (applying Rule 3, the summation rule: $[2 \times (+3)] + [3 \times (-2)] = 0$).

Redox is redefined here in terms of changes in oxidation number.

Oxidation: an increase in oxidation number (oxidation state)

Reduction: a decrease in oxidation number (oxidation state)

Study Example 5.4 in the text to review how we use oxidation numbers to identify oxidation and reduction.

Example 5.1 Using oxidation numbers

The following is a balanced net ionic equation for an oxidation reduction reaction:

$$3SO_3^2 + H_2O + 2MnO_4 \longrightarrow 3SO_4^{2-} + 2MnO_2 + 2OH$$

Determine the oxidation numbers of all elements in the reaction in order to determine which substance is oxidized and which is reduced. Also determine which substance is the oxidizing agent and which is the reducing agent.

Analysis:

The key to this problem is the determination of the oxidation numbers and then finding out which oxidation numbers increase and which decrease.

Assembling the Tools:

We will use the rules for assigning oxidation numbers as shown in the text.

Solution:

The hydrogens and oxygens will be +1 and +2 respectively in all cases. In the MnO_4 we can write an equation to calculate the oxidation number of Mn following rule #3.

$$Mn + 4 \times O = -1$$

Since oxygen is –2 we get

$$Mn + (4x - 2) = -1$$

Solving this for Mn we get +7 for the oxidation number of manganese.

In a similar calculation we can determine that the Mn in MnO_2 must be +4. In SO_3^{2-} the oxidation number of sulfur is +4 and in the sulfate ion, SO_4^{2-}, the oxidation number is +6 using the same method.

Since the oxidation number on manganese is reduced from +7 to +4, we say that the Mn (also the permanganate ion) is reduced. Since the oxidation number on the sulfur increases from +4 to +6, the sulfur (also the sulfite ion) is oxidized. The permanganate ion is the oxidizing agent since it causes the oxidation of the sulfite ion. Similarly the sulfite ion is the reducing agent since it causes the permanganate to be reduced.

Is The Answer Reasonable?

The first test is to be sure that something gets oxidized *and* something else is reduced. This does occur. Conveniently, the substance that is *reduced* undergoes a *reduction* in oxidation number. That is true in this case and we are confident that the rest of the assignments are correct.

Self–Test

1. Consider the reaction of magnesium with fluorine to give the ionic compound MgF_2.

$$Mg + F_2 \longrightarrow MgF_2$$

   (a) Write separate equations showing the gain and loss of electrons.

   _____

   _____

   (b) Which substance is oxidized?   _____

   (c) Which substance is reduced?   _____

   (d) Which reactant is the oxidizing agent?   _____

   (e) Which reactant is the reducing agent?   _____

2. Assign oxidation numbers to each atom in the following:

   (a) NO_3^- _____

   (b) $SbCl_5$ _____

   (c) $CaHAsO_4$ _____

(d) ClF_3 _____

(e) I_3^- _____

(f) S_8 _____

3.   Determine whether the following changes are oxidation, reduction, or neither oxidation nor reduction.

(a) SO_3^{2-} to SO_4^{2-} _____

(b) Cl_2 to ClO_3^- _____

(c) N_2O_4 to NH_3 _____

(d)   PbO to $PbCl_4^{2-}$ _____

(e)   Ag to Ag_2S _____

4.   Consider the balanced equation,

$$3H_2O + 3Cl_2 + NaI \longrightarrow 6HCl + NaIO_3$$

(a)   Which substance is oxidized?   _____

(b)   Which substance is reduced?   _____

(c)   Which substance is the oxidizing agent?   _____

(d)   Which substance is the reducing agent?   _____

5.2   Balancing Redox Equations

Learning Objective

You will learn to balance equations for oxidation–reduction reactions in acidic or basic solutions.

Review

In applying the ion–electron method, we divide a redox equation into two half–reactions, balance the half–reactions separately, and then combine the balanced half–reactions to give the balanced overall net ionic equation. The method is not difficult to apply if you proceed in a stepwise fashion, without skipping any of the steps. If you can't obtain a balanced equation, it is probably because you haven't remembered to do things in sequence. Be sure you learn the seven steps summarized in Section 5.2 in the text for reactions that occur in acidic solutions.

In applying the ion–electron method, one of the most frequent causes for error is not remembering to write the correct charges on the formulas. You need these charges to get the correct numbers of electrons, so if you forget to write the charges on the ions, you will surely get wrong answers.

Another common problem students have is computing the net charge on each side of a half–reaction. Suppose we have reached the following stage in balancing a half–reaction.

$$6H_2O + N_2H_5^+ \longrightarrow 2NO_3^- + 17H^+$$

To obtain the charge contributed by each substance, multiply its coefficient by its charge.

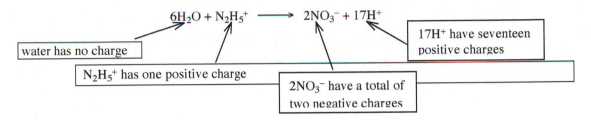

Thus, on the right side of this half–reaction the net charge is

$$[2 \times (1-)] + [17 \times (1+)] = 15+$$

The 2 is the coefficient of the NO_3^- and the 1– is the charge on the NO_3^-. Similarly, 17 is the coefficient of H^+ and 1+ is the charge on H^+.

In determining the number of electrons that must be added to balance a half–reaction, be especially careful when the charges on opposite sides of the half–reaction have *opposite* algebraic signs. For example, consider the following half–reaction just before we add electrons to it.

Net charge is 5+   Net charge is 1–

$$7H^+ + SO_3^{2-} \rightarrow HS^- + 3H_2O$$

The net charge on the left is 5+, the net charge on the right is 1–. The number of electrons we must add equals the *algebraic difference* between them.

$$\text{number of electrons to be added} = (5+) - (1-) = 6$$

The electrons are always added to the more positive (or less negative) side, so the half–reaction properly balanced is

$$6e^- + 7H^+ + SO_3^{2-} \longrightarrow HS^- + 3H_2O$$

Notice that the net charge is the same on both sides.

Reactions in basic solutions

To balance an equation for a reaction in basic solution, first balance it as if the solution were acidic. Then follow a three–step conversion to basic solution. First add hydroxide ions equal to the number of hydrogen ions; then add the same number of hydroxide ions to the other side of the equation (we must keep the equation balanced). Second, combine any pairs of H^+ and OH^- ions to make water molecules. Finally cancel any like terms. Notice that the result of the change over is to convert all the H^+ to H_2O and to add an equal number of OH^- to the other side. For example, when we add $7OH^-$ to both sides we get

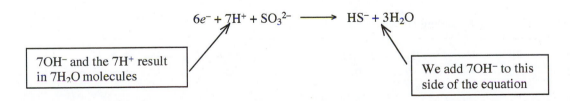

This gives

$$6e^- + 7H_2O + SO_3{}^{2-} \longrightarrow HS^- + 3H_2O + 7OH^-$$

Finally, we delete three water molecules from each side to give the balanced equation (a half–reaction, in this instance).

$$6e^- + 4H_2O + SO_3{}^{2-} \longrightarrow HS^- + 7OH^-$$

Example 5.2   Balancing REDOX reactions

Balance the following reaction, it occurs in acid solution

$$S_2O_3{}^{2-} + IO_3{}^- \longrightarrow I_2 + S_4O_6{}^{2-}$$

Analysis:

We need a balanced equation as described in Section 2.4.   After a few tries we may see that the inspection method does not work and we need a more precise approach such as the ion–electron method.

Assembling the Tools:

We will use the ion–electron method as our tool.

Solution:

The first step is to divide the reaction into two half–reactions. We can see that the obvious pairs are

1st Half–Reaction	2nd Half–Reaction
$IO_3{}^- \longrightarrow I_2$	$S_2O_3{}^{2-} \longrightarrow S_4O_6{}^{2-}$

We next balance all atoms except hydrogen and oxygen (notice where the coefficients were added below).

$$2IO_3{}^- \longrightarrow I_2 \qquad\qquad 2S_2O_3{}^{2-} \longrightarrow S_4O_6{}^{2-}$$

The next step is to balance oxygen atoms by adding one water molecule for each oxygen needed.

$$2IO_3{}^- \longrightarrow I_2 + \mathbf{6H_2O} \qquad\qquad 2S_2O_3{}^{2-} \longrightarrow S_4O_6{}^{2-}$$

Hydrogen atoms are balanced by adding H^+ where needed.

$$12H^+ + 2IO_3{}^- \longrightarrow I_2 + \mathbf{6H_2O} \qquad\qquad 2S_2O_3{}^{2-} \longrightarrow S_4O_6{}^{2-}$$

charges   12+   2–   0   0   4–   2–

We now count the charge on each side of the two half–reactions. As shown, the left side of the iodine half–reaction has twelve positive charges and two negative charges for a total of ten positive charges and zero charge on the right. The $2S_2O_3^{2-}$ half–reaction has four negative charges on the left and two negative charges on the right. (Failing to count charges properly is the most common error encountered in balancing half reactions.)

Now we balance the charge by adding electrons to bring the side with the more positive charge down to the value of the less positive (or perhaps negative) charge.

$$12H^+ \ + \ 2IO_3^- \ + \ 10e \longrightarrow I_2 \ + \ 6H_2O \qquad\qquad 2S_2O_3^{2-} \longrightarrow S_4O_6^{2-} \ + \ 2e$$

We equalize the electrons in both half–reactions by multiplying the right half–reaction by 5 before adding the half–reactions.

$$12H^+ \ + \ 2IO_3^- + \ 10e \longrightarrow I_2 \ + \ 6H_2O$$

$$10S_2O_3^{2-} \longrightarrow 5S_4O_6^{2-} \ + \ 10e$$

$$12H^+ \ + \ 2IO_3^- \ + \ 10S_2O_3^{2-} \longrightarrow I_2 \ + \ 6H_2O + + \ 5S_4O_6^{2-}$$

Add the two half–reactions and cancel electrons and any other IDENTICAL species. Check to see if the coefficients can be simplified and recheck that the equation is balanced.

Is the Answer Reasonable?

The ultimate check of a balanced equation is whether or not everything is balanced. Checking the charge balance is often sufficient. We count 22 negative charges and 12 positive charges for a total of 10 negative charges on the left. There are 10 negative charges on the right and the charges are balanced

	$12H^+$	$+$	$2IO_3^-$	$+$	$10S_2O_3^{2-}$	$\longrightarrow$	I_2	$+$	$6H_2O$	$+$	$+$	$5S_4O_6^{2-}$
charges	12+		2–		20–		0		0			10–

Self–Test

5. Balance the following half–reactions for an acidic solution.

   (a) $NO \longrightarrow NO_3^-$ _____

   (b) $Br_2 \longrightarrow BrO_3^-$ _____

   (c) $P_4 \longrightarrow HPO_3^{2-}$ _____

6. Balance the following half–reactions for a basic solution.

   (a) $Cl_2 \longrightarrow OCl^-$ _____

   (b) $AsO_4^{3-} \longrightarrow AsH_3$ _____

   (c) $S_2O_4^{2-} \longrightarrow SO_4^{2-}$ _____

7. Balance the following reaction that occurs in an acidic solution.

$$H_2SeO_3 + I^- \longrightarrow I_2 + Se$$

_____

8. Balance this reaction for a basic solution.

$$SeO_3^{2-} + I^- \longrightarrow I_2 + Se$$

_____

5.3   Acids as Oxidizing Agents

Learning Objective

To illustrate at the molecular level how a metal reacts with an acid.

Review

Every acid releases hydrogen ions, H^+ (which, of course, actually exist in solution as H_3O^+). Hydrogen ion is a mild oxidizing agent and can oxidize many metals. The products are the metal ion and H_2 gas. For example, tin reacts with H^+ to give H_2 and Sn^{2+}.

$$Sn(s) + 2H^+(aq) \longrightarrow Sn^{2+}(aq) + H_2(g)$$

Many metals dissolve in acids such as HCl. Some examples are iron, zinc, tin, aluminum, and magnesium. These metals are said to be *more active* than H^+. (In the next section, you will learn more about how you can tell whether a given metal will dissolve in acids.)

There are some metals that are not attacked by H^+, and they will not dissolve in acids that have H^+ as the strongest oxidizing agent. Such acids are called *nonoxidizing acids*. An example is HCl, which gives H^+ and Cl^- in solution. The Cl^- ion cannot gain any more electrons, so it cannot act as an oxidizing agent. Solutions of HCl therefore have H^+ as the only oxidizing agent.

Metals that won't dissolve in nonoxidizing acids often will dissolve in *oxidizing acids* such as HNO_3. Copper is the example that's given in the text Section 5.3. Notice that when the nitrate ion serves as the oxidizing agent, hydrogen gas is *not* among the products. Instead, the reduction product comes from the NO_3^- ion, which shows that NO_3^- is the oxidizing agent. Also, you should remember that when concentrated nitric acid is used as an oxidizing agent, the reduction product is usually NO_2, and when dilute nitric acid is used, the reduction product is usually NO. When a very strong reducing agent such as zinc reacts with NO_3^-, the nitrogen can be reduced all the way to the -3 oxidation state.

Hot concentrated H_2SO_4 is also an oxidizing agent in which SO_4^{2-} undergoes reduction to SO_2.

Self–Test

9. Hydrobromic acid, HBr, is a nonoxidizing acid. Write the chemical equation for its reaction with magnesium.

   HBr + Mg$_{(s)}$ → MgBr + H$_2$

10. Cadmium reacts with dilute solutions of sulfuric acid with the evolution of hydrogen. Write a chemical equation for the reaction.

   _____

11. In the reaction described in Question 10, which is the oxidizing agent and which is the reducing agent?

   Oxidizing agent _____

   Reducing agent _____

12. Mercury dissolves in concentrated nitric acid, with the evolution of a reddish–brown gas, just as in the reaction shown in Figure 5.4 for copper. The oxidation state of the mercury after reaction is +2. Write a balanced net ionic equation for the reaction.

   _____

13. Construct the balanced molecular equation for the reaction described in Question 12.

   _____

5.4   Redox Reactions of Metals

Learning Objective

To explain the activity series of metals and use it to predict the product of a redox reaction involving a metal.

Review

If one metal is more active than another, then the more active metal will be able to reduce the ion of the less active metal. In the reaction, the more active metal is oxidized. This is what happens when metallic zinc is placed into a solution that contains a soluble copper salt. The zinc (which is the more active metal) is oxidized to zinc ion and the copper ion is reduced to metallic copper.

The activity series, Table 5.3, lists metals in order of increasing activity (that is, in order of increasing ability to be oxidized and to serve as a reducing agent). Remember that any metal in this table is able to displace the ion of any metal above it from compounds. For instance, if you turn to Table 5.3 you will see that aluminum can reduce the ions of manganese, zinc, chromium, iron, and ions of all the other metals above it in the table.

Table 5.3 can also be used to determine if a given metal will dissolve in a nonoxidizing acid. If the metal is *below* hydrogen in the table, a nonoxidizing acid will react with it to give the metal ion and hydrogen gas. If a metal is above hydrogen in this table, an oxidizing acid such as HNO_3 or hot, concentrated H_2SO_4 must be used to dissolve it.

Example 5.3 Creating an activity series

Construct an activity series based on the fact that the reactions below are known to occur. List the best oxidizing agent first.

$$Cu + 2Ag^+ \longrightarrow Cu^{2+} + 2Ag$$
$$Fe + 2Ag^+ \longrightarrow Fe^{2+} + 2Ag$$
$$Sn + Cu^{2+} \longrightarrow Sn^{2+} + Cu$$
$$Fe + Sn^{2+} \longrightarrow Fe^{2+} + Sn$$
$$Sn + 2Ag^+ \longrightarrow Sn^{2+} + 2Ag$$

Analysis:

A substance that is reduced is an oxidizing agent. That means we want to focus on the reactants that are ions since they are reduced to their metals. Ions written as reactants must be stronger oxidizing agents than ions written as products, otherwise the reaction would have been written in the reverse direction. For the first reaction we can say that $Ag^+ > Cu^{2+}$ or the silver ion is a stronger oxidizing agent than the copper ion. We continue in this manner to get a list as shown in the solution.

Assembling the Tools:

The activity series of the metals is the tool needed to answer this question.

Solution:

From the reactions we can write the following list of oxidizing strengths:

$$Ag^+ > Cu^{2+}$$
$$Ag^+ > Fe^{2+}$$
$$Cu^{2+} > Sn^{2+}$$
$$Sn^{2+} > Fe^{2+}$$
$$Ag^+ > Sn^{2+}$$

Ag^+ is apparently a stronger oxidizing agent than all of the other three ions and we list Ag^+ first on our list and cross off reactions 1, 2, and 5 from the list, leaving

$$Cu^{2+} > Sn^{2+}$$
$$Sn^{2+} > Fe^{2+}$$

From this short list we see that $Cu^{2+} > Sn^{2+}$ and then $Sn^{2+} > Fe^{2+}$ which leads to the conclusion that Fe^{2+} must also be a weaker oxidizing agent than Cu^{2+}. Our list should be

$$Ag^+ > Cu^{2+} > Sn^{2+} > Fe^{2+}$$

Is the Answer Reasonable?

We can check our work and reasoning, but it is difficult to discern an error if there was one. A better check is to consult the activity series, Table 5.3 in Section 5.4. We see that silver metal is the least active reducing agent of the metals given, but that means that the silver ion is among the most active oxidizing agents. The other ions follow in the order we determined.

Self–Test

14. Write molecular equations for any reaction that will occur when the following reactants are combined:

(a)   Manganese metal added to a solution of $CoCl_2$.

_____

(b)   Iron metal added to a solution of $AuCl_3$.

_____

(c)   Cadmium metal added to a solution of $MgCl_2$.

_____

15. Which metal ions will be reduced if an excess amount of iron powder is added to a solution that contains a mixture of $Ca(NO_3)_2$, $Cu(NO_3)_2$, $AgNO_3$, KNO_3, $Pb(NO_3)_2$, and $Hg(NO_3)_2$?

_____

16. Which of the following metals will dissolve in a solution of HBr: aluminum, tin, cobalt, gold, mercury, manganese?

_____

5.5   Molecular Oxygen as an Oxidizing Agent

Learning Objective

To describe the reaction of oxygen with organic compounds, metals, and nonmetals.

Review

Molecular oxygen, O_2, is a very good oxidizing agent. In this section you learn about several kinds of reactions that are brought about by O_2. The purpose is to enable you to make reasonable predictions of the outcome of combustion reactions.

1. **Reactions of O_2 with organic compounds**. Organic compounds normally contain carbon, hydrogen, oxygen, and perhaps several other elements. When these compounds are burned in a plentiful supply of O_2, the carbon forms CO_2, the hydrogen forms H_2O, and if any sulfur is present in the compound, it is changed to SO_2. Any oxygen in the compound becomes incorporated in the other oxygen–containing products.

    If the combustion takes place in a limited supply of O_2, hydrogen still is changed to H_2O, but the carbon is only oxidized to CO. In extremely limited supplies of O_2, elemental carbon (soot) is formed.

2. **Reactions of O_2 with metals**. Many metals combine with oxygen directly. Aluminum, iron, and magnesium are given as examples in the text. The products are metal oxides. When the reactions are slow, they are described by the term corrosion or tarnishing. If the reaction is rapid, it's combustion.

3. **Reactions of O_2 with nonmetals**. These reactions give nonmetal oxides. An important nonmetal that does not react with O_2 at atmospheric pressure is nitrogen.

Example 5.4 Technique for balancing combustion reactions

Write and balance the chemical equation for the combustion of pentanol ($C_5H_{11}OH$).

Analysis:

We are informed this will be a combustion reaction and one reactant is given. For combustion reactions we know that the other reactant is O_2 and that the products are CO_2 and water. The specific technique for balancing combustion equations is then used.

Assembling the Tools:

We consider using tools for balancing combustion reactions to answer this question

Solution:

The reactants are butanol and molecular oxygen, O_2. The products, unless otherwise indicated, are assumed to be CO_2 and H_2O. After writing the reaction, balancing is done by balancing everything except the O_2, which is balanced last.

The unbalanced reaction is written as
$$C_5H_{11}OH \ + \ O_2 \ \longrightarrow \ CO_2 \ + \ H_2O$$

Carbon atoms are balanced first to get:
$$C_5H_{11}OH \ + \ O_2 \ \longrightarrow \ 5CO_2 \ + \ H_2O$$

Hydrogen atoms are balanced next:
$$C_5H_{11}OH \ + \ O_2 \ \longrightarrow \ 5CO_2 \ + \ 6H_2O$$

The O_2 is balanced last. There are 16 oxygens on the product side and one oxygen in the pentanol; leaving 15 oxygen atoms to be balanced by O_2. With a little math we see that 15/2 (or 7.5) O_2 molecules are needed to balance the atoms.

$$C_5H_{11}OH \ + \ 7.5O_2 \ \longrightarrow \ 5CO_2 \ + \ 6H_2O$$

We do not want fractional coefficients. If we multiply *all coefficients* by 2 we will get

$$2C_5H_{11}OH \ + \ 15O_2 \ \longrightarrow \ 10CO_2 \ + \ 12H_2O$$

This is the best answer for this reaction since simpler, whole–number coefficients cannot be found.

Is the Answer Reasonable?

The only reasonable answer for balancing an equation is if it is indeed balanced. We count 10 carbon atoms, 24 hydrogens, and 32 oxygen atoms on either side of the arrow. Since 15 is an odd number and the other coefficients are even, the coefficients cannot be simplified. (Note: If the sum of half the hydrogens, plus the oxygen and sulfur atoms in the formula, is an odd number, balancing will often be quicker if the coefficient of 2 is given to the organic compound before starting the balancing process.)

Self–Test

17. Complete and balance the following equations:

(a) ___C_2H_6 + ___O_2 ⟶ _____ $CO_2 + H_2O$ _____

(b) ___$C_6H_{12}O_6$ + ___O_2 ⟶ _____

(c) ___C_3H_6O + ___O_2 ⟶ _____

(d) ___C_2H_5SH + ___O_2 ⟶ _____

(e) 4 Al + __3__ O_2 ⟶ _____ 2 AlO_3 _____

(f) ___S + _____O_2 ⟶ _____ SO_3 _____

(g) ___Ca + _____O_2 ⟶ _____ CaO_2 _____

18. When burned in an excess supply of oxygen, phosphorus forms the compound P_4O_{10}. Write a chemical equation for the reaction.

_____

19. Write a balanced chemical equation for the combustion of benzene, C_6H_6, when there is a severely limited supply of O_2.

_____

5.6 Stoichiometry of Redox Reactions

Learning Objective

To perform calculations involving stoichiometry of a redox reaction.

Review

The general principles of stoichiometry discussed earlier apply to redox reactions as well. However, redox reactions are usually more complex than metathesis and their equations are similarly more complex.

The use of potassium permanganate, $KMnO_4$, as a reactant in redox titrations is discussed. The advantage of this substance is that the MnO_4^- ion is deeply colored, but its reduction product is almost colorless. Permanganate ion serves as its own indicator in a titration. When the last of the reducing agent has reacted, the next drop of the MnO_4^- solution causes the reaction mixture to take on a pink color. This signals the end point in the titration. Study the Analyzing and Solving Multi-Concept Problem at the end of Chapter 5. To test your understanding, review the following example and take the Self–Test below.

Example 5.5 Stoichiometry using a redox reaction

A sample of tin ore with a mass of 1.225 g was dissolved in acid and all the tin converted to Sn^{2+}. The solution required 24.33 mL of 0.0200 M $KMnO_4$ to reach an end point in a titration in which the Sn^{2+} was converted to Sn^{4+}. What was the percentage of SnO_2 in the original ore sample?

Analysis:

We are asked for the percentage of SnO_2 and we can write an equation for it.

$$\% \ SnO_2 \ = \ \frac{g \ SnO_2}{g \ sample} \ x \ 100$$

Since we know the mass of the sample, this tells us that we need to calculate the g SnO_2. It seems that we can get the g SnO_2 if we solve a stoichiometry problem. We will need to determine the molar mass of SnO_2 and also find the balanced equation. We see that one half–reaction is $Sn^{2+} \rightarrow Sn^{4+}$. The second half–reaction involves permanganate as a reactant. Permanganate has different products depending on the acidity of the solution. In this case the solution appears to be acid (we dissolved the ore in acid) and we will use the acid product, Mn^{2+}, and write the $KMnO_4$ as the permanganate ion MnO_4 to set up the other half–reaction. In our calculation we will calculate Sn^{4+} but also note that one Sn^{4+} is equivalent to one SnO_2 based on the chemical formula.

Assembling the Tools:

We need to balance the redox chemical equation with the ion–electron method (Section 5.2). Stoichiometry tools in Section 3.5 and titration tools in Section 4.8 are also needed here.

Solution:

Now we will put it all together. First we balance the chemical equation. The half–reactions are

$$MnO_4 + 8H^+ + 5e^- \longrightarrow Mn^{2+} + 4H_2O \quad \text{and} \quad Sn^{2+} \longrightarrow Sn^{4+} + 2e^-$$

These half–reactions are combined to obtain the balanced reaction

$$2MnO_4 + 16H^+ + 5Sn^{2+} \longrightarrow 2 \ Mn^{2+} + 8H_2O + 5Sn^{4+}$$

We now ask how many grams of SnO_2 are equivalent to 24.33 mL of $KMnO_4$ and set up the calculation.

$$24.33 \ mL \ MnO_4^- \Leftrightarrow ? \ g \ SnO_2$$

We will now use the appropriate conversion factors to reveal the answer.

$$24.33 \ mL \ MnO_4^- \left(\frac{0.0200 \ mol \ MnO_4^-}{1000 \ mL \ MnO_4^-} \right) \left(\frac{5 \ mol \ Sn^{4+}}{2 \ mol \ MnO_4^-} \right) \left(\frac{1 \ mol \ SnO_2}{1 \ mol \ Sn^{4+}} \right) \left(\frac{150.7 \ g \ SnO_2}{1 \ mol \ SnO_2} \right) \Leftrightarrow 0.183 \ g \ SnO_2$$

Now we can take the 0.183 g SnO_2 and calculate the percentage in the ore as

$$\% \ SnO_2 = \frac{0.183 \ g \ SnO_2}{1.225 \ g \ sample} \ x \ 100 \ = \ 14.9\% \ SnO_2$$

Is the Answer Reasonable?

First, the answer is less than 100%, which is a good indication. We can check our balanced equation and check to be sure all units cancel properly in our problem. Finally, we can estimate the answer by noting that 24.33 x 0.02 is approximately 0.5; multiplying that by 5/2 is a little more than 1. That leaves us with 150.7 divided by 1000, which is approximately 0.15 g, a number that is very close to our answer of 0.183 g.

Self–Test

20. (a) Write a balanced net ionic equation for the reaction of sulfurous acid with permanganate ion in an acidic solution. The products of the reaction are sulfate ion and manganese(II) ion.

    _____

    (b) How many milliliters of 0.150 M $KMnO_4$ are required to react with a solution that contains 0.440 g of H_2SO_3?

    _____

21. The reaction of MnO_4^- with Sn^{2+} in acidic solution follows the equation:

    $$2MnO_4^- + 5Sn^{2+} + 16H^+ \longrightarrow 2Mn^{2+} + 5Sn^{4+} + 8H_2O$$

    (a) How many grams of $KMnO_4$ must be used to react with 35.0 g of $SnCl_2$?

    _____

    (b) How many milliliters of 0.0500 M $KMnO_4$ solution would be needed to react with all the tin in 25.0 mL of 0.250 M $SnCl_2$ solution?

    _____

    (c) A 0.500 g sample of solder, which is an alloy containing lead and tin, was dissolved in acid and all the tin was converted to Sn^{2+}. The solution was then titrated with 0.0200 M $KMnO_4$ solution. The titration required 27.73 mL of the $KMnO_4$ solution. What is the percentage by mass of tin in the solder?

    _____

22. Sodium chlorite, $NaClO_2$, is used as a bleaching agent. A 2.00 g sample of a bleach containing this compound was dissolved in an acidic solution and treated with excess NaI. The net ionic equation for the reaction that took place is

    $$4H^+ + ClO_2^- + 6I^- \longrightarrow Cl^- + 2I_3^- + 2H_2O$$

    The solution containing the I_3^- was then titrated with 28.80 mL of 0.200 M $Na_2S_2O_3$ solution. The reaction that took place during the titration was

    $$I_3^- + 2S_2O_3^{2-} \longrightarrow 3I^- + S_4O_6^{2-}$$

    What is the percentage by mass of $NaClO_2$ in the bleach?

    _____

Answers to Self–Test Questions

1. (a) $Mg \longrightarrow Mg^{2+} + 2e^-$; $F_2 + 2e^- \longrightarrow 2F^-$, (b) Mg, (c) F_2, (d) F_2, (e) Mg

2. (a) $N, +5$; $O, -2$ (b) $Sb, +5$; $Cl, -1$ (c) $Ca, +2$; $H, +1$; $As, +5$; $O, -2$
   (d) $Cl, +3$; $F, -1$ (e) $I, -1/3$ (f) S, zero

3. (a) oxidation, (b) oxidation, (c) reduction, (d) neither, (e) oxidation

4. (a) NaI, (b) Cl_2, (c) Cl_2, (d) NaI

5. (a) $2H_2O + NO \longrightarrow NO_3^- + 4H^+ + 3e^-$
   (b) $6H_2O + Br_2 \longrightarrow 2BrO_3^- + 12H^+ + 10e^-$
   (c) $12H_2O + P_4 \longrightarrow 4H_3PO_3 + 12H^+ + 12e^-$

6. (a) $4OH^- + Cl_2 \longrightarrow 2OCl^- + 2H_2O + 2e^-$
   (b) $8e^- + 7H_2O + AsO_4^{3-} \longrightarrow AsH_3 + 11OH^-$
   (c) $8OH^- + S_2O_4^{2-} \longrightarrow 2SO_4^{2-} + 4H_2O + 6e^-$

7. $4H^+ + 4I^- + H_2SeO_3 \longrightarrow Se + 2I_2 + 3H_2O$

8. $3H_2O + 4I^- + SeO_3^{2-} \longrightarrow Se + 2I_2 + 6OH^-$

9. $Mg + 2HBr \longrightarrow MgBr_2 + H_2$

10. $Cd + H_2SO_4 \longrightarrow CdSO_4 + H_2$

11. Cd is the reducing agent, H^+ is the oxidizing agent.

12. $Hg + 2NO_3^- + 4H^+ \longrightarrow Hg^{2+} + 2NO_2 + 2H_2O$

13. $Hg + 4HNO_3 \longrightarrow Hg(NO_3)_2 + 2NO_2 + 2H_2O$

14. (a) $Mn + CoCl_2 \longrightarrow MnCl_2 + Co$, (b) $2AuCl_3 + 3Fe \longrightarrow 3FeCl_2 + 2Au$,
    (c) $Cd + MgCl_2 \longrightarrow$ no reaction

15. $Cu^{2+}, Ag^+, Pb^{2+}, Hg^{2+}$

16. Al, Sn, Co, Mn

17. (a) $2C_2H_6 + 7O_2 \longrightarrow 4CO_2 + 6H_2O$ (b) $C_6H_{12}O_6 + 6O_2 \longrightarrow 6CO_2 + 6H_2O$
    (c) $C_3H_6O + 4O_2 \longrightarrow 3CO_2 + 3H_2O$
    (d) $2C_2H_5SH + 9O_2 \longrightarrow 4CO_2 + 6H_2O + 2SO_2$
    (e) $4Al + 3O_2 \longrightarrow 2Al_2O_3$ (f) $S + O_2 \rightarrow SO_2$
    (g) $2Ca + O_2 \longrightarrow 2CaO$

18. $4P + 5O_2 \longrightarrow P_4O_{10}$

19. $2C_6H_6 + 3O_2 \longrightarrow 12C + 6H_2O$

20. (a) $5H_2SO_3 + 2MnO_4^- \longrightarrow 5SO_4^{2-} + 2Mn^{2+} + 4H^+ + 3H_2O$ (b) 14.3 mL

21. (a) 23.3 g $KMnO_4$, (b) 50.0 mL, (c) 32.9% Sn

22. 6.51% $NaClO_2$

Tools for problem solving

In this chapter you learned to apply the following concepts as tools in solving problems. Study each one carefully so that you know what each is used for. When faced with solving a problem, recall what each tool does and consider whether it will be helpful in finding a solution. This will aid you in selecting the tools you need.

You might want to remove these pages to use while solving problems in this chapter.

Oxidizing and reducing agents (Section 5.1)

The substance reduced is the oxidizing agent; the substance oxidized is the reducing agent.

Assigning oxidation numbers (Section 5.1)

The rules permit us to assign oxidation numbers to elements in compounds and ions.

Oxidation and reduction reactions (Section 5.1)

In an oxidation reaction, the substance loses electrons; in a reduction reaction, the reactant gains electrons. Oxidation and Reduction reactions always occur together.

Ion–electron method for acidic solutions, (Section 5.2)

Use this method when you need to obtain a balanced net ionic equation for a redox reaction in an acidic solution.

Ion–electron method for basic solutions, (Section 5.2)

Use this method when you need to obtain a balanced net ionic equation for a redox reaction in a basic solution.

Oxidizing and nonoxidizing acids (Table 5.2, Section 5.3)

Nonoxidizing acids will react with metals below hydrogen in Table 5.3 to give H_2 and the metal ion.

Activity series of metals (Table 5.3, Section 5.4)

A metal in the table will reduce the ion of any metal above it in the table, leading to a single replacement reaction.

Combustion reactions of hydrocarbons with oxygen (Section 5.5)

Hydrocarbon combustion with plentiful supply of oxygen

$$\text{hydrocarbon} + O_2 \rightarrow CO_2 + H_2O \text{ (plentiful supply of } O_2)$$

Hydrocarbon combustion with limited supply of oxygen

$$\text{hydrocarbon} + O_2 \rightarrow CO + H_2O \text{ (limited supply of } O_2)$$

Hydrocarbon combustion with extremely limited supply of oxygen

$$\text{hydrocarbon} + O_2 \rightarrow C + H_2O \text{ (very limited supply of } O_2)$$

Combustion of compounds containing C, H, and O (Section 5.5)

Complete combustion gives CO_2 and H_2O:

$$(\text{C, H, O compound}) + O_2 \rightarrow CO_2 + H_2O \text{ (complete combustion)}$$

Combustion of organic compounds containing sulfur (Section 5.5)

If an organic compound contains sulfur, SO_2 is formed in addition to CO_2 and H_2O when the compound is burned.

Summary of Useful Information

Rules for Assigning Oxidation Numbers

1.   The oxidation number of any free element is zero, regardless of how complex its molecules might be.

2.   The oxidation number of any simple, monatomic ion is equal to the charge on the ion.

3.   The sum of all the oxidation numbers of the atoms in a molecule or ion must equal the charge on the particle.

4.   In its compounds, fluorine has an oxidation number of -1.

5.   In its compounds, hydrogen has an oxidation number of $+1$.

6.   In its compounds, oxygen has an oxidation number of -2.

Ion–Electron Method — Acidic Solution

Step 1.   Divide the equation into two half–reactions.

Step 2.   Balance atoms other than H and O.

Step 3.   Balance O by adding H_2O.

Step 4.   Balance H by adding H^+.

Step 5.   Balance net charge by adding e^-.

Step 6.   Make e^- gain equal e^- loss; then add half–reactions.

Step 7.   Cancel anything that's the same on both sides.

Additional Steps for Basic Solution

Step 8.   Add to *both* sides of the equation the same number of OH^- as there are H^+.

Step 9.   Combine H^+ and OH^- to form H_2O.

Step 10. Cancel any H_2O that you can.

Notes:

Chapter 6

Energy and Chemical Change

We turn from the materials budgets of chemical reactions (Chapters 3, 4, and 5) to their energy budgets. Some chemical reactions will not occur until the reactants have received energy—usually heat energy but sometimes electrical energy, and sometimes other forms. Other reactions occur and spontaneously give off energy as heat, light, sound, electricity, or as mechanical energy (as from an explosion). We introduce here the important concepts needed to describe such energy changes. We also introduce some very important topics in thermodynamics.

6.1   Energy: The Ability to Do Work

Learning Objectives

Explain the difference between potential and kinetic energy and the law of conservation

of energy.

Review

Energy is the ability to do work and supply *heat*. *Work* is motion accomplished against an opposing force. For example, work is accomplished when you lift an object against the opposing force of gravity, or when you squeeze a spring that offers resistance.

The energy of a moving object is called its *kinetic energy* (KE) and is defined by the equation $KE = (1/2)mv^2$, where m is the mass and v is the velocity. *Potential energy* is stored energy. Potential energy exists in a stretched spring, a compressed spring, or in a book raised above the desk. These are examples that illustrate some of the ways by which potential energy can be stored. There are two important situations where we can tell that changes in potential energy have occurred.

1.   Potential energy increases by pulling apart objects that attract each other. (An example is lifting a book upward against the force of gravitational attraction.)

2.   Potential energy increases by pushing together objects that push back or repel each other (for example, pushing objects together that are held apart by a spring, which involves compressing the spring).

The existence of positively and negatively charged particles in atoms, ions, and molecules means that there are repelling and attracting forces in substances. Motions of these particles lead to changes in both kinetic and potential energies.

The SI unit of energy is the *joule (J)*, although we will often express energies in the larger unit, *kilojoule (kJ)*. The base SI units for the joule are kg m^2 s^{-2}. Older energy units are the calorie (cal) and kilocalorie (kcal). The nutritional Calorie unit is really the kilocalorie. Study the definitions in the text to be sure you understand how to apply the conversions between joules and calories, and between kilojoules and kilocalories. Be sure to learn the following conversion relationships between the energy units.

$$4.184 \text{ J} = 1 \text{ cal}$$

$$4,184 \text{ kJ} = 1 \text{ kcal}$$

The original definition of the calorie was based on how the absorption of heat affects the temperature of water; 1 cal is enough heat to raise the temperature of 1 g of water by 1 °C.

Heat is energy and it is spontaneously transferred from hotter objects to colder ones. The heat energy is transferred by the warmer object losing kinetic energy while the cooler object's molecules increase their kinetic energy. Heat and temperature are not the same thing. Temperature is directly related to the average kinetic energy of the particles in an object.

Example 6.1 Calculating kinetic energy for an object

What is the kinetic energy, in joules, of a 145 gram baseball thrown by a pitcher at 97 miles per hour?

Analysis:

We need to identify the equation that allows us to calculate kinetic energy and then we need to be sure we have the necessary information, with the correct units. The mass needs to be converted from g to kg and the velocity needs to be converted from miles per hour to m s^{-1} units.

Assembling the Tools:

We need the kinetic energy equation (KE = ½mv^2) for our tool.

Solution:

Converting g to kg results in 0.145 kg for the mass of the baseball. The conversion for the speed is

$$\frac{97 \text{ miles}}{1 \text{ hour}} \times \frac{1 \text{ km}}{0.62 \text{ miles}} \times \frac{1000 \text{ m}}{1 \text{ km}} \times \frac{1 \text{ hour}}{60 \text{ min}} \times \frac{1 \text{ min}}{60 \text{ s}} = \frac{43 \text{ m}}{\text{s}}$$

Now we can solve the equation for KE.

KE = ½(0.145 kg)(43 m s^{-1})^2 = 134 J = 1.3 x 10^2 J (properly rounded)

Is the Answer Reasonable?

First we check our calculations. Cancel the units above to satisfy yourself that the setup is correct. We could also estimate that the numerator is approximately 100,000 and the denominator is approximately 2,000, resulting in an approximate answer of 50 m s^{-1}, which is close to the speed calculated. We can do a similar estimate with the numbers for the KE and see that the value is reasonable. Finally, the units for joules are kg m^2 s^{-1}, which is what we have.

Self-Test

1. Describe how the kinetic energy and the potential energy of a stone thrown upward change during its upward motion, at the top of its trajectory, and during its downward flight.

---

2. Which actions decrease the potential energies of the systems?

    (a) Two electrons move away from each other.

    (b) A stretched rubber band is released.

    © Coal burns.

    (d) All of the above.   _____ b.

3. How much heat, measured in calories, is needed to raise the temperature of 38.7 g of water from 25.4 °C to 48.7 °C?

    _____

4. Express the answer to the preceding question in units of joules.   _____

5. How many kilocalories correspond to an energy of 77.3 kJ?   _____

6. How many joules are possessed by an 18.7 kg object traveling at a speed of 45.2 m/s?

    _____

6.2   Heat, Temperature, and Internal Energy

Learning Objective

Explain the connection between temperature, energy, and the concept of a state function.

Review

There are several key concepts introduced in this section. One is that the atoms and molecules of a substance are in constant motion and possess molecular kinetic energy. The sum of all the kinetic and potential energies of the particles in a substance is its internal energy, E. We use the Greek letter Δ (delta) to signify a change, so a change in internal energy is given the symbol ΔE. Changes in quantities are defined as "final" minus "initial." Therefore,

$$\Delta E = E_{final} - E_{initial} \quad \text{or} \quad \Delta E = E_{products} - E_{reactants}$$

The algebraic sign of the internal energy change defines the direction of energy flow. If ΔE is positive for a chemical reaction, then the products have more internal energy than the reactants. then

Another important concept is that the temperature of an object is related to the average molecular kinetic energy of its particles (atoms, molecules, or ions). Figure 6.4 is particularly important because it illustrates how the molecular kinetic energies are distributed among the molecules, and how this distribution changes with temperature. Notice how the shapes of the curves change with changing temperature. We will use these graphs in later chapters to help us understand properties of liquids and solids and how temperature affects the speeds of chemical reactions.

The concept of a *state function* is introduced in this section as well. The change in the value of a state function depends only on the initial and final conditions or *state*, not on how the change occurs. For example, temperature qualifies as a state function because a change in the temperature of an object depends only on its initial and final temperatures. If the final temperature is 30 °C and the initial temperature is 25 °C, the temperature changes, Δt, equals +5 °C. The quantity Δt doesn't depend on how the temperature went from 25 to 30 °C. State functions are useful quantities because we don't have to worry about the path from start to finish, only on the values in the initial and final states.

Finally, you learn in this section how molecular kinetic energy accounts for the transfer of heat by collisions between molecules.

Self-Test

7. Define internal energy.

_____

8. Considering the equation for kinetic energy, why does kinetic energy never have a negative value?

_____

9. Why is it impossible for a system to have a temperature lower than 0 K?

_____

10. When a substance cools, what happens to the average kinetic energy of its atoms, molecules, or ions?

_____

11. What is a state function? Give two examples.

_____

_____

12. Suppose that in a chemical reaction, the products have 35.6 kJ more energy than the reactants. What is the value (with appropriate algebraic sign) of ΔE?

_____

13. On a separate sheet of paper, make a graph that illustrates the distribution of molecular kinetic energies (i.e., how molecular kinetic energy varies with the fraction of molecules with that kinetic energy) for two different temperatures. Indicate on the graph the average kinetic energy for each distribution and indicate which distribution corresponds to the higher temperature. (Refer to Figure 6.4 to check your answer.)

6.3   Measuring Heat

Learning Objectives

Determine an amount of heat exchanged from the temperature change of an object.

Review

Two important terms that you must be sure to understand are *system* and *surroundings*, because almost everything we do in thermochemistry and thermodynamics relates to these concepts. The *system* consists of that part of the universe we are studying. All the rest is the surroundings. Separating the system and the surroundings is a *boundary*, which can be visible (like the walls of a beaker) or invisible (like the arbitrary boundary that separates the earth's atmosphere from outer space).

Be sure to understand the differences among open, closed, and isolated systems given in the the text. In this chapter we will be most interested in closed systems and will study the flow of energy between systems and their surroundings.

Heat flow is specified by the symbol q and is given a positive sign if heat enters a system and a negative sign if heat leaves. When heat moves across the boundary between system and surroundings, the sign of q for the system is just the opposite of the sign of q for the surroundings. Thus, if the system absorbs 30 J of heat, it must absorb the heat from the surroundings and the surroundings will lose 30 J of heat. For the system, $q = +30$ J and for the surroundings, $q = -30$ J. (When energy flows across the boundary between the system and surroundings, the amount of energy absorbed by one is always equal in magnitude to the amount of energy released by the other. The signs are opposite to indicate the direction of energy flow.)

The amount of heat that enters or leaves an object is directly proportional to the temperature change.

$$q \propto \Delta t$$

The proportionality constant is called the *heat capacity* and is given the symbol C. Inserting this constant into the expression above gives the equation

$$q = C\Delta t$$

This is Equation 6.5 in the text. Study Example 6.1 in the text.

Heat capacity is a property that depends on the size (mass) of the sample. If we divide the heat capacity C by the mass m (both extensive properties), we obtain a quantity that is independent of sample size. We call it the specific heat, s. (This is similar to dividing mass by volume to obtain the intensive property density.)

$$s = C/m$$

Solving for C gives $C = s\,m$, and substituting into the equation for heat gives

$$q = sm\Delta t$$

This is Equation 6.7 in the textbook. Study Example 6.2 in the text to see how we use specific heat in a calculation.

Example 6.2 Determining heat capacity and specific heat

A 26.35 g sample of a metal is heated to 100 °C in boiling water, dried quickly and then dropped into 225 mL of distilled water at 22.15 °C in a foam cup. The temperature of the water increases to 23.33 °C. What is the heat capacity of this sample and specific heat of this metal?

Analysis:

We need to find the heat capacity and the specific heat of this sample. We need to know the amount of heat, q, transferred. We'll calculate q from the increase in temperature of the water itself, and ignore the mass of the foam cup.

Assembling the Tools:

We need the tools for the heat capacity (Equation 6.5 in text) and specific heat (Equation 6.7 in text). These are $C = q/\Delta t$ and $s = q/(m\Delta t)$ respectively.

Solution:

First we calculate the heat exchanged to the water from the equation $q = ms\Delta t$.

$$q = 225 \text{ g H}_2\text{O } (4.184 \text{ J g}^{-1} \,^{\circ}\text{C})(23.33 \,^{\circ}\text{C} - 22.15 \,^{\circ}\text{C})$$

$$= 1.11 \times 10^3 \text{ J}$$

The heat gained by the water is equal to the heat lost by the metal. Therefore q for the metal is -1.11×10^3 J. Solving the two equations gives us

$$C = -1.11 \times 10^3 \text{ J}/(23.33 \,^{\circ}\text{C} - 100 \,^{\circ}\text{C})$$

$$= -1.11 \times 10^3 \text{ J}/-76.67 \,^{\circ}\text{C}$$

$$= 14.5 \text{ J} \,^{\circ}\text{C}^{-1}$$

$$s = -1.11 \times 10^3 \text{ J}/(26.35 \text{ g})(23.33 \,^{\circ}\text{C} - 100 \,^{\circ}\text{C})$$

$$= -1.11 \times 10^3 \text{ J}/(-2020 \text{ g} \,^{\circ}\text{C})$$

$$= 0.549 \text{ J g}^{-1} \,^{\circ}\text{C}^{-1}$$

Is the Answer Reasonable?

We check that the units are correctly written and result in the expected units for each part of the problem, and they do. Next we can make some approximate calculations. For q we see that 225 times 4.184 is approximately 1000 and the temperature change is close to one degree. Therefore q should be about 1000 J, and it is. The equation for C divides 1000 by approximately 100 and the answer should be in the neighborhood of 10, and it is. Since s is C divided by the mass of the sample, we see that if we round 14.5 to 15 and 26.35 to 25, that ratio is 15/25 or 3/5 or 0.6, which is close to the 0.549 we calculated.

Self-Test

14. What is the difference between an open and a closed system?

_____

15. Suppose that a change occurs to a beaker of water and $q = -85$ J.

    (a) Will the temperature of the water increase or decrease? _____

    (b) What is the value of q for the surroundings? _____

16. A 52.9 g sample of nickel required 446 J to raise its temperature from 35.8 °C to 49.6 °C.

    (a) What is the heat capacity of the nickel sample (with correct units)? _____

    (b) What is the specific heat of nickel (with correct units)? _____

    (c) How many joules would be needed to raise the temperature of 255 g of nickel from 35.6 °C to 55.9 °C?

    _____

    (d) If 455 J were removed from a 78.2 g sample of nickel initially at 63.2 °C, what would the temperature change to?

    _____

6.4   Energy of Chemical Reactions

Learning Objectives
Describe the energy changes in exothermic and endothermic reactions.

Review

The net attractive force that exists between two atoms in a molecule is called a *chemical bond*. The nature of these attractions will be studied in detail later. For now, our attention is on the potential energy associated with chemical bonds, because when bonds form and break, potential energy is lost and gained, respectively. This kind of potential energy, which we associate with chemical bonds in molecules, is often called *chemical energy*. Remember the following:

- Breaking a chemical bond requires energy, so it is accompanied by an increase in the potential energy (chemical energy) of the atoms.

- Forming a chemical bond releases energy, so it is accompanied by a decrease in the potential energy (chemical energy) of the atoms.

When a chemical reaction involves both the breaking and forming of bonds, there is almost always a net change in potential energy. Overall, if there is a net increase in attractions, there will be a lowering of the potential energy. Let's see what this does to the temperature of the system.

If the potential energy decreases, it can't just disappear; that would violate the law of conservation of energy. Instead, the PE that's lost changes to molecular KE. This causes the *total* molecular KE to rise and also leads to an increase in the *average molecular KE*. As we noted earlier, when the average molecular KE increases, so does the temperature. The conclusion, then, is that forming stronger bonds (with greater attractions between atoms) leads to a rise in temperature. This is what is described for the reaction of methane with oxygen in the text.

When the temperature of a reacting system rises, we can view it as resulting from an increase in the amount of heat in the system. This heat comes from the reaction that's taking place, so we can view the heat as a product of the reaction. Chemical reactions in which heat is released by the chemicals are said to be *exothermic*; those in which heat is consumed and changed to potential energy are said to be *endothermic*.

It is easy to tell whether a reaction is exothermic or endothermic. If the reaction mixture becomes warmer as the reaction occurs, the reaction is exothermic. If the mixture becomes cooler, the reaction is endothermic.

Self-Test

17. How does the average molecular kinetic energy of the molecules of a system change during an exothermic reaction among them?

18. When the products of a reaction have less potential (chemical) energy than the reactants, will the reaction be exothermic or endothermic?

19. When potassium iodide dissolves in water, it dissociates and the ions become hydrated. We could write the dissociation reaction as

$$KI(s) + nH_2O \longrightarrow K^+(aq) + I^-(aq)$$

where the water that surrounds the ions is written as a reactant. Forming a solution of KI in water is endothermic.

(a) Does the mixture become warmer or cooler as the KI dissolves?

(b) Rewrite the equation showing heat as a reactant or product, whichever is appropriate.

(c) Overall, do the attractions increase or decrease as the dissociation takes place?

(d) What happens to the molecular KE and chemical energy during the dissociation process (increase or decrease)?

6.5 Heat, Work, and the First Law of Thermodynamics

Learning Objectives
State the first law of thermodynamics and explain how it applies to chemistry.

Review

This section reviews the determination of heat, q, released or absorbed in a chemical process. Work is presented as another form of energy that can accompany a chemical process. Work is most easily measured as $P\Delta V$, and the units of pressure (force/area) multiplied by the units of volume give us the units for work.

Pressure is a ratio of force to area. The gases of the atmosphere exert a pressure (atmospheric pressure) equal to approximately 14.7 lb in.$^{-2}$. This pressure is very close to the pressure units called the *standard atmosphere* (*atm*) and the *bar*.

Heats of reaction measured at constant volume are indicated by q_v and those measured at constant pressure are indicated by q_p. Study Example 6.4, which illustrates that q_p may not equal q_v, especially for reactions that involve gases.

The difference between q_p and q_v arises from pressure–volume work, which is given by Equation 6.9. Pressure–volume work is another way for a system to transfer energy. According to the *first law of thermodynamics*, the change in internal energy of a system is a function of the heat, q, added to the system and the work, w, done on the system. (Compressing a gaseous system is an example of work done on or to a system.)

$$\Delta E = q + w$$

It's important to keep the algebraic signs of q and w straight. Both are positive when energy is being added to the system (i.e., when heat is added or the system has work done on it). Be sure to study the sign conventions about q and w given in Section 6.5 in the text.

The values of q and w depend on how a change is carried out, as the example of the discharging battery of Figure 6.8 in the text explains. Therefore, q and w are not state functions. Nevertheless, their algebraic sum, ΔE, is the difference between two state functions, $E_{final} - E_{initial}$.

Self-Test

20. A certain process gives off 204 J of heat to the surroundings while the energy needed to compress the system requires 68 J of energy. What is ΔE for this process? _____

21. What *PV* work is done when a gas is allowed to expand at a constant temperature from 2.50 L (and 4.00 atm pressure) to 10.0 L against an opposing pressure of 1.00 atm? _____

22. What is the total *PV* work when a gas is expanded at a constant temperature in the following two steps? First, the gas expands from 2.50 L (and 4.00 atm pressure) to 5.00 L against an opposing pressure of 2.00 atm. Second, the gas from step 1 expands from 5.00 L to 10.0 L with an opposing force of 1.00 atm. _____

23. Consider Questions 21 and 22. (a) Do the two systems both have the same initial and final states? (b) Is the *PV* work the same for both systems? (c) Can you deduce whether or not *PV* work is a state function?

_____

_____

6.6   Heats of Reaction

Learning Objectives

Explain the difference between a heat of reaction obtained at constant pressure and the heat of reaction at constant volume.

Review

The heat absorbed or given off during a chemical reaction is called the heat of reaction and is measured using a calorimeter either under conditions of constant volume (using a rigid container that cannot change in volume) or at constant pressure.

Heats of reaction measured at constant volume are indicated by q_v and those measured at constant pressure are indicated by q_p. Study Example 6.3, which illustrates that q_p may not equal q_v, especially for reactions that involve gases.

If a system cannot change in volume during a reaction, then the heat of reaction at constant volume, q_v, equals ΔE for the change.

$$\Delta E = q_v \text{ (where } q_v \text{ is the heat of reaction at constant volume)}$$

Study Example 6.5, which describes measuring q_v using a bomb calorimeter. Solve Practice Exercises 11 and 12.

Enthalpy and enthalpy changes

The "corrected" internal energy of a system at constant pressure is called the system's *enthalpy*. Like internal energy, enthalpy is a state function. By definition, $H = E + PV$, so when P is a constant, a change in H can be represented by

$$\Delta H = \Delta E + P\Delta V$$

When the only kind of work a system can do is expansion work (i.e., pressure–volume work) against the opposing atmospheric pressure, then ΔH equals the heat of reaction at constant pressure, q_p.

$$\Delta H = q_p \text{ (where } q_p \text{ is the heat of reaction at constant pressure)}$$

For most reactions, particularly those that do not involve gases, the difference between ΔE and ΔH is very small and can be neglected. However, the difference can be significant for reactions involving gases.

As with the internal energy, a change in enthalpy is the difference between the enthalpies of the initial and final states.

$$\Delta H = H_{final} - H_{initial}$$

or

$$\Delta H = H_{products} - H_{reactants}$$

The value of ΔH is positive for endothermic reactions and negative for exothermic ones. Study the Analyzing and Solving Multi-Concept Problems exercise in this section of the text.

Example 6.3 Calculating the heat of combustion using a bomb calorimeter

When 0.1000 mol of aspirin (acetylsalicylic acid) was burned in a bomb calorimeter, the temperature of the system rose from 24.000 °C to 27.332 °C. The heat capacity of the calorimeter was 118.0 kJ °C^{-1}. The equation for the reaction is

$$C_9H_8O_4(s) + 9O_2(g) \longrightarrow 9CO_2(g) + 4H_2O(l)$$

(a)   How many kilojoules were liberated by the combustion of this sample?

(b)   How many kilojoules were liberated per mole of aspirin?

(c)   What is the heat of combustion for aspirin?

Analysis:

We need a relationship between temperature change and energy. From our discussion we see that we can use the equation for calculating the heat, q, using the heat capacity.

For part (b), we see that the amount of heat we calculate in part (a) is for 0.1000 mol of sample. A simple conversion can result in the heat from 1.000 mol of aspirin.

Assembling the Tools:

We need Equation 6.5 and our stoichiometry tools to determine the heat per mole of aspirin.

Solution:

Solving for the heat energy we get

$$q = 118.0 \text{ kJ °C}^{-1}(\ 27.332 \text{ °C} - 24.000 \text{ °C}) = 393.2 \text{ kJ}$$

Since the 393.2 kJ was produced by 0.1000 mol of aspirin we can calculate

$$393.2 \text{ kJ}/(0.1000 \text{ mol aspirin}) = 3932 \text{ kJ mol}^{-1} \text{ for aspirin}$$

Finally, the last answer is the same as the heat of combustion.

Is the Answer Reasonable?

We check that we have the correct equations. Then the math is approximately 100 x 3.3 for an estimate of 330 for the first answer, and we got 393, which is close enough. Dividing our first answer by 0.1 is the same as multiplying by 10. Multiplying by 10 gives us the answer we calculated and we have confidence that the answer is correct.

Self-Test

24. To what state functions do q_p and q_v each belong? _____

25. Give the definition of enthalpy in terms of internal energy, pressure, and volume.

    _____

26. When a reacting system in an insulated container becomes warmer, does its total energy change? Explain.

    _____

27. ΔH_{system} is defined as

    (a) $H_{reactants} - H_{products}$

    (b) $H_{products} - H_{reactants}$

    (c) $\Delta H_{products} - \Delta H_{reactants}$

    (d)   the change in the heat capacity of the system.

28. If we had some way of knowing that $H_{products}$ = 400 kcal and that $H_{reactants}$ = 600 kcal, then the $\Delta H_{reaction}$ is

    (a)   1000 kcal

    (b)   200 kcal

    (c)   –200 kcal

    (d)   –1000 kcal

29. If the value of ΔH_{system} = –200 J, and the system's temperature before and after the change is the same, then the value of $\Delta H_{surroundings}$ is

    (a)   –200 J

    (b)   200 J

    (c)   0

    (d)   not measurable

30. An 8.72 kg mass of water in a well-insulated vat was warmed from 19.75 °C to 19.50 °C. How much energy entered the water? Give the answer in J, kJ, cal, and kcal. The specific heat of water in this temperature range is 4.179 J g^{-1} °C^{-1}.

---

31. Hydrochloric acid, HCl(aq), can be made by bubbling hydrogen chloride, HCl(g), into water. When HCl(g) was bubbled into 225.00 g of water at 25.00 °C until the solution had a mass of 226.16 g, the temperature rose to 27.51 °C. Assume that the specific heat of the water is 4.184 J g^{-1} °C^{-1} throughout the reaction and the change in temperature. Calculate $\Delta H_{\text{solution}}$ for this change in kJ/mol HCl.

---

32. The heat of the reaction between KOH and HBr solutions was determined using a coffee cup calorimeter. The reaction was

$$\text{KOH}(aq) + \text{HBr}(aq) \longrightarrow \text{KBr}(aq) + \text{H}_2\text{O}$$

A solution of 45.0 mL of 1.00 M HBr at 23.5 °C was mixed with a solution of 45.0 mL of 1.00 M KOH, also at 23.5 °C. The temperature quickly rose to 30.2 °C. Assume that the specific heats of each solution is 4.18 J g^{-1} °C^{-1}, that the densities of the solutions are 1.00 g mL^{-1}, and that the system loses no heat to its surroundings. Calculate the heat of reaction in kilojoules per mole of HCl.

6.7   Thermochemical Equations

Learning Objective

Describe the assumptions and utility of thermochemical equations.

Review

The major concept in this section is the *thermochemical equation,* what it is and how to write one.

The value of ΔH for a given reaction depends on both the temperature and the pressure at which the system is kept. To compare ΔH values for different reactions, they must be measured under the same conditions. For enthalpy experiments, the *standard conditions* are 25 °C and 1 atm of pressure (and 1 M concentrations of reactants in solution). Values of ΔH measured under these conditions are called *standard heats of reaction* (or *standard enthalpies of reaction*), and the associated symbol is $\Delta H°$.

The actual value of $\Delta H°$ depends on the *scale* of the reaction—the actual mole quantities of substances as specified by the coefficients in the balanced equation. An equation that includes $\Delta H°$ is called a *thermochemical equation*. The following are all valid thermochemical equations for the reaction of hydrogen and oxygen that gives water.

(1) $\qquad 2\text{H}_2(g) + \text{O}_2(g) \longrightarrow 2\text{H}_2\text{O}(l) \qquad\qquad \Delta H° = -571.5 \text{ kJ}$

(2) $\qquad 3\text{H}_2(g) + 3/2\text{O}_2(g) \longrightarrow 3\text{H}_2\text{O}(l) \qquad\qquad \Delta H° = -857.3 \text{ kJ}$

(3) $\qquad 4\text{H}_2(g) + 2\text{O}_2(g) \longrightarrow 4\text{H}_2\text{O}(l) \qquad\qquad \Delta H° = -1143 \text{ kJ}$

The value of $\Delta H°$ differs for each because, as the coefficients indicate, the scale changes.   Always be careful to notice the physical states of the substances involved (*s, l,* or *g*), because enthalpy data depend on these, too. Notice also that the coefficients in a thermochemical equation do not have to be whole numbers and that they are not necessarily the smallest set of coefficients needed to balance the equation.   This is because the coefficients specify numbers of *moles*, not individual molecules, and it is possible to have fractions of moles.

Self-Test

33.   What are the experimental conditions to which standard enthalpy data refer?

34.   The thermochemical equation for the combustion of 2 mol of $CO(g)$ is

$$2CO(g) + O_2(g) \longrightarrow 2CO_2(g) \qquad \Delta H° = -566 \text{ kJ}$$

Write the thermochemical equation for the combustion of 6 mol of $CO(g)$.

6.8   Hess's Law

Learning Objective

Use Hess's law to determine the enthalpy of a reaction.

Review

Because enthalpy is a state function, it doesn't matter how we proceed from an initial state to a final state. The enthalpy change, ΔH, will be the same along any path as long as we begin and end at the same initial and final states. This interesting fact allows us to calculate an unknown ΔH for a reaction by combining known ΔH values for a different set of reactions that take us from the same reactants to the same products.

Suppose, for example, that we wished to determine $\Delta H°$ for the reaction

$$2NO(g) + O_2(g) \longrightarrow 2NO_2(g)$$

We could use the following two thermochemical equations to accomplish this.

(1) $\qquad\qquad N_2(g) + O_2(g) \longrightarrow 2NO(g) \qquad \Delta H° = +180.7 \text{ kJ}$

(2) $\qquad\qquad N_2(g) + 2O_2(g) \longrightarrow 2NO_2(g) \qquad \Delta H° = +67.6 \text{ kJ}$

An *enthalpy diagram* can be used to obtain a graphical solution to the problem, as illustrated below.

The vertical axis is standard enthalpy, $H°$, as indicated by the heavy arrow pointing upward at the left. The line at the bottom represents the enthalpy of $N_2(g)$ plus $2O_2(g)$, which we've written as $O_2(g) + O_2(g)$. This will make it a bit easier to see what's happening.

Combining $N_2(g)$ with $O_2(g)$ to form $2NO(g)$ produces an enthalpy increase of $+180.7$ kJ, so the enthalpies of these products are shown higher on the enthalpy scale. (Recall that a positive value of ΔH means the products have a higher enthalpy than the reactants.) Notice that we've carried along an extra mole of O_2 that hasn't reacted. We'll use that shortly.

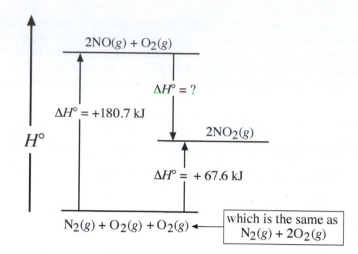

The second reaction combines $N_2(g)$ with $2O_2(g)$ to give $2NO_2(g)$, which gives an enthalpy increase of $+67.6$ kJ. Therefore, the $2NO_2(g)$ is shown higher on the enthalpy scale, too [but not as high as $2NO(g) + O_2(g)$].

Examining the diagram, we can see that to go from $2NO(g) + O_2(g)$, at the top of the diagram, to $2NO_2(g)$ we will move downward on the enthalpy scale. This means $H°$ is decreasing, so $\Delta H°$ will be negative. The *size* of the difference is also easy to determine from the graph; we just take the difference between 180.7 kJ and 67.6 kJ, or 113.1 kJ. Therefore, for the reaction $2NO(g) + O_2(g) \rightarrow 2NO_2(g)$, $\Delta H° = -113.1$ kJ.

Combining thermochemical equations without an enthalpy diagram

Hess's law, given in this section tells us we can combine Equations 1 and 2 above to obtain the same results as with the enthalpy diagram. To accomplish this, we set up the given thermochemical equations so that when we add them together we obtain the desired equation (let's call it the *target equation*). *If the sum of the chemical equations gives the target equation, then the sum of their $\Delta H°$ values will equal the $\Delta H°$ of the target reaction.*

Before we can add the equations, we have to adjust them so that, when added together, the sum is the target equation. We will use the rules for manipulating thermochemical equations given in the text. In this case, we will reverse Equation 1, which will change the sign of $\Delta H°$. Let's do that and then add the two equations.

(1 reversed)	$2NO(g) \rightarrow N_2(g) + O_2(g)$	$\Delta H° = -180.7$ kJ
(2)	$N_2(g) + 2O_2(g) \rightarrow 2NO_2(g)$	$\Delta H° = +67.6$ kJ

$$2NO(g) + \cancel{N_2(g)} + \cancel{2}O_2(g) \rightarrow \cancel{N_2(g)} + \cancel{O_2(g)} + 2NO_2(g) \qquad \Delta H° = -113.1 \text{ kJ}$$

Notice that we've canceled $N_2(g)$ as well as one $O_2(g)$ from each side of the final equation. What's left is our target equation. Also notice that we've obtained the same value of $\Delta H°$ as when we used the enthalpy diagram.

Applying Hess's law can sometimes seem like a daunting project, but it's not so bad if you proceed slowly. The following example illustrates a strategy you can follow. We suggest you read through the solution from beginning to end and then go back and study it carefully.

Example 6.4 Manipulating thermochemical equations

Calculate $\Delta H°$ for the reaction of copper(II) sulfide, $CuS(s)$, with gaseous hydrogen to give metallic copper and hydrogen sulfide.

$$CuS(s) + H_2(g) \longrightarrow Cu(s) + H_2S(g)$$

Use the thermochemical equations below.

(1) $2CuS(s) + 3O_2(g) \longrightarrow 2CuO(s) + 2SO_2(g)$ $\Delta H° = -807.18 \text{ kJ}$

(2) $2CuO(s) + C(s) \longrightarrow 2Cu(s) + CO_2(g)$ $\Delta H° = -83.05 \text{ kJ}$

(3) $3H_2(g) + SO_2(g) \longrightarrow H_2S(g) + 2H_2O(l)$ $\Delta H° = -292.32 \text{ kJ}$

(4) $C(s) + O_2(g) \longrightarrow CO_2(g)$ $\Delta H° = -393.51 \text{ kJ}$

(5) $2H_2(g) + O_2(g) \longrightarrow 2H_2O(l)$ $\Delta H° = -571.70 \text{ kJ}$

Analysis:

We have two goals as we proceed. The first is to rearrange the listed equations so that we obtain each substance in the target equation (the one we want to work toward) on the correct side of the arrow with the correct coefficient. The second goal is to be sure anything that should *not* be in the target equation will cancel when we add the adjusted equations.

Before you get experience with these kinds of problems, the biggest difficulty is deciding where to start. There aren't any pat rules, but probably the best advice is to pick one of the compounds in the target equation *that appears in only one of the listed equations.* If we have more than one such choice, then just select one of them and begin.

Assembling the Tools:

Adding equations, covered in Section 5.2, and the process of manipulating thermochemical equations are two tools needed for this problem and Hess's law is another.

Solution:

There are many paths to a solution of a problem like this. To illustrate one, first scan the equations to find compounds in the target equation that appear in only one of the listed equations. The formulas $CuS(s)$, $Cu(s)$, and $H_2S(g)$ fit this requirement. Notice, however, that $H_2(g)$ does not; it appears in both Equations (3) and (5), so we have no way of knowing which equation to adjust and in what way. We will have to leave $H_2(g)$ until near the end.

Let's begin with $CuS(s)$. In the target, it appears on the left with a coefficient of 1. In Equation (1) $CuS(s)$ is on the left, but its coefficient is 2. We'll divide all the coefficients of Equation (1) by 2 and we'll divide its $\Delta H°$ by 2 as well. We suggest you follow along by listing the equations on a piece of scrap paper as we adjust them. Then, in the end, you'll have them arranged so you can simply add them.

(1-adjusted) $\boxed{CuS(s)} + \frac{3}{2}O_2(g) \longrightarrow CuO(s) + SO_2(g)$ $\Delta H° = -403.59 \text{ kJ}$

We've placed a box around the CuS so we know we have it where we want it with the correct coefficient.

Next, let's work on $Cu(s)$. This appears in the second equation, but its coefficient has to be divided by 2 also. Therefore, we divide all the coefficients of Equation (2) by 2, including the value of $\Delta H°$.

(2-adjusted) $\quad CuO(s) + \frac{1}{2}C(s) \longrightarrow \boxed{Cu(s)} + \frac{1}{2}CO_2(g)$ $\qquad \Delta H° = -41.52$ kJ

Now we'll work on $H_2S(g)$, which is in Equation (3). This compound is on the correct side and its coefficient is what we want, so we don't have to do anything to this equation. Let's just rewrite it, placing a box around the H_2S.

(3) $\quad 3H_2(g) + SO_2(g) \longrightarrow \boxed{H_2S(g)} + 2H_2O(l)$ $\qquad \Delta H° = -292.32$ kJ

Let's combine these three equations, ignoring their $\Delta H°$ for the time being, so we can see where we stand.

$$\boxed{CuS(s)} + \frac{3}{2}O_2(g) \rightarrow \cancel{CuO(s)} + \cancel{SO_2(g)}$$
$$\cancel{CuO(s)} + \frac{1}{2}C(s) \rightarrow \boxed{Cu(s)} + \frac{1}{2}CO_2(g)$$
$$3H_2(g) + \cancel{SO_2(g)} \rightarrow \boxed{H_2S(g)} + 2H_2O(l)$$
$$\overline{\boxed{CuS(s)} + \frac{3}{2}O_2(g) + \frac{1}{2}C(s) + 3H_2(g) \rightarrow \boxed{Cu(s)} + \frac{1}{2}CO_2(g) + \boxed{H_2S(g)} + 2H_2O(l)}$$

Notice that the SO_2 and CuO cancel, but there is still work to do. We need to cancel $\frac{3}{2}O_2(g)$, $\frac{1}{2}C(s)$, and $2H_2(g)$ from the left, and $\frac{1}{2}CO_2(g)$ and $2H_2O(l)$ from the right. We'll do this by using the remaining thermochemical equations.

We can eliminate the $\frac{1}{2}C(s)$, $\frac{1}{2}O_2(g)$, and $\frac{1}{2}CO_2(g)$ by reversing Equation (4) and dividing its coefficients by 2.

(4-adjusted) $\qquad \frac{1}{2}CO_2(g) \longrightarrow \frac{1}{2}C(s) + \frac{1}{2}O_2(g)$ $\qquad \Delta H° = +196.76$ kJ

Notice that this places $\frac{1}{2}CO_2(g)$ on the left, so it will cancel with the $\frac{1}{2}CO_2(g)$ on the right. We similarly will be able to cancel $\frac{1}{2}C(s)$ and $\frac{1}{2}O_2(g)$. This still leaves $\frac{2}{2}O_2(g)$ to be removed, as well as $2H_2(g)$. We can finally clear these up by reversing Equation (5).

(5-reversed) $\qquad 2H_2O(l) \longrightarrow 2H_2(g) + O_2(g)$ $\qquad \Delta H° = +571.70$ kJ

Let's check to be sure everything cancels to give the target equation. Then we can add the $\Delta H°$ values to get $\Delta H°$ for the target equation.

$$\boxed{CuS(s)} + \cancel{\tfrac{3}{2}O_2(g)} + \cancel{\tfrac{1}{2}C(s)} + \boxed{\overset{1}{\cancel{3}H_2(g)}} \rightarrow \boxed{Cu(s)} + \cancel{\tfrac{1}{2}CO_2(g)} + \boxed{H_2S(g)} + \cancel{2H_2O(l)}$$
$$\cancel{\tfrac{1}{2}CO_2(g)} \rightarrow \cancel{\tfrac{1}{2}C(s)} + \cancel{\tfrac{1}{2}O_2(g)} \qquad \boxed{\begin{array}{l}\text{These combine}\\\text{to give }\frac{3}{2}O_2(g)\end{array}}$$
$$\cancel{2H_2O(l)} \rightarrow \cancel{2H_2(g)} + \cancel{O_2(g)}$$

Target equation $\qquad CuS(s) + H_2(g) \rightarrow Cu(s) + H_2S(g)$

Spend some time to study this so you can see how everything cancels except the formulas of the compounds in the target equation.

Equation Used	$\Delta H°$
(1) adjusted by dividing coefficients by 2	–403.59 kJ
(2) adjusted by dividing coefficients by 2	–41.52 kJ
(3) as is	–292.32 kJ
(4) reversed and coefficients divided by 2	+196.76 kJ
(5) reversed	+571.70 kJ

$$\text{Net } \Delta H° \;=\; +31.03 \text{ kJ}$$

Therefore, for the target equation we can write

$$CuS(s) + H_2(g) \longrightarrow Cu(s) + H_2S(g) \qquad \Delta H° = 31.02 \text{ kJ}$$

Is the Answer Reasonable?

The best check is to note that when we add the adjusted equations, we obtain the target equation. We can check to be sure we've followed the rules for adjusting the $\Delta H°$ values when manipulating thermochemical equations. Once we've done that, we need to check to be sure we've done the arithmetic correctly. After all that, we can feel confident in our answer.

The example we've just worked out is about as complicated as they get. If you're careful and follow the procedure we've outlined above, you will get correct answers. Just be sure the adjusted equations add to give the target equation, and keep your eye on the target as you make decisions as to how to manipulate the equations.

Self-Test

35. The direct reaction of methane (CH_4) with oxygen to give carbon monoxide is hard to accomplish without also producing carbon dioxide. However, we can use a Hess's law calculation to calculate $\Delta H°$, anyway. Calculate $\Delta H°$ in kJ for the following reaction from the thermochemical equations given.

$$2CH_4(g) + 3O_2(g) \longrightarrow 2CO(g) + 4H_2O(l)$$

Use the following thermochemical equations:

$$(1) \quad 2C(s) + O_2(g) \longrightarrow 2CO(g) \qquad \Delta H° = -221.08 \text{ kJ}$$

$$(2) \quad CH_4(g) + 2O_2(g) \longrightarrow CO_2(g) + 2H_2O(l) \qquad \Delta H° = -890.4 \text{ kJ}$$

$$(3) \quad C(s) + O_2(g) \longrightarrow CO_2(g) \qquad \Delta H° = -393.51 \text{ kJ}$$

36. Acetic acid can be made from acetylene by the following overall reaction:

$$2C_2H_2(g) + 2H_2O(l) + O_2(g) \longrightarrow 2HC_2H_3O_2$$

acetylene   acetic acid

Calculate $\Delta H°$ for the formation of *one* mole of acetic acid by this reaction using the following thermochemical equations:

(1) $C_2H_4(g) \longrightarrow H_2(g) + C_2H_2(g)$ $\Delta H° = 174.464$ kJ
    ethylene

(2) $C_2H_5OH(l) \longrightarrow C_2H_4(g) + H_2O(l)$ $\Delta H° = 44.066$ kJ
    ethyl alcohol

(3) $C_2H_5OH(l) + O_2(g) \longrightarrow HC_2H_3O_2(l) + H_2O(l)$ $\Delta H° = -495.22$ kJ

(4) $2H_2(g) + O_2(g) \longrightarrow 2H_2O(l)$ $\Delta H° = -571.70$ kJ

6.9   Standard Heats of Reaction

Learning Objective

Determine and use standard heats of formation to solve problems.

Review

Thermochemical data is available in several formats, including heats of combustion. The standard heat of combustion, $\Delta H_c°$, is the energy released in the complete combustion of 1 mol of a compound in oxygen, all substances being in their standard states. For example, for butane, $\Delta H_c°$ corresponds to $\Delta H°$ for the reaction

$$C_4H_{10}(g) + \frac{13}{2}O_2(g) \longrightarrow 4CO_2(g) + 5H_2O(l)$$

Study Example 6.9 to see how we can use $\Delta H_c°$ in calculations.

The standard enthalpy of formation (also called the standard heat of formation), $\Delta H_f°$, is the amount of heat absorbed or released in the formation of 1 mol of a compound at 25 °C and a pressure of 1 bar from its elements in their standard states. The key phrase here is "from its elements in their standard states," which means from the elements as we normally find them at 25 °C and ordinary atmospheric pressure. Be sure to examine the thermochemical equations which illustrate this point. Notice that $\Delta H_f°$ has units of energy per mole (e.g., kJ mol–1).

Heats of formation are useful for calculating heats of reaction using the Hess's law equation (Equation 6.14 in the text).

$$\Delta H° = (\text{sum of } \Delta H_f° \text{ of products}) - (\text{sum of } \Delta H_f° \text{ of reactants})$$

Study Example 6.11 in this Section of the text to see how we apply the Hess's law equation. Problems of this kind are not difficult as long as you are careful about the algebraic signs of the energy terms. Take your time to avoid errors.

Example 6.5 Using $\Delta H_f°$ to calculate $\Delta H°$ of a reaction

Calculate $\Delta H°$ for the target reaction

$$CuS(s) + H_2(g) \longrightarrow Cu(s) + H_2S(g)$$

Use the following $\Delta H_f°$ data.

$$CuS(s) \quad \Delta H_f° = -48.53 \text{ kJ mol}^{-1}$$

$$H_2S(g) \quad \Delta H_f° = -17.506 \text{ kJ mol}^{-1}$$

Analysis:

We need to find the heats of formation for each of the reactants and products and combine them as discussed in this section. The first simplification is that all elements at standard state have heats of formation equal to zero. We don't have to look up $H_2(g)$ and $Cu(s)$. The problem gives us the heats of formation for the other two substances so we don't have to look any further for the data to solve the problem.

Assembling the Tools:

Hess's law is the tool we need for this problem.

Solution:

It seems obvious that Hess's law equation is applicable in this situation and we have seen that it is summarized as

$$\Delta H° = (\text{sum of } \Delta H_f° \text{ of products}) - (\text{sum of } \Delta H_f° \text{ of reactants})$$

$$\Delta H° = (-17.506 \text{ kJ mol}^{-1} \times 1 \text{ mol } H_2S) - (-48.53 \text{ kJ mol}^{-1} \times 1 \text{ mol CuS})$$

$$= (-17.506 \text{ kJ mol}^{-1} \times 1 \text{ mol } H_2S) + (+48.53 \text{ kJ mol}^{-1} \times 1 \text{ mol CuS})$$

$$= +31.02 \text{ kJ}$$

Is the Answer Reasonable?

This is a simple addition/subtraction problem. We have correctly combined the two minus signs to make the second term a positive number. Since 48 is larger than 17 we expect an answer that is positive (endothermic).

Self-Test

37. Any substance in its most stable chemical form at 25 °C and 1 bar is said to be in its

(a) thermochemical state   (b)   standard state

(c) STP state   (d)   zero H state   _____

38. Write the thermochemical equations together with ΔH_f° values (Table 6.2 in the textbook) for the formation of each of the following compounds from its elements. Use units of kJ/mol.
    (a) $H_2SO_4(l)$ (b) $Al_2O_3(s)$ (c) $CO(NH_2)_2(s)$

    _____

    _____

39. Write the thermochemical equations, including correct values of ΔH° in kJ, for the formation of the specified amount of each of the following compounds from its elements. Use Table 6.2 in the textbook for necessary data.

    (a)   2 mol of Fe_2O_3

    (b)   3 mol of $C_2H_6(g)$

    _____

40. Draw an enthalpy diagram for the formation of one mole of H_2O_2 from its elements. Use data in Table 6.2 of the textbook in units of kJ as needed.

41. Use the Hess's law equation and ΔH_f° data in Table 6.2 of the textbook to calculate ΔH° for the reaction of methane with oxygen given in Problem 35. (Having done 35 the long way you'll see the powerful advantage of the Hess's law equation.)

42. Use the Hess's law equation, data in Table 6.2 of the textbook, and the results of the calculations in Problem 36 to calculate ΔH_f° for acetic acid in kJ mol^{-1}.

43. Calculate $\Delta H^\circ_{combustion}$ for the combustion of acetylene, $C_2H_2(g)$, in kJ mol^{-1} from ΔH_f° data from Table 6.2 of the textbook. The equation is

$$2C_2H_2(g) + 5O_2(g) \longrightarrow 4CO_2(g) + 2H_2O(l)$$

Answers to Self-Test Questions

1. On the upward flight, kinetic energy changes to potential energy. At the top, where the stone is motionless, all of the initial kinetic energy has become potential energy. On the way down, potential energy changes to kinetic energy as the stone picks up speed.

2. d

3. 902 cal

4. 3.77×10^3 J

5. 18.5 kcal

6. 3.82×10^4 J

7. E equals the sum of all KE and PE of atoms, molecules, and ions in a sample.

8. KE = $1/mv^2$; mass (m) is never negative and even if velocity (v) were negative, the square of the velocity must be a positive number. Hence, KE is always positive (or zero). (Note: As you may have already studied in a physics course, velocity can be negative because it is a vector quantity.)

9. Because there is no such thing as negative kinetic energy and because 0 K corresponds to the cessation of all motions associated with molecular kinetic energy.

10. The average molecular kinetic energy decreases.

11. A change in a state function depends only on the initial and final states, not on the path taken by the system as it changes from one state to the other. Examples are temperature and internal energy.

12. ΔE = 35.6 kJ

13. See Figure 6.4 in the textbook.

14. Open system: exchanges matter and energy with surroundings. Closed system: only exchanges energy.

15. (a) decrease, (b) +85 J

16. (a) 32.3 J °C^{-1}, (b) 0.610 J g^{-1} °C^{-1}, (c) 3.16×10^3 J, (d) 53.7 °C

17. It increases.

18. exothermic

19. (a) cooler, (b) heat + KI(s) + nH$_2$O $\longrightarrow$ K$^+$(aq) + I$^-$(aq), (c) decrease, (d) KE decrease, PE increase

20. ΔE = −136 kJ

21. 7.5 L atm

22. 5.0 L atm + 5.0 L atm = 10.0 L atm

23. *PV* work depends on the path and is not a state function.

24. q_v = ΔE at constant volume; q_p = ΔH at constant pressure

25. $H = E + PV$

26. No. Law of conservation of energy.

27. b

28. c

29. b

30. 27×10^3 J, 27 kJ, 6.5×10^3 cal, 6.5 kcal

31. 74.9 kJ/mol HCl

32. 55 kJ/mol HCl

33. 25 °C and 1 atm

34. 6CO(g) + 3O$_2$(g) $\rightarrow$ 6CO$_2$(g) $\Delta H°$ = 1698 kJ

35. −1214.9 kJ

36. −427.90 kJ
37. b
38. (a) $H_2(g) + S(s) + 2O_2(g) \longrightarrow H_2SO_4(l)$ $\Delta H_f^\circ = -811.32$ kJ mol^{-1}

    (b) $2Al(s) + \frac{3}{2}O_2(g) \longrightarrow Al_2O_3(s)$ $\Delta H_f^\circ = -1669.8$ kJ mol^{-1}

    © $C(s) + \frac{1}{2}O_2(g) + N_2(g) + 2H_2(g) \longrightarrow CO(NH_2)_2(s)$ $\Delta H^\circ = -333.19$ kJ mol^{-1}

39. (a) $4Fe(s) + 3O_2(g) \longrightarrow 2Fe_2O_3(s)$ $\Delta H^\circ = -1644$ kJ

    (b) $6C(s) + 9H_2(g) \longrightarrow 3C_2H_6(g)$ $\Delta H^\circ = -254.00$ kJ

40.

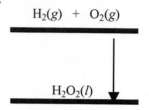

$\Delta H^\circ = -187.6$ kJ   Because ΔH° is negative, the product is placed lower than the reactants.

41. −1214.9 kJ
42. −487.05 kJ
43. 1299.7 kJ mol^{-1}

Tools for problem solving

In this chapter you learned to apply the following concepts as tools in solving problems. Study each one carefully so that you know what each is used for. When faced with solving a problem, recall what each tool does and consider whether it will be helpful in finding a solution.

You might want to tear these pages out to use along with solving problems in this chapter.

Kinetic energy (Section 6.1)

$$KE = \frac{1}{2}mv^2$$

Potential energy change. (Section 6.1)

You should know how potential energy varies when the distance changes between objects that attract or repel.

Heat capacity and *q* (Section 6.3)

$$q = C\Delta t$$

Specific heat capacity (Section 6.3)

Also called specific heat, *s*, it is an intensive property. When mass, *m*, and temperature change, Δt, are known, *q* is calculated by the equation

$$q = ms\Delta t$$

Heat transfer (Section 6.3)

When heat is transferred between two objects, the size of *q* is identical for both objects, but the algebraic signs of *q* are opposite.

$$q_1 = -q_2$$

First law of thermodynamics (Section 6.5)

$$\Delta E = q + w$$

The algebraic sign of ΔH (Section 6.6)

The sign of ΔH indicates the direction of energy flow.

For an endothermic change: ΔH is positive.

For an exothermic change: ΔH is negative.

Thermochemical equations (Section 6.7)

These equations show enthalpy changes ($\Delta H°$) for reactions where the amounts of reactants and products in moles are represented by the coefficients.

Enthalpy diagrams (Section 6.8)

We use enthalpy diagrams to provide a graphical picture of the enthalpy changes associated with a set of thermochemical equations that are combined to give some net reaction.

Hess's law (Section 6.8)

This law allows us to combine thermochemical equations to give a final desired equation and its associated $\Delta H°$.

Manipulating thermochemical equations (Section 6.8)

Changing the direction of a reaction changes the sign of $\Delta H°$. When the coefficients are multiplied by a factor, the value of $\Delta H°$ is multiplied by the same factor. Formulas can be canceled only when the substances are in the same physical state. These rules are used in applying Hess's law.

Hess's law equation (Section 6.9)

$$\Delta H° = (\text{sum of } \Delta H_f° \text{ of products}) - (\text{sum of } \Delta H_f° \text{ of reactants})$$

Summary of Important Equations

Heat capacity:

$$\text{Heat capacity} = \frac{\text{heat absorbed}}{\Delta t}$$

Specific heat:

$$\text{Specific heat} = \frac{\text{heat capacity}}{\text{mass}(g)}$$

Using specific heat:

$$\text{Heat energy} = \text{specific heat} \times \text{mass} \times \Delta t$$

Hess's law equation:

$$\Delta H° = (\text{sum of } \Delta H_f° \text{ of products}) - (\text{sum of } \Delta H_f° \text{ of reactants})$$

Chapter 7

The Quantum Mechanical Atom

In Chapter 0 you learned that atoms are not the simplest particles of matter. They are composed of still simpler parts called protons, neutrons, and electrons. The protons and neutrons of an atom are found in the nucleus and the electrons surround the nucleus in the atom's remaining volume. Because the nucleus is so small and so far from the outer parts of an atom, it has very little direct influence on ordinary chemical and physical properties. Its indirect role is in determining the number of electrons a neutral atom will have. As you will learn in this chapter, it is the number of electrons and their energy distribution (which we call the atom's *electronic structure*) that is the primary factor in determining an atom's chemical and physical properties.

An important fact about electrons in atoms is that they do not obey the classical laws of physics, which would predict that an orbiting electron should quickly crash into the nucleus. In fact, tiny particles like the electron sometimes behave like particles and at other times they behave as waves, a phenomenon we call *wave/particle duality*. The understanding of atomic structure came about through the development of *quantum theory*, also called *quantum mechanics*. We introduce you to some of the concepts and results of the theory, but (mercifully!) we cannot delve too deeply into it because of its highly mathematical nature.

You will find that much of this chapter deals with theory, and you may find that a lot of it seems remote from everyday experience. If you find it difficult to understand at first, don't be discouraged. Reread the discussions, study the worked examples, and do the exercises. It may take some time, but it should all fit together eventually.

7.1   Electromagnetic Radiation

Learning Objective

Describe light as both a particle and a wave, and use the equations describing the wave and particle nature of light.

Review

Electromagnetic energy is energy that is carried by *electromagnetic radiation* (light waves). These waves are characterized by their *frequency*, ν (the number of oscillations per second), which is given in units called hertz (Hz). Remember,

$$1 \text{ Hz} = 1 \text{ s}^{-1}$$

(s^{-1} means *second* raised to the minus one power, which is the same as 1/second).

$$1 \text{ s}^{-1} = \frac{1}{s} = \frac{1}{\text{second}}$$

A wave is also characterized by its *wavelength*, λ, which is the distance between any two successive peaks. In the SI, wavelength is expressed in meters (or submultiples of meters, such as nanometers).

The product of wavelength and frequency is the speed of the wave, and for electromagnetic radiation traveling through a vacuum, this speed is a constant called the *speed of light* (symbol, c). Be sure you know the equation

$$\lambda \times \nu = c$$

Check with your teacher to find out whether you are expected to know the value of c, 3.00×10^8 m/s (or 3.00×10^8 m s^{-1}). Study Examples 7.1 and 7.2 to be sure you can change wavelength to frequency and frequency to wavelength.

The electromagnetic spectrum

Electromagnetic radiation comes in a range of frequencies or wavelengths that we call the electromagnetic spectrum. Visible light constitutes only a narrow band of wavelengths ranging from approximately 400 to 700 nm. Infrared, microwaves, and radio and TV broadcasts are waves of longer wavelength (lower frequency) than visible light. Ultraviolet, X rays, and gamma radiation are waves of shorter wavelength (higher frequency) than visible light. The absorption of electromagnetic radiation, particularly in the infrared, visible, and ultraviolet, by molecules and atoms allows us to characterize them based on patterns of wavelengths absorbed.

The energy of light

Light often behaves as though it consists of a stream of tiny packets (also called *quanta*) of energy that we call *photons*. The energy of a photon is proportional to the radiation's frequency. The equation is

$$E = h\nu$$

where E is the energy of the photon, ν is its frequency, and h is a proportionality constant called Planck's constant. Ask your teacher whether you are expected to memorize the value of Planck's constant ($h = 6.62 \times 10^{-34}$ J s). The relationship between energy and frequency tells us that light with high frequencies (short wavelengths) consists of photons of high energy, whereas light with low frequencies (long wavelengths) has photons of lower energy.

Example 7.1 Calculating the energy of photons

How many microwave photons, each with a wavelength of 1.05 cm, must be absorbed to raise the temperature of a cup of tea (240 mL) from room temperature (25 °C) to 42 °C? Assume the tea has a density of 1.05 g/mL and a specific heat of 4.18 J g^{-1} °C^{-1}.

Analysis:

We can calculate the energy of one photon from the equation $E = h\nu$, but first we must convert the wavelength to frequency with the equation $\lambda\nu = c$. (Alternatively, we can combine these two equations to give $E = hc/\lambda$.) To find the number of such photons required, we have to calculate the total energy needed. We obtain this from the product of the specific heat of the tea (4.18 J g^{-1} °C^{-1}), the mass of the tea (252 g), and the temperature change that the tea undergoes (17 °C).

specific heat × mass × temp. change = energy

Once we know the total amount of energy needed (in joules), we divide by the energy per photon to calculate the number of photons needed. (The answer is 3.04×10^{31} photons.)

Assembling the tools:

From previous chapters we need the tools for calculating heat; conversion factors from the English system to the SI units, and the wavelength, frequency, and energy relationships in this section.

Solution:

The energy on one microwave photon is

$$E = hc/\lambda = (6.63 \times 10^{-34} \text{ J s})(3.00 \times 10^{8} \text{ m s}^{-1})/(1.05 \times 10^{-2} \text{ m}) = 1.89 \times 10^{-23} \text{ J (per photon)}$$

The energy needed to heat the cup of tea is

$$q = (4.184 \text{ J g}^{-1} \text{ °C}^{-1})(240 \text{ mL} \times 1.05 \text{ g mL}^{-1})(42 \text{ °C} - 25 \text{ °C}) = 1.79 \times 10^{4} \text{ J}$$

The number of photons will be $q/E = 1.79 \times 10^{4} \text{ J}/(1.89 \times 10^{-23} \text{ J}) = 9.47 \times 10^{26}$ photons.

Is the Answer Reasonable?

Because photons carry a very small amount of energy, the large number of photons we calculated is expected and reassures us the answer is correct. We can check calculations (ignoring units for the time being). In the first calculation we combine the exponents $(-34 + 8 - (-2)) = -24$ and our exponent is -23, so that calculation is most likely correct. In the second calculation we see that the first two terms are approximately 1000 and the 17 degree temperature change gives a product of 17,000, which is very close to 1.79×10^4. For the last calculation, again the exponents are $4 - (-23) = 27$, which is close to the 26 of the answer.

Self-Test

1. Radio station KROQ in Los Angeles broadcasts at a frequency of 106.7 MHz on the FM radio band. What is the wavelength in meters of these radio waves?

   _____

2. A certain compound is found to absorb infrared radiation strongly at a wavelength of 2.5×10^{-6} m. What is the frequency of this radiation?

   _____

3. What is the wavelength in nanometers of electromagnetic radiation that has a frequency of 3.3×10^{15} Hz?

   _____

4. Calculate the energy in joules of a photon of red light that has a wavelength of 646 nm.

   _____

5. What is the energy in kilojoules of one mole of photons having a wavelength of 285 nm?

   _____

6. Arrange the following kinds of electromagnetic radiation in order of decreasing energy (highest → lowest): X rays, radio waves, infrared radiation, ultraviolet light, visible light, microwaves.

7.2   Line Spectra and the Rydberg Equation

Learning Objective

Calculate the wavelength of a line in the spectrum of hydrogen and relate it to the energy levels of an electron.

Review

Atomic spectra

A *continuous spectrum* (Figure 7.7a) contains all wavelengths of light and is produced by the glow of a hot object, such as the filament in a light bulb, or the sun. The spectrum emitted by energized (*excited*) atoms is called an *atomic spectrum* or *emission spectrum* and is not continuous. It contains only a relatively few wavelengths, as illustrated in Figures 7.7b and c.

The atomic spectrum is different for each element, and can be used to identify the element; it's like the element's "fingerprint." Atomic spectra suggest that the energy of an electron in an atom is quantized, which means the electron can have only certain specific amounts of energy. These energies correspond to a characteristic set of *energy levels* possessed by atoms of the element. When an electron goes from one energy level to a lower one the difference in energy, ΔE, appears as light having a frequency determined by the equation $\Delta E = h\nu$.

The Rydberg equation is an *empirical equation*, which means it was derived from data obtained by experimental measurements. It can be used to calculate the wavelengths of the lines in the spectrum of hydrogen as illustrated in Example 7.3 in the text. (You probably don't need to memorize the Rydberg equation, but check with your instructor to be sure.)

Self-Test

7. Use the Rydberg equation to calculate the wavelength in nanometers of the spectral line produced when an electron drops from the sixth Bohr orbit to the second. What color is this spectral line?

7.3   The Bohr Theory

Learning Objective

Using the Bohr model of hydrogen, explain how a line spectrum of hydrogen is generated.

Review

Atomic spectrum of hydrogen and the Bohr model of the atom

Bohr developed his model of the atom to explain the spectrum of hydrogen. You should know the basic features of the model:

1. The electron was believed to travel in circular orbits.

2. The energy and size of an orbit can have only certain values, which are related to the value of the quantum number n.

3. When the atom absorbs energy, the electron is raised to a higher, more energetic orbit.

4. When the electron drops from a higher to a lower orbit, a photon is emitted whose energy equals the energy difference between the two orbits. The frequency of the photon is determined by the relationship $\Delta E = h\nu$.

Bohr was able to use his model to derive an equation that matched the Rydberg equation almost exactly. This equation was $E = -b/n^2$ (where $b = 2.18 \times 10^{-18}$ J). This was the theory's greatest success.

Although Bohr's model of the atom was later shown to be incorrect, he was the first to recognize the existence of quantized energy levels in atoms. He was also the first to introduce the idea of a *quantum number*—an integer related to the energy of an electron in an atom.

Example 7.2 Determining the energy of electron transitions

How much energy, in joules, is needed to excite an electron from the third energy level to the sixth energy level in a hydrogen atom?

Analysis:

We need to determine the energy difference between these two levels in the hydrogen atom. This can be done if we can determine the energy of each level and then take the difference between the two. Equation 7.3 in the text, $E = -b/n^2$ (where $b = 2.18 \times 10^{-18}$ J) gives us the needed equation to calculate the energy for each level to solve this problem.

Assembling the Tools:

Equation 7.3 in the text, $E = -b/n^2$ (where $b = 2.18 \times 10^{-18}$ J) gives us the needed equation to calculate the energy for each level to solve this problem. Another tool that could be used is the Rydberg equation.

Solution:

The energy at $n = 3$, we will call it E_3, is

$$E_3 = -(2.18 \times 10^{-18} \text{ J})/9 = -2.42 \times 10^{-19} \text{ J}$$

Similarly the energy at the sixth level is

$$E_6 = -(2.18 \times 10^{-18} \text{ J})/36 = -6.06 \times 10^{-20} \text{ J}$$

Recall that whenever we calculate a difference in chemistry it is always the initial state subtracted from the final state or

$$\Delta E = E_{\text{difference}} = E_6 - E_3 = -6.06 \times 10^{-20} \text{ J} - (-2.42 \times 10^{-19} \text{ J}) = 1.81 \times 10^{-19} \text{ J}$$

Is the Answer Reasonable?

The best check is that when an electron is excited, energy must be put into the system and the sign for the energy must be positive in that case (endothermic), and it is. Next, since we are calculating the energy change of one electron, we expect the number to be very small, and the 10^{-19} is a very small number as expected.

Self-Test

8. According to Bohr's model of the hydrogen atom, what was the value of the quantum number for the lowest-energy orbit?

   _____

9. In terms of b in Equation 7.3, what is the energy of the electron in the first Bohr orbit ($n = 1$)?

   _____

   What is the energy of the electron in the second Bohr orbit ($n = 2$)?

   _____

   In terms of b, how much energy is emitted when the electron drops from the orbit with $n = 2$ to the orbit with $n = 1$?

   _____

7.4   The Wave Mechanical Model

Learning Objective

Describe the main features of the wave mechanical model of an atom.

Review

De Broglie (pronounced De Broy) proposed that particles have wave properties, each with a wavelength, λ, that is inversely proportional to the product of the particle's mass and velocity.

$$\lambda = \frac{h}{mv}$$

The proportionality constant, h, is Planck's constant. Only very light particles, such as the electron, proton, and neutron, have wavelengths large enough to be observed experimentally. Diffraction—the constructive and destructive interference of waves—can be used to demonstrate the wave nature of these particles.

In this section, we examine two types of waves, traveling waves and standing waves (ones in which the positions of the peaks, troughs, and nodes don't move). The discussion about notes on a guitar string shows that standing waves lead naturally to integer relationships concerning the wavelengths that can exist on a string. Bringing this together with the de Broglie relationship for an electron on a string shows that the energy of the electron, behaving like a wave, is restricted in its values. In other words, the energy of the electron is quantized. The significant point here is that this quantized behavior is a direct result of the wave characteristics of the electron.

7.5   Quantum Numbers of Electrons in Atoms

Learning Objective

Define and use the three major quantum numbers and their possible values.

Review

Wave mechanics (quantum mechanics) is the name we give to the theory about the electronic structures of atoms and molecules. In this theory, the electron in an atom is considered to be a three-dimensional standing wave.   Electron waves are described by a mathematical expression called a wave function, usually represented by the symbol ψ. Each individual wave function describes an *orbital*, which is another term used to describe an electron wave.   Each orbital is identified by a set of values for three quantum numbers. You should learn the names of the quantum numbers and their allowed values (including the restrictions on their values).

The *principal quantum number*, n, can only have integer values that range from 1 to ∞.

$$n = 1 \text{ or } 2 \text{ or } 3 \text{ or } ...\infty$$

In categorizing the energies of the orbitals, n identifies the shell. An orbital with $n = 1$ is in the first shell, and so forth. The energy and size of an orbital is determined in large measure by its value of n.

The *secondary quantum number*, l, divides the shells into subshells. This quantum number determines the shapes of the orbitals, and to some degree, their energies as well. For a given n, allowed values of l range from 0 to $(n-1)$. Remember the letter designations for the subshells.

value of l	0	1	2	3
letter	s	p	d	f

The *magnetic quantum number*, m_l, divides the subshells into individual orbitals and determines the orientations of the orbitals relative to each other. In a given subshell, the values of m_l range from $+l$ to $-l$. These values determine the number of orbitals in a given kind of subshell.

type of subshell	s	p	d	f
number of orbitals	1	3	5	7

These relationships are illustrated in Figure 1 on the next page, which should be studied in conjunction with Table 7.1 of the text. The relative energies of the various subshells and orbitals are described by Figure 7.18 of the text. You need not memorize this figure, although we will use it in Section 7.7 to determine how the electrons in an atom are distributed among the various orbitals.

In Figure 7.18 of the text, there are several points to note. First, we see that every shell has an *s* subshell. Shells above the first shell each have a *p* subshell. Any shell above the second shell has a *d* subshell as well, and beyond the third shell, each also has an *f* subshell. Second, notice that the energies of the shells increase with increasing value of n. Third, notice that within a shell the energies of the subshells are in the order $s < p < d < f$, and that all orbitals of a given subshell are of equal energy. Finally, notice how the subshells of one shell begin to overlap with subshells of other shells as n becomes larger.

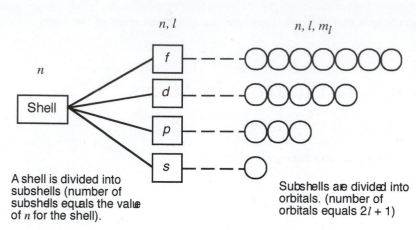

n, l   *n, l, m_l*

Figure 1
The way the quantum numbers for electrons correspond to shells, subshells, and individual orbitals in an atom. Each circle represents an orbital.

A shell is divided into subshells (number of subshells equals the value of *n* for the shell).

Subshells are divided into orbitals. (number of orbitals equals $2l + 1$)

Self-Test

10. What do the terms *in phase* and *out of phase* mean?

_____

_____

11. What would be the subshell designation (e.g., 1*s*) corresponding to the following sets of values of *n* and *l*?

(a) $n = 2, l = 1$ _____

(b) $n = 4, l = 0$ _____

(c) $n = 3, l = 2$ _____

(d) $n = 5, l = 3$ _____

12. Which of the following sets of quantum numbers are <u>unacceptable</u>?

(a) $n = 2, l = 1, m_l = 0$ (b) $n = 2, l = 2, m_l = 1$

(c) $n = 2, l = 1, m_l = -2$ (d) $n = 3, l = 2, m_l = -2$

(e) $n = 0, l = 0, m_l = 0$ _____

13. What subshells are found in the fourth shell?

_____

14. Which subshell is higher in energy?

(a)   3*s* or 3*p*   _____

(b)   4*p* or 4*d*   _____

(c)   3*p* or 4*p*   _____

7.6   Electron Spin

Learning Objective

Describe electron spin and apply it to explain paramagnetism and diamagnetism along with the Pauli exclusion principle.

Review

The electron behaves like a tiny magnet. A way of explaining this is to imagine that it spins like a top. Two directions of spin are possible, which are identified by the values of the spin quantum number, m_s. The Pauli exclusion principle states that no two electrons in the same atom can have the same values for all four of their quantum numbers. The result is that no more than two electrons can occupy any one orbital. (This means that no more than two electrons can have the same waveform.) When all of the electrons in an atom are not *paired*, a residual magnetism occurs and the atom is paramagnetic. Paramagnetic substances are attracted weakly to a magnetic field. On the other hand, if all the electrons are paired, the atom is diamagnetic. Diamagnetic substances are often said to have no magnetic properties but, in fact, they are very weakly repelled by a magnetic field.

Self-Test

15.   How many electrons can occupy

    (a)   a 2*p* subshell?   _____

    (b)   a 4*f* subshell?   _____

    (c)   the shell with $n = 4$?   _____

    (d)   the shell with $n = 2$?   _____

    (e)   the subshell with $n = 4$ and $l = 1$?   _____

7.7   Energy Levels and Ground State Electron Configurations

Learning Objective

Write the ground state electron configuration of an atom.

Review

The distribution of electrons among an atom's orbitals is called the electronic structure, or electron configuration of the atom. Electrons always fill orbitals of lowest energy first (Figure 7.18). When filling a set of equal-energy orbitals, Hund's rule requires that the electrons spread out over the orbitals of a subshell as much as possible and that the number of unpaired electrons in the subshell be a maximum. This gives the ground state (lowest energy) configuration. In Section 7.8 you will see that the periodic table can also be used in place of Figure 7.18 to obtain an atom's electron configuration.

Learn to draw orbital diagrams. When indicating the orbitals of a *p*, *d*, or *f* subshell, be sure to show *all* of the orbitals of that subshell, even if some of them are unoccupied. For example, for boron we should write

B   1*s*   2*s*   2*p*   (correct)

The following is *incorrect*, because only one of the three 2*p* orbitals is shown.

B   1*s*   2*s*   2*p* (incorrect)

Self-Test

16. Use Figure 7.18 to write the electron configuration for

    (a) P   _____

    (b) Se   _____

    (c) Co   _____

17. Construct orbital diagrams for

    (a) phosphorus

    _____

    (b) silicon

    _____

7.8   Periodic Table and Ground State Electron Configurations

Learning Objective

Explain the connection between the periodic table and electron configurations of atoms.

Review

Figure 7.20 illustrates how we can use the periodic table as a guide in writing electron configurations. Following the aufbau principle, we assign electrons to subshells beginning with the 1*s* (corresponding to

Period 1 in the periodic table). Then, as we cross the periodic table row after row, we let the table tell us which subshells become occupied.

When we cross through Groups 1A and 2A, we fill an *s* subshell; across Groups 3A through Group 8A we fill a *p* subshell. For the *s* and *p* subshells, the value of *n* that goes with the *s* and *p* designations equals the period number. Crossing a row of transition elements corresponds to filling a *d* subshell with *n* equal to *one less than the period number*. Crossing a row of inner transition elements (the lanthanides or actinides) corresponds to filling an *f* subshell with *n* equal to *two less than the period number*. Be sure to study Example 7.4 in the text.

Basis for the periodic recurrence of properties

The main point of this discussion is that atoms of a given group in the periodic table have similar electron configurations for their outer shells. Thus, all elements in Group 1A have an outer shell with one electron in an *s* orbital. Similarly, the elements in Group 2A each have two electrons in their outer shell *s* orbital.

Abbreviated configurations

In the shorthand electron configuration of an element, we only show subshells above those that are filled in an atom of the preceding noble gas. For example, for potassium (atomic number $Z = 19$), we only show orbitals beyond those that are filled in argon, the preceding noble gas. Argon ($Z = 18$) has completed $1s$, $2s$, $2p$, $3s$, and $3p$ subshells ($2 + 2 + 6 + 2 + 6 = 18\ e^-$). The next subshell is the $4s$, so for potassium we write

$$K\ [Ar]\ 4s^1$$

where [Ar] stands for the "argon core" of a potassium atom. We write the abbreviated orbital diagram for potassium as

$$K \quad [Ar] \quad 4s$$

$$\uparrow$$

Following the same procedure, for the element nickel we write

$$Ni\ [Ar]\ 3d\,^8 4s^2$$

and give its orbital diagram as

$$Ni \quad [Ar] \qquad 3d \qquad\qquad 4s$$

$$\uparrow\downarrow \; \uparrow\downarrow \; \uparrow\downarrow \; \uparrow \; \uparrow \qquad \uparrow$$

Valence shell configurations

In writing the valence shell configuration for an element, we only list subshells that are in the *outer shell* of the atom. (Valence shell configurations are only of interest to us for the representative elements.) The valence shell configuration for the element sulfur, for example, is

$$S \qquad 3s^2 3p^4$$

Unexpected electron configurations sometimes occur for transition and inner transition elements

Half-filled subshells and (especially) filled subshells have extra stability that causes the electron configurations of some elements such as chromium, copper, silver, and gold to deviate from the configurations that we would predict by following the procedures described above. You should learn the electron configurations of these elements as exceptions.

Example 7.3 Determining magnetic properties of substances

Determine which of the following are paramagnetic and which are diamagnetic. (Justify your reasoning.) (a) a sodium atom, (b) a calcium ion, (c) a copper atom, and (d) a chloride ion.

Analysis:

We look to the definitions of paramagnetic and diamagnetic and find that the distinguishing characteristic is whether or not there are one or more unpaired electrons in the substance.

Assembling the Tools:

We will need our procedures for drawing orbital diagrams, which show electron spins, along with Hund's rule, which allows us to determine the number of unpaired electrons.

Solution:

Unpaired electrons will appear in the outermost electrons and therefore we can use the abbreviated format for the electron configurations. They are:

(a) sodium Na $[Ne] 3s^1$ (b) calcium ion Ca^{2+} [Ar] or $[Ne] 3s^2 3p^6$

(c) copper Cu $[Ar] 4s^1 3d^{10}$ (d) chloride ion Cl^- [Ar] or $[Ne] 3s^2 3p^6$

From these we see an odd number of electrons for sodium and copper, and they are classified as paramagnetic. The calcium ion and chloride ion are diamagnetic because their subshells are completely filled. It is important to remember that an even number of electrons does not guarantee that a substance will be diamagnetic. For example, chromium has 24 electrons but there are six unpaired electrons.

Is the Answer Reasonable?

We can recheck our electron configurations (Appendix A in the text) and also our logic in determining the noble gas used to represent the core electrons.

Self-Test

18. What is the shorthand electron configuration of tellurium?

_____

19. What is the shorthand electron configuration of nickel?

_____

20. How many unpaired *d* electrons are found in an atom of nickel?

_____

21. Use the periodic table to write the valence shell electron configuration of

(a) oxygen   _____

    (b)   carbon   _____

    (c)   sulfur   _____

    (d)   lead   _____

22. Use the periodic table to predict the shorthand electron configuration for

    (a)   Tc   _____

    (b)   Fe   _____

    (c)   Gd   _____

23. From what you learned in this section, predict the electron configuration of europium, Eu ($Z = 63$).

7.9   Atomic Orbitals: Shapes and Orientations

Learning Objective

Draw the shapes and orientations of s, p, and d orbitals in an atom.

Review

The Heisenberg uncertainty principle comes from a mathematical proof that it is impossible to know or predict exactly where an electron is in an atom at any given moment. Wave mechanics, however, provides us with a way to predict the probability of finding the electron at any given location around the nucleus. This probability, which varies as we move from point to point, is given by the square of the wave function, ψ^2. The notion of a probability distribution for the electron in the space around the nucleus leads to the concept of the electron behaving as a blur or cloud with a greater *electron density* in some places than in others.

In general, the size of a given kind of orbital increases with increasing n. The s orbitals are all spherical in shape, meaning that if we examine a surface on which the probability of finding the electron is everywhere the same, the surface is a sphere. Study Figure 7.22, which illustrates the electron density distribution in a $1s$ orbital. In Figure 7.23, we see that the $2s$, $3s$, and higher s orbitals are also spherical, but spherical nodes, too. Nodes are a consequence of wave behavior, although they are not really important for our discussions.

The p orbitals are sometimes described as being dumbbell-shaped. Each p orbital has *two* regions of electron density located on opposite sides of the nucleus (Figures 7.24 and 7.25) and the three p orbitals of a p subshell are oriented perpendicular to each other (Figure 7.26).

The d orbitals are more complex, still, than the p orbitals (Figure 7.27). Four of them in a given subshell have the same shape, each consisting of four lobes of electron density. The fifth consists of a pair of lobes along the (arbitrary) z axis, plus a donut-shaped ring of electron density in the xy plane.

Self-Test

24. On a separate piece of paper, practice sketching the shapes of *s* and *p* orbitals. Indicate where you expect to find spherical nodes and nodal planes.

25. On the axes below, sketch the shapes of d_{xy}, d_{xz}, d_{yz}, and $d_{x^2-y^2}$ orbitals.

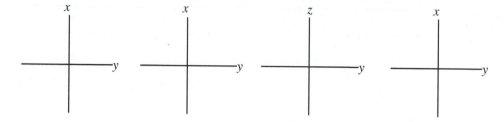

7.10   Periodic Table and Properties of the Elements

Learning Objective

Describe how the electron configurations of the elements and the periodic table help predict the physical and chemical properties of the elements.

Review

The key to understanding the way the properties discussed in this section vary within the periodic table is understanding the concept of effective nuclear charge. The notion here is that the outer electrons do not feel the full effect of the charge on the nucleus because the charge of the electrons in shells below the outer shell partially offset the nuclear charge. The positive charge that the outer electrons do feel is called the *effective nuclear charge*.

Sizes of atoms

Although, in a strict sense, atoms and ions have no true outer boundary, they often behave as if they have nearly constant sizes. The size of an atom or ion is generally given in terms of its radius and is expressed in units of picometers, nanometers, or angstroms. Even though the angstrom isn't an SI unit, it is still widely used for expressing small distances. Remember: 1 Å = 0.1 nm = 100 pm.

Within the periodic table, size *decreases* from left to right in a period and from bottom to top in a group. Going across a period, the amount of positive charge felt by the outer electrons increases because electrons in the same shell are not very effective at shielding each other from the nuclear charge. Going from top to bottom in a group, the outer-shell orbitals feel a relatively constant effective nuclear charge and become larger as *n* becomes larger. As a result, atoms become larger as we descend a group. Study Figure 7.30.

Sizes of ions

When ions are formed from atoms, their sizes increase as electrons are added and decrease as electrons are removed. Negative ions are always larger than the atoms from which they are formed. Positive ions are always smaller than the atoms from which they are formed.

Ionization energy

The *ionization energy* (IE) is the energy needed to pull an electron from an isolated atom or ion in its ground state. It is a measure of how tightly held the electrons are. Atoms with more than one electron have a series of ionization energies corresponding to the removal of the electrons one by one. For any given atom, successive ionization energies increase.

You should remember that, in general, the ionization energy increases from left to right across a period and decreases from top to bottom within a group. The increase across a period occurs because of the increasing effective nuclear charge felt by the outer electrons. The decrease down a group occurs because the outer electrons become farther from the nucleus (from which they are well shielded) and are held less tightly. You should also note the special stability of the noble gas configuration. It is also helpful to remember that as atomic radius becomes *smaller*, IE becomes *larger*.

Electron affinity

Energy is normally released when an electron is added to an isolated gaseous atom to form a negative ion. This energy is the *electron affinity* (EA). Most values are given with a negative sign because the process is usually exothermic.

The changes in EA within the periodic table parallel the changes in IE, and for the same reasons. Both increase from left to right in a period and from bottom to top in a group.

When more than one electron is added to an atom (i.e., when an ion with a charge of 2– or 3– is formed) the attachment of the second and third electron is always endothermic. Overall, formation of a negative ion with a charge larger than 1– is endothermic.

Self-Test

26. Which is the larger ion, Fe^{2+} or Fe^{3+}? _____

27. Explain your answer to Question 26. _____

_____

28. The following particles each have the same number of electrons: N^{3-}, O^{2-}, and F^-. Their atomic numbers increase from N to O to F. How would you expect their sizes to vary? Explain.

_____

_____

29. In each pair, choose the species with the larger IE.

    (a) Na or Mg _____   (d) Co or Co^{2+} _____

    (b) Rb or K _____   (e) Ar or Al _____

    (c) Si or N _____

30. Which elements, metals or nonmetals, tend to have the smaller ionization energies?

    _____

31. Choose the species with the more exothermic EA.

    (a) P or O _____   (c) Te or I _____

    (b) Se or Cl _____   (d) P or P^- _____

32. From which ion, O^{2-} or O^-, would you expect it to be easier to remove an electron?

    _____

Answers to Self-Test Questions

1. 2.81 m
2. 1.2×10^{14} Hz
3. 91 nm
4. 3.07×10^{-19} J
5. 419 kJ
6. X rays, ultraviolet light, visible light, infrared radiation, microwaves, radio waves
7. 410 nm, violet
8. $n = 1$
9. $E = -b$ for $n = 1$, $E = -b/4$ for $n = 2$, change in energy is $0.75b$
10. *In phase*—amplitudes add to give a new wave with a larger amplitude.
    Out of phase—amplitudes cancel to give a new wave with a smaller, or even zero, amplitude.
11. (a) $2p$, (b) $4s$, (c) $3d$, (d) $5f$
12. (b), (c), (e) are unacceptable.
13. $4s, 4p, 4d, 4f$
14. (a) $3p$, (b) $4d$, (c) $4p$
15. (a) 6, (b) 14, (c) 32, (d) 8, (e) 6
16. (a) $1s^2 2s^2 2p^6 3s^2 3p^3$ (b) $1s^2 2s^2 2p^6 3s^2 3p^6 3d^{10} 4s^2 4p^4$
    (c) $1s^2 2s^2 2p^6 3s^2 3p^6 3d^7 4s^2$
17. (a)

    (b)

18. Te = [Kr] $4d^{10}5s^25p^4$

19. Ni = [Ar] $3d^84s^2$

20. two

21. (a) $2s^22p^4$, (b) $2s^22p^2$, (c) $3s^23p^4$, (d) $6s^26p^2$

22. (a) Tc = [Kr] $4d^55s^2$, (b) Fe = [Ar] $3d^64s^2$, (c) Gd = [Xe] $4f^75d^16s^2$

23. Eu [Xe] $6s^24f^7$ (the $4f$ subshell becomes half-filled)

24. Check your sketches against Figures 7.24 and 7.26.

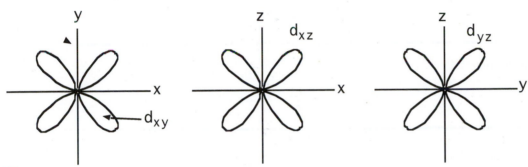

25.

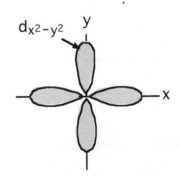

($d_{x^2-y^2}$ and d_{xy} are concentrated in the xy plane, the d_{xz} is concentrated in the xz plane, and the d_{yz} is concentrated in the yz plane.)

26. Fe^{2+}

27. In Fe^{3+} there is one less e^- than in Fe^{2+} and therefore less inter-electron repulsion, allowing the e^- in Fe^{3+} to be pulled closer to the nucleus.

28. $N^{3-} > O^{2-} > F^-$. As the nuclear charge increases, the electrons are pulled closer to the nucleus and the size decreases.

29. (a) Mg (b) K (c) N (d) Co^{2+} (e) Ar

30. Metals

31. (a) O (b) Cl (c) I (d) P

32. O^{2-}

Tools for problem solving

In this chapter you learned to apply the following concepts as tools in solving problems. Study each one carefully so that you know what each is used for. When faced with solving a problem, recall what each tool does and consider whether it will be helpful in finding a solution.

You might want to tear these pages out to use along with solving problems in this chapter.

Wavelength-frequency relationship (Section 7.1)

$$c = \frac{\lambda}{\nu}$$

Energy of a photon (Section 7.1)

$$E = h\nu = h\frac{c}{\lambda}$$

Rydberg equation (Section 7.2)

$$\frac{1}{\lambda} = R_H \left(\frac{1}{n_1^2} \frac{1}{n_2^2} \right)$$

Periodic table and chemical properties (Section 7.8)

Valence electrons define chemical properties of the elements. Elements within a group (family) have similarities in their valence electrons and thus similar chemical properties.

Periodic trends in atomic and ionic size (Section 7.10)

The relative sizes of atoms and ions increase moving down the periodic table and decrease moving across the periodic table.

Periodic trends in ionization energy (Section 7.10)

The ionization energy of an atom increases moving up the periodic table and across the periodic table.

Periodic trends in electron affinity (Section 7.10)

The electron affinity of an atom is more exothermic moving up and across the periodic table.

Summary of Important Equations

The most important equations in this chapter involve the wavelength–frequency relationship and Planck's equation for the energy of a photon.

Wavelength–frequency relationship:

$$\lambda \times \nu = c = 3.00 \times 10^8 \text{ m s}^{-1}$$

Energy of a photon of frequency ν:

$$E = h\nu \quad \text{(where } h = \text{Planck's constant } (6.626 \times 10^{-34} \text{ J s}))$$

Your teacher may also ask you to learn the following:

The Rydberg equation, where $R_H = 109{,}687 \text{ cm}^{-1}$

$$\frac{1}{\lambda} = R_H \left(\frac{1}{n_1^2} \frac{1}{n_2^2} \right)$$

De Broglie's equation for the wavelength of a matter wave:

$$\lambda = \frac{h}{mv}$$

where h is Planck's constant, m is the mass of the particle, and v is the velocity of the particle.

Notes:

Chapter 8

The Basics of Chemical Bonding

This is the first of two chapters that deal with the attractions that hold atoms to each other in chemical compounds. In this chapter we examine chemical bonds on a rather elementary level. Yet, even these simple explanations provide us with many useful tools for understanding chemical and physical properties.

8.1 Energy Requirements for Bond Formation

Learning Objective

Describe the necessary condition for bond formation that produces a stable compound.

Review

There are two broad categories of chemical bonds—*ionic bonds*, which occur when atoms transfer electrons between them, and *covalent bonds*, which occur when atoms share electrons. An ionic "bond" is really just the attraction that exists between oppositely charged ions.

For a bond to be formed between atoms, there must be a net lowering of the potential energy of the particles. This is a simple but important principle that will be explored in depth in the rest of this chapter. This concept will be used in later chapters also.

Self-Test

1. Can you observe potential energy directly, if so what instrument would you use to measure it?

_____

_____

2. What are two observations that would indicate that a bond has been formed?

(a) _____

(b) _____

8.2   Ionic Bonding

Learning Objective

Explain the factors involved in ionic bonding.

Review

There are two broad categories of chemical bonds—*ionic bonds*, which occur when atoms transfer electrons between them, and *covalent bonds*, which occur when atoms share electrons. An ionic "bond" is really just the attraction that exists between oppositely charged ions.

For a bond to be formed between atoms, there must be a net lowering of the potential energy of the particles. For ionic bonding, the three most important factors that contribute to the overall potential energy change are (1) the ionization energy of the element that forms the cation, (2) the electron affinity of the element that forms the anion, and (3) the *lattice energy*, which is the potential energy lowering that is produced by the attractions between the ions. For most elements, (1) and (2) taken together produce a net increase in potential energy, so it is the stabilizing (energy-lowering) influence of the lattice energy that enables ionic compounds to exist. But the lattice energy can do this *only* if it is larger than the net PE increase caused by (1) and (2). This restriction leads to certain generalizations about the kinds of elements that form ionic compounds and the charges on the ions that are created in the reaction:

> Ionic compounds tend to be formed between metals (low IE) and nonmetals (relatively large exothermic EA).

The Born–Haber cycle described in this section is another method to account for energy changes that occur in a reaction. We recall that the enthalpy, H, is a state function. Therefore any way we can get from the reactants to the products will result in the same ΔH_f for the reaction. We can write and measure the heat of reaction as we did in Chapter 7. We could also study the logical steps leading to the compound. These often include vaporization, ionization, and then the combination of the ions to release the lattice energy. All of the steps are directly measurable except the lattice energy, which is calculated.

Self-Test

3.   What is the lattice energy and why is it important?

_____

_____

4.   For the formation of the mineral fluorite, CaF_2, calcium metal and fluorine gas can react directly. Give the steps of the second pathway to form fluorite from calcium metal and fluorine gas.

_____

8.3   Octet Rule and Electron Configurations of Ions

Learning Objective

Write electron configurations of ions.

Review

Ions are formed when atoms lose or gain electrons. Cations have positive charges and are the result of an atom losing one or more electrons. Anions are formed by gaining one or more electrons and they have negative charges.

The tendency of certain elements to achieve a noble gas configuration gives rise to the octet rule, which states that *many elements tend to gain or lose electrons until they have achieved a valence shell that contains eight electrons*.

1.   The representative metals of Groups 1A and 2A plus aluminum lose electrons until they have achieved an electron configuration corresponding to that of a noble gas.

2.   Nonmetals gain electrons until they have also achieved an electron configuration that is the same as that of a noble gas.

The transition metals and the metals that follow the transition elements in Periods 4, 5, and 6 (the post-transition metals) do not follow any particular rule. They lose electrons, but often more than one cation is possible, depending on conditions. The charges on the ions of the common transition metals were given in Table 2.3 in the text; if necessary, review them.

When electrons are lost by an atom, they always come first from the shell with largest n. For a transition element, the shell with largest n is an s subshell. After the s subshell is emptied, electrons are lost from the d subshell below the outer shell.

When electrons are lost from a given shell, they come from the highest energy subshell first. Because the energies of subshells vary in the order $s < p < d < f$, post-transition metals lose electrons from the outer shell p subshell before they lose them from the outer shell s subshell.

It is often convenient to rearrange the electron configuration of an element from the energy or aufbau ordering to the energy level ordering. This simplifies the process of removing or adding electrons.

Self-Test

5.   Write in words describing what happens to the electron configurations of calcium and bromine atoms when they form ions. _____

_____

_____

6. What happens to the electron configuration of a manganese atom when it forms (a) the Mn^{2+} ion and (b) the Mn^{3+} ion?

_____

_____

7. When Rb and Cl_2 react, why doesn't the compound $RbCl_2$ form? What compound does form?

_____

_____

_____

8. Write the abbreviated electron configuration of

   (a) Pb^{2+} _____   (b) Pb^{4+} _____

9. Show what happens to the valence shells of Ba and I when these elements react to form an ionic compound. What is the formula of this compound?

_____

_____

_____

8.4   Lewis Symbols: Keeping Track of Valence Electrons

Learning Objective

Write Lewis symbols for atoms and ions.

Review

An element's Lewis symbol is constructed by placing one dot for each valence electron around the chemical symbol for the element. The number of valence electrons, and therefore the number of dots in the Lewis symbol, is equal to the element's group number. (In general, Lewis symbols are only used for the representative elements.)

One way to draw Lewis symbols is as follows. First write the chemical symbol. Then imagine a diamond with the symbol at its center. Place a dot at one of the corners of the diamond (it doesn't matter where you start). Move around the diamond from corner to corner as you place additional dots, until the required number are shown. For example, for sulfur (Group VIA), there must be six dots around the symbol S.

S · S · S · ·S· ·S· ·S :

| imagine a diamond around S | 1st dot | 2nd dot | 3rd dot | 4th dot | 5th dot | 6th dot |

The Lewis symbol for sulfur is ·S: (or any other arrangement such as ·S: that shows two pairs of dots and two single dots).

Example 8.4 in the text illustrates how Lewis symbols can be used to diagram the transfer of electrons that takes place during the formation of an ionic substance. Notice that when we write the Lewis symbol for an anion, we enclose it in brackets to show that all the electrons belong to the ion.

Self-Test

10. Write Lewis symbols for the following:

   (a)   P   _____   (c)   Cl   _____

   (b)   Te   _____   (d)   B   _____

11. Diagram the reaction that occurs when Li reacts with Se to form the compound Li_2Se.

8.5   Covalent Bonds

Learning Objective

Describe covalent bonds through the energetics of bond formation, octet rule, and multiple bonds.

Review

Covalent bonds are formed when ionic bonds are not energetically favored. In the formation of a covalent bond, there is also a potential energy lowering, but it is achieved by a different method than in ionic bonding.

When atoms approach each other to form a covalent bond, there is an attraction of the electrons of each atom toward both nuclei, which leads to an overall lowering of the potential energy. The two nuclei also repel each other because they are of the same charge, so the atoms cannot approach each other too closely. The *bond length* is determined by the balance of these attractive and repulsive forces, and the net decrease in PE that occurs when the bond is formed is called the *bond energy*, which is usually expressed in units of kilojoules per mole of bonds formed. (See Figure 8.7.) The bond energy is also the energy that would be necessary to break the bond.

One of the most important features of covalent bonding is the pairing of electrons that occurs; a covalent bond is the sharing of a *pair* of electrons with their spins in opposite directions. When Lewis symbols are used to represent a molecule, a shared pair of electrons is shown as a pair of dots or a dash between the chemical symbols. One dash stands for *two* paired electrons.

When counting electrons in the valence shells of atoms attached to each other by a covalent bond, we count both electrons of the shared pair as belonging to both atoms joined by the bond. When applied to covalent bonds, the octet rule states that *atoms tend to share sufficient electrons so as to obtain an outer shell with eight electrons.* An exception to the octet rule is hydrogen, which completes its valence shell when it has a share of two electrons (i.e., when it forms one covalent bond).

Atoms can share one, two, or three pairs of electrons to give single, double, and triple bonds. Notice how the numbers of electrons needed by atoms of carbon, nitrogen, and oxygen to achieve octets are related to the number of covalent bonds these atoms tend to form.

Self-Test

12. What happens to the electron density between the two atoms when they form a covalent bond?

_____

_____

13. On a separate piece of paper, sketch a diagram that shows how the potential energy varies with the distance between the nuclei of a pair of atoms becoming joined by a covalent bond. On the diagram indicate the bond energy and the bond length. Refer to Figure 8.7 to check your answer.

14. Predict the formulas of the simplest compound that might be formed between hydrogen and each of the following:

    (a) Ge   _____

    (b) Te   _____

    (c) I   _____

15. Draw the Lewis structures for each of your answers to Question 15.

    (a)   (b)   (c)

16. Formaldehyde, used in preserving biological specimens, has the formula H_2CO. A molecule of this substance has two hydrogen atoms and an oxygen atom bonded to the carbon. Use Lewis symbols to write a structural formula for H_2CO. (Hint: There's a double bond in the structure.)

17. Count the number of electrons around the atoms in the following molecule (called urea).

$$\begin{array}{ccccc} H & & :\!\overset{..}{O}\!: & & H \\ | & & \| & & | \\ H\!-\!\overset{..}{N}\!-\!&\!C\!-\!&\overset{..}{N}\!-\!H \\ \overset{..}{} & & \overset{..}{} & & \overset{..}{} \end{array}$$

8.6   Bond Polarity and Electronegativity

Learning Objective

Explain how electronegativity affects the polarity of covalent bonds and the reactivity of the elements.

Review

Electronegativity
Because different atoms have different attractions for electrons, the electrons in a covalent bond can be shared unequally. When this happens, the electron density in the bond is shifted toward that atom with the greater attraction for electrons, so this atom acquires a partial negative charge, δ^-. The atom at the other end of the bond loses some electrical charge and carries a partial positive charge, $\delta+$. This produces an *electric dipole*, which is two equal but opposite electric charges separated by some distance. Bonds in which electrons are shared unequally are *polar bonds* and the bond is a *dipole* (two poles of equal but opposite charges separated by the bond distance). The degree of polarity of a bond is expressed quantitatively by the bond's *dipole moment* μ, which is the product of the charge on either end of the bond multiplied by the distance between the charges. Molecules as a whole can be polar and have dipole moments. Dipole moments are measured in units called *debeys* (symbol D).

Electronegativity is the attraction an atom has for the electrons in a bond. The greater the electronegativity *difference* between two atoms that are bonded to each other, the more polar the bond is and the greater is the partial negative charge on the more electronegative of the two atoms.

It is important to remember that there is no sharp dividing line between covalent and ionic bonding. If we arranged all sorts of bonds in order of increasing polarity, we would find a gradual transition from nonpolar covalent bonds to bonds that are essentially 100% ionic. All degrees of ionic character (polarity) are possible.

You should be sure to learn how electronegativity varies in the periodic table. This is illustrated in Figure 8.9 in the text. Metals, in the lower left corner of the table, have low electronegativities and nonmetals, in the upper right, have high electronegativities.

Self-Test

18. Use the data in Table 8.2 of the textbook to calculate the amount of charge, in electronic charge units, on opposite ends of the NO molecule. (1 D = 3.34×10^{-30} C m and the electronic charge unit = 1.60×10^{-19} C)

19. Use Figures 8.9 and 8.10 to arrange the following bonds in order of increasing percent ionic character: $Li-I, Ba-F, N-O, As-Cl$.

The reactivities of metals and nonmetals

For metals, reactivity refers to how easily oxidized the element is. Because oxidation involves the removal of electrons from a substance, a metal that is easily oxidized is one that does not hold its electrons very tightly. This corresponds to a metal with a low electronegativity. By this reasoning, we come to the generalization that the smaller the electronegativity of a metal, the more easily oxidized it should be. This correlation is reflected in Figure 8.11, where we see that the metals with the lowest electronegativities (Groups IA and IIA) are most easily oxidized. Study Figure 8.11 and note where in the periodic table the least reactive metals are found.

The reactivity of a nonmetal generally is associated with its ability to serve as an oxidizing agent and is directly proportional to the nonmetal's electronegativity, which increases from left to right in a period and decreases from top to bottom in a group.

Self-Test

20. In each pair, select the element with the specified property.

    (a)   The better reducing agent: K or Be   _____

    (b)   The more easily oxidized: Mg or Fe   _____

    (c)   The more reactive metal: Ti or Au   _____

    (d)   The better oxidizing agent: P or Cl   _____

    (e)   The more easily reduced: O or Se   _____

    (f)   The more easily oxidized: Ga or Se   _____

8.7   Lewis Structures

Learning Objective

Draw Lewis structures for covalent molecules and calculate formal charges for individual atoms in a molecule

Review

Although the octet rule is often a useful tool in constructing the Lewis structure for a molecule or polyatomic ion, there are some instances in which the rule is not obeyed. Whenever an atom is bonded to more than four other atoms, the octet rule cannot be obeyed. Examples are PCl_5 and SF_6 illustrated in the text. Remember that only atoms below Period 2 are able to exceed an octet; their valence shells are able to accommodate more than eight electrons. (Period 2 elements never go beyond eight electrons in the outer shell because the second shell has only *s* and *p* subshells, which together can hold a maximum of $8e^-$.) Beryllium and boron are unusual elements because they sometimes have *less* than an octet in compounds.

Procedure for drawing Lewis structures

To draw a Lewis structure for a molecule or ion, follow the guidelines in Figure 8.12 in the text. The first step in the procedure is to decide on the skeletal structure.

Remember:

- In oxoacids, hydrogen is bonded to oxygen, which in turn is bonded to the other nonmetal.
- The central atom in a molecule is usually the least electronegative atom. In the formulas for most simple molecules and polyatomic ions, the central atom is written first.
- When all else fails, the most symmetrical arrangement of atoms is the "best guess."

Next, count all of the valence electrons. Remember that the number of the group in which a representative element (A-group element) is found is the same as the number of valence electrons that it contributes. Sulfur, for example, is in Group VI, so a sulfur atom contributes six valence electrons. Also remember that if the species whose Lewis structure you are working on is a negative ion, add one electron for each negative charge. If it is a positive ion, subtract one electron for each positive charge.

Once you know how many electrons are to be represented in the structure, place them in the structure as described in Figure 8.12. Study Examples 8.6 through 8.7, work the Practice Exercises, and then answer the Self-Test questions below.

Example 8.1 Drawing Lewis structures

What is the Lewis structure for dimethyl ether, CH_3OCH_3?

Analysis:
To draw the Lewis structure, we need to count all available valence electrons and distribute them, if possible, in octets. Note that most organic substances will obey the octet rule. After counting the valence electrons we need a skeleton structure of how the atoms are arranged. Once we have the skeleton structure, we can add electrons.

Assembling the Tools:
To draw the Lewis structure, we will need the tools for the octet rule and how to use Lewis symbols to represent covalent bonds. The tool for how to draw Lewis structures is also helpful.

Solution:
We count the valence electrons,

6 hydrogen atoms have	$6e^-$
2 carbon atoms have	$8e^-$
1 oxygen atom has	$6e^-$
total e⁻	$20e^-$

The formula given in the statement of the problem suggests the structure contains a $C - O - C$ structure with hydrogen atoms connected only to the carbon atoms. We start by writing a structure of

```
       H     H
    H  C  O  C  H
       H     H
```

Now we add a pair of electrons between each pair of atoms to create a bond. Counting electrons we see that we have used 16 of the 20 available electrons. We also note that each carbon has an octet of electrons but oxygen does not. We can easily make an octet on oxygen by adding the remaining 4 electrons.

```
       H          H
       ••         ••
   H : C : Ö : C : H
       ••    ••   ••
       H          H
```

Is the Answer Reasonable?

We can check our result by seeing if our atoms have octets of electrons. Recall that an "octet" for hydrogen is 2. Let's circle the octets. It looks reasonable.

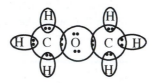

The bond length between a given pair of atoms *decreases* as the bond order increases (i.e., on going from a single to a double to a triple bond). At the same time, the bond energy increases. These trends, along with experimentally measured values of these bond properties, help us "fine tune" our descriptions of the bonding in various molecules. One of the observations we can make is that the best Lewis structure for a molecule or ion is one in which the *formal charges* on the atoms are a minimum.

To determine the formal charge on an atom in a Lewis structure, add up the number of bonds it forms plus the number of unshared electrons. Subtract this number from the number of electrons an isolated atom of the element has. The difference is the formal charge. (This is what Equation 8.3 tells us to do.) To select the preferred ("best") Lewis structure for a molecule or ion, we seek to minimize the number of formal charges. Study Examples 8.8 through 8.9 as well as the additional worked example below.

Example 8.2   Using formal charges to select Lewis structures

What is the preferred Lewis structure for chlorous acid, $HClO_2$?

Analysis:

We know that the preferred Lewis structure is the one that has the minimum number of formal charges. Therefore, the first step is to construct the Lewis structure. Then we calculate the formal charge on each atom. Since the best Lewis structure is the one with the minimum number of formal charges, we attempt to reduce the formal charges by moving electrons.

Assembling the Tools:

To draw the Lewis structure, we will need the tools for the octet rule and how to use Lewis symbols to represent covalent bonds. The tool for drawing Lewis structures, following the general procedure given in Figure 8.12, is useful. We then use the tools for determining formal charges and optimizing them.

Solution:

Below we have the Lewis structure for $HClO_2$ drawn according to the procedure outlined earlier. Notice that we have counted the number of bonds and unshared electrons, which we need to know to calculate the formal charges according to the formula

$$\text{Formal charge} = \begin{pmatrix} \text{number of electrons} \\ \text{an isolated atom} \\ \text{of the element has} \end{pmatrix} - \begin{pmatrix} \text{number of bonds} & + & \text{number of} \\ \text{to the atom} & & \text{unshared } e^- \end{pmatrix}$$

$$H-\overset{..}{\underset{..}{O}}-Cl-\overset{..}{\underset{..}{O}}:$$

2 bonds plus   1 bond plus
4 unshared e^- 6 unshared e^-

For the oxygen between the H and Cl, the sum of 2 bonds and 4 unshared e^- gives a total of 6. An isolated oxygen atom has 6 valence electrons, so the formal charge on this oxygen is zero. For the chlorine, the sum of bonds and unshared e^- is also 6, and when subtracted from 7 (the number of valence e^- an isolated Cl atom has) we obtain +1. This is the formal charge on the Cl. For the oxygen at the right, the sum of bonds and unshared e^- is 7. When 7 is subtracted from 6 (the number of valence e^- an oxygen atom has), we get −1. This is the formal charge on the oxygen at the right. Placing these formal charges in circles alongside the atomic symbols gives us

$$H-\overset{..}{\underset{..}{O}}-\overset{..}{\underset{..}{Cl}}^{\oplus}-\overset{..}{\underset{..}{O}}^{\ominus}:$$

The next step is to attempt to reduce formal charges to obtain a better structure. We do this by shifting an unshared pair of electrons on the more negative atom into the bond it forms to the more positive atom.

$$H-\overset{..}{\underset{..}{O}}-\overset{..}{\underset{..}{Cl}}^{\oplus}-\overset{..}{\underset{..}{O}}^{\ominus}: \dashrightarrow H-\overset{..}{\underset{..}{O}}-\overset{..}{Cl}=\overset{..}{\underset{..}{O}}:$$

A lone pair of electrons is moved from the oxygen into the Cl—O bond. This effectively transfers an electron from O to Cl, thereby reducing the formal charges to zero

This gives a Lewis structure with no formal charges, so we select it as the preferred Lewis structure for the molecule.

Is the Answer Reasonable?

First, we can check the initial Lewis structure to be sure we've included all the valence electrons. We can also check our calculation of formal charges by adding them up algebraically. Their sum must equal the net charge on the molecule, which it does. When rearranging electrons, we should move them from the atom with the negative formal charge toward the atom with the positive formal charge, and we've done that as well. Having done all these steps correctly gives us confidence that our answer is correct.

When one atom donates a pair of electrons to another atom in order to form a single bond, we call the bond a *coordinate covalent bond*. Once formed, it is really no different than any other single bond because electrons can't "remember" where they came from. This is really just a bookkeeping device. When we want to note the origin of the electrons of a coordinate covalent bond, we use an arrow instead of a dash to represent the bond. The arrow points from the donor to the acceptor of the electrons.

Self-Test

21. If an atom forms more than four bonds, it definitely does not obey the octet rule. Why?

   _____

22. What is the maximum number of electrons that can be held in the orbitals in the valence shell of

   (a)   nitrogen?   _____

   (b)   phosphorus?   _____

   (c)   What kind of orbitals does the valence shell of phosphorus have that the valence shell of nitrogen does not?

   _____

23. How many valence electrons are in each of the following?

   (a) PO_4^{3-} _____   (d) IF_5 _____

   (b) IF_4^- _____   (e) CO_3^{2-} _____

   (c) NO_2^+ _____   (f) C_2^{2-} _____

24. Write the Lewis structures for (a) $HClO_2$ and (b) H_2CO_3 following the procedure in Figure 8.12.

   (a)   (b)

25. Construct Lewis structures for each substance in Question 23 following the procedure in Figure 8.12.

    (a)   (b)   (c)

    (d)   (e)   (f)

26. Phosphorous acid has the Lewis structure shown at the left below when constructed as described in the preceding section. Assign formal charges to the atoms in this structure and, if possible, draw a "better" Lewis structure for the molecule.

$$
\begin{array}{c}
:\!\overset{\displaystyle ..}{\underset{\displaystyle ..}{O}}\!: \\[2pt]
| \\[2pt]
H-\overset{..}{\underset{..}{O}}-P-\overset{..}{\underset{..}{O}}-H \\[2pt]
| \\[2pt]
H
\end{array}
$$

27. Boron trifluoride, BF_3, can react with a fluoride ion, F^-, to form the tetrafluoroborate ion, BF_4^-. Diagram this reaction using Lewis symbols to show the formation of a coordinate covalent bond.

28. BF_3 reacts with organic chemicals called ethers to form addition compounds. Use Lewis formulas to diagram the reaction of BF_3 with dimethyl ether. The structure of dimethyl ether is

$$
\begin{array}{c}
CH_3 \\
| \\
:\!\overset{}{\underset{}{O}}\!: \\
| \\
CH_3
\end{array}
$$

8.8 Resonance Structures

Learning Objective

Draw and explain resonance structures.

Review

For some molecules and ions, a single Lewis structure cannot be drawn that adequately explains experimental bond lengths and bond energies. In these instances, we view the true structure of the particle as a sort of average of two or more Lewis structures, and we call this true structure a *resonance hybrid* of the contributing structures.

When you have a choice as to where to place a double bond in a Lewis structure, as when writing the Lewis structure for the HCO_2^- ion, the number of resonance structures you should draw is equal to the number of choices that you have. This rule works in most cases.

A resonance hybrid is more stable than any of its individual structures would be, if they were to actually exist. The extra stability is called the resonance energy.

Self-Test

29. The oxalate ion, $C_2O_4^{2-}$, has the skeletal structure shown below.

$$O \qquad O$$
$$C \quad C$$
$$O \qquad O$$

Draw the resonance structures for this ion in the space below.

30. How would the C—O bond lengths and bond energies in the $C_2O_4^{2-}$ ion (Question 26) compare to those in the following?

$$\overset{\displaystyle H}{\underset{\displaystyle H}{H-\overset{|}{\underset{|}{C}}-\ddot{O}-H}} \quad \text{and} \quad \ddot{O}=C=\ddot{O}$$

8.9   Covalent Compounds of Carbon

Learning Objective

Classify organic compounds and identify functional groups.

Review

There are so many carbon–containing compounds and reactions that an entire discipline, organic chemistry, is devoted to their study. We will look at just a small group of organic compounds that will help us when we describe their properties later.

Hydrocarbons are compounds that contain only carbon and hydrogen. Saturated hydrocarbon molecules have carbon atoms held together with covalent bonds. The carbon atoms can form a single chain of one carbon after another (called a straight-chain molecule). Carbon atoms also can branch off from the straight chain to make branched molecules. These all have the general formula. Sometimes the carbon atoms bond to make rings. Once the carbon chain is defined, hydrogen atoms are added so that each carbon has four bonds. Unsaturated hydrocarbon molecules have double or triple bonds between one or more pairs of carbon atoms. They can be straight-chain, branched, or form rings.

We consider naming organic compounds in Chapter 23. However, the first ten straight-chain hydrocarbons are named methane, ethane, propane, butane, pentane, hexane, heptane, octane, nonane, and decane. Notice that after butane, the remaining hydrocarbons use the prefixes in Section 2.6 in the text that we have used for naming hydrates and molecular compounds.

Structural isomers occur because the carbon atoms can often be bonded together in different ways so that the same number of carbon and hydrogen atoms will form two or more distinctly different molecules. For example, CH_4, CH_3CH_3, and $CH_3CH_2CH_3$ can have only one isomer each. The four–carbon hydrocarbon can exist as one of the two possible isomers.

In addition to isomers and double and triple bonds, organic compounds can have functional groups attached to hydrocarbons. You should be able to identify alcohols (ROH), ketones (RC=OR), aldehydes (RC=OH), amines (RNH_2, RR'NH or RR'R"N) or carboxylic acids (RCOOH). Names of the simpler compounds are given in the text.

Self-Test

31. Write the functional group that matches the following.

(a) alcohol   _____

(b) ketone   _____

(c) acid   _____

(d) aldehyde   _____

32. How many different alcohols can have four carbon atoms? Write out their names.

_____

_____

_____

Answers to Self-Test Questions

1. No. Other measurements or observations allow us to infer that the potential energy has decreased. No instrument will do this.

2. We could observe that heat is given off or that work was done (expansion), both of which are a result of lowering potential energy.

3. Lattice energy is the energy released when ions come together in the formation of an ionic compound. This energy is important since the formation of ionic compounds would not occur.

4. In the formation of calcium fluoride we can imagine calcium metal being evaporated into the gas phase and then the removal of two electrons to form Ca^{2+}. Fluorine is already a gas, but the F_2 molecule must be divided into fluorine atoms and then an electron must be added to form the F^- ion. Two fluoride ions and one calcium ion combine to form CaF_2.

5. Calcium ions lose their $4s^2$ electrons, resulting in the same electron configuration identical to argon. Bromine atoms gain an electron to form the bromide ion and end up with an electron configuration that is the same as krypton. Both krypton and argon have octets of electrons.

6. (a) $Mn ([Ar] 3d^5 4s^2) \longrightarrow Mn^{2+} ([Ar] 3d^5) + 2e^-$
   (b) $Mn^{2+} ([Ar] 3d^5) \longrightarrow Mn^{3+} ([Ar] 3d^4) + e^-$

7. Removal of an electron from Rb^+ requires too much of a PE increase. The compound that does form contains Rb^+ and has the formula $RbCl$.

8. (a) Pb^{2+} $[Xe] 4f^{14}5d^{10}6p^2$ (b) Pb^{4+} $[Xe] 4f^{14}5d^{10}$

9. $Ba ([Xe] 6s^2) \longrightarrow Ba^{2+} ([Xe]) + 2e^-$; $I (5s^2 5p^5) + e^- \longrightarrow I^- (5s^2 5p^6)$
   The compound is BaI_2.

10.
   (a) $:\!\overset{\bullet}{\underset{\bullet}{P}}\!\bullet$ (b) $:\!\overset{\bullet}{\underset{\bullet}{Te}}\!\bullet$ (c) $:\!\overset{\bullet\bullet}{\underset{\bullet\bullet}{Cl}}\!\bullet$ (d) $\bullet\,\overset{\bullet}{B}\,\bullet$

11.
   $Li\!\cdot\curvearrowright\cdot\overset{\bullet}{\underset{\bullet\bullet}{Se}}\cdot\curvearrowleft\cdot Li \longrightarrow 2Li^+ \ [:\!\overset{\bullet\bullet}{\underset{\bullet\bullet}{Se}}\!:]^{2-}$

12. It shifts toward the region between the two atoms.
13. See Table 7.2.
14. (a) GeH_4 (b) H_2Te (c) HI

15.
   (a) $H\!-\!\underset{\displaystyle H}{\overset{\displaystyle H}{Ge}}\!-\!H$ (b) $H\!-\!\overset{\bullet\bullet}{\underset{\bullet\bullet}{Te}}\!-\!H$ (c) $H\!-\!\overset{\bullet\bullet}{\underset{\bullet\bullet}{I}}\!:$

16.

Hydrogen can complete its valence shell by forming one bond, carbon by four bonds, and oxygen by forming two bonds. This is the only arrangement of atoms that satisfies this condition.

17. Two around each H, eight around each N, C, and O.

18. Charge = $0.029e^-$

19. $N-O < As-Cl < Li-I < Ba-F$

20. (a) K, (b) Mg, (c) Ti, (d) Cl, (e) O, (f) Ga

21. Two electrons have to be in each bond. Since $2 \times 4 = 8$, more than four bonds means more than an octet.

22. (a) eight (b) eighteen (c) A phosphorus atom has d orbitals in its empty $3d$ subshell.

23. (a) 32 (b) 36 (c) 16 (d) 42 (e) 24 (f) 10

24.

25.

26.

27.

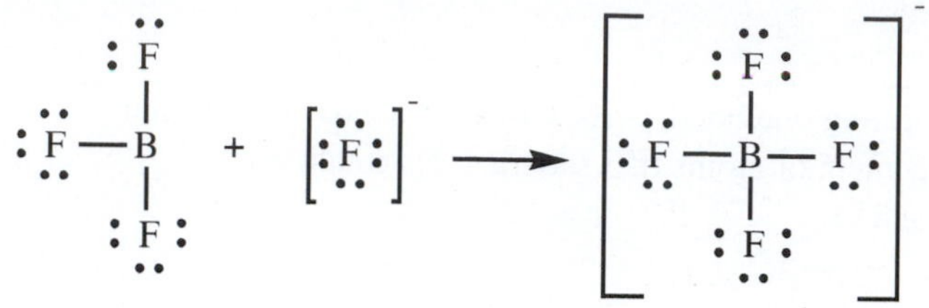

28.

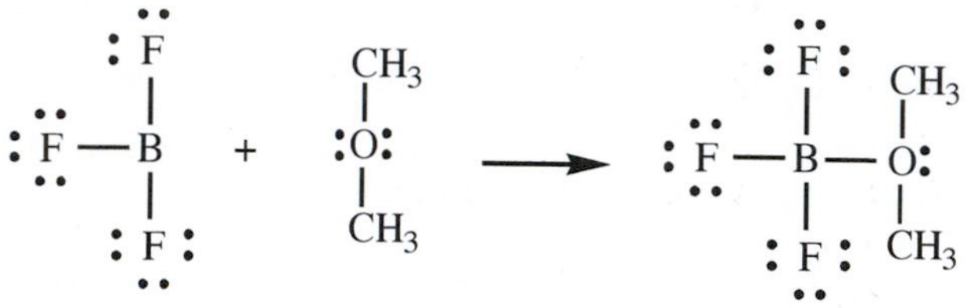

29.

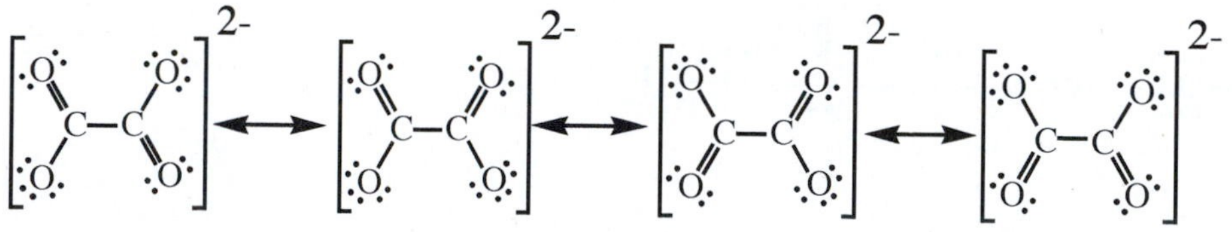

30. The C—O bond length decreases: $CH_3OH > C_2O_4^{2-} > CO_2$
    The C—O bond energy increases: $CH_3OH < C_2O_4^{2-} < CO_2$
31. Alcohol (ROH), ketone (RC=OR), carboxylic acid (RCOOH), aldehyde (RC=OH)
32. Butane-1-ol, butane-2-ol, 2-methylpropan-1-ol, 2-methylpropan-2-ol

Tools for problem solving

In this chapter you learned to apply the following concepts as tools in solving problems. Study each one carefully so that you know what each is used for. When faced with solving a problem, recall what each tool does and consider whether it will be helpful in finding a solution.

   You might want to tear these pages out to use along with solving problems in this chapter.

Octet rule (Section 8.3)

Atoms tend to gain or lose electrons until they have achieved an outer shell that contains an octet of electrons.

Order in which electrons are lost from an atom (Section 8.3)

Electrons are lost first from the shell with largest n. For a given shell, electrons are lost from subshells in the p subshell before the s subshell. If additional electrons are lost, they come from an available $(n - 1)$ d orbital.

Lewis symbols (Section 8.4)

Lewis symbols are a bookkeeping device that we use to keep track of valence electrons (the outermost s and p electrons) in atoms and ions. For a neutral atom of the representative elements, the Lewis symbol consists of the atomic symbol surrounded by dots equal in number to the number of valence electrons.

Octet rule and covalent bonding (Section 8.5)

The octet rule helps us construct Lewis structures for covalently bonded molecules. Elements in Period 2 never exceed an octet in their valence shells.

Dipole moment (Section 8.6)

The dipole moment (μ), in debye units, of a diatomic molecule is calculated as the charge on an end of the molecule, q, multiplied by the bond length, r.

$$\mu = q \times r$$

Periodic trends in electronegativity (Section 8.6)

Electronegativity increases across a period and decreases down a group in the periodic table and can be used to estimate the degree of polarity of bonds and to estimate which atoms in a bond are the most electronegative.

Reactivity of metals and the periodic table (Section 8.6)

A knowledge of where the most reactive and least reactive metals are located in the periodic table gives a qualitative feel for how reactive a metal is by locating it in the periodic table.

Reactivity of nonmetals and the periodic table (Section 8.6)

Oxidizing ability increases from left to right across a period and from bottom to top in a group.

Drawing Lewis structures (Section 8.7)

The method described in Figure 8.12 yields Lewis structures in which the maximum number of atoms obey the octet rule.

Correlation between bond properties and bond order (Section 8.7)

The correlations between increasing bond energy and decreasing bond length with bond order allow us to compare experimental covalent bond properties with those predicted by theory.

Calculating formal charges (Section 8.7)

$$\begin{bmatrix} \text{Formal} \\ \text{charge} \end{bmatrix} = \begin{bmatrix} \text{Number of } e^- \text{ in valence} \\ \text{shell of the isolated atom} \end{bmatrix} - \begin{bmatrix} \text{Number of bonds} \\ \text{to the atom} \end{bmatrix} + \begin{bmatrix} \text{Number of} \\ \text{unshared } e^- \end{bmatrix}$$

Selecting the best Lewis structures (Section 8.7)

A valid Lewis structure having the smallest formal charges is preferred.

Resonance structures (Section 8.8)

Expect resonance structures when there is more than one equivalent option for assigning the location of a double bond. The number of equivalent positions for the double bond is the number of resonance structures that can be drawn.

Summary of Useful Information

Rules for drawing Lewis structure (see also Figure 8.12.):

1. Decide which atoms are bonded to each other and construct the skeleton structure. (Remember, H cannot be a central atom.)

2. Count valence electrons. (Remember: add an electron for each negative charge on an ion or subtract an electron for each positive charge.)

3. Place two electrons in each bond.

4. Complete octets of atoms (except H) that surround the central atom.

5. Place any remaining electrons on the central atom *in pairs*.

6. Form multiple bonds if the central atom has less than an octet.

   (Rule 6 is omitted for the elements Be and B.)

Chapter 9

Theories of Bonding and Structure

There are two principal goals in this chapter. One of them is to introduce you to the variety of three-dimensional shapes molecules can have. You will learn to sketch the shapes and also how to use Lewis structures to predict them. Knowledge of molecular shape is important because many of the physical properties of substances depend on the shapes of their molecules. As we point out in the introduction to the chapter, molecular shape is a prime factor in determining the behavior of biologically active molecules.

In this chapter we will also present more advanced views of covalent bonding that are based on the results of quantum theory. Our aim is to enable you to understand how the theories *explain* covalent bonding in terms of the way the orbitals of atoms interact.

9.1   Five Basic Molecular Geometries

Learning Objective

Draw diagrams of the five basic molecular geometries, including bond angles.

Review

This section describes five basic geometric shapes that form the basis for most molecular structures you will encounter. The shapes are: *linear*, *planar triangular*, *tetrahedral*, *trigonal bipyramidal*, and *octahedral*. It is important that you develop the ability to visualize these shapes in three dimensions. It is also wise to learn to sketch them. Even if you don't consider yourself much of an artist, you should make the effort, because these three-dimensional concepts are likely to become important tools in other science courses you take.

The linear and planar triangular structures can be drawn on a two-dimensional surface, so you should have no difficulty with them. Simplified representations of the trigonal bipyramidal and octahedral structures are shown in Figures 9.2 and 9.3 in the text. Instructions for drawing the trigonal bipyramidal and octahedral structures are given in Study Guide Figure 9.1.

You should know the bond angles in the various structures. Notice that except for the trigonal bipyramid, all the other structures have equal angles between their bonds. Thus, in the tetrahedron, the bond angles are all 109.5°. In the octahedron, the bond angles are all 90°. In the trigonal bipyramid, the axial bonds make 90° angles with the equatorial bonds; the equatorial bonds are at 120° angles to each other.

Self-Test

1.   On a separate sheet of paper, make sketches of the five basic molecular shapes described in this section. Indicate the bond angles on the drawings. (Continue to practice this until the structures you draw make three-dimensional sense to *you*.)

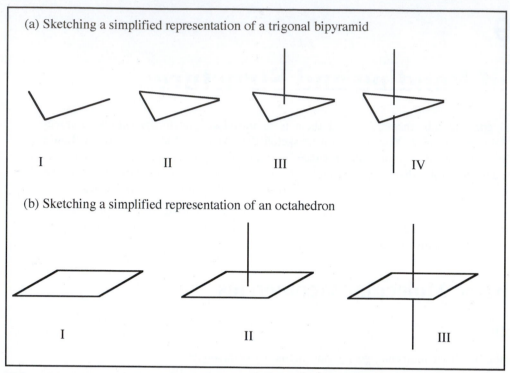

Figure 9.1 *(a) To draw a trigonal bipyramid, begin by making a check mark (I), then draw a line across the top (II). Next, draw a line from the center of the triangle upward, representing an axial bond (III). Finally, draw a line downward (projecting from below the triangular plane), which represents the other axial bond (IV). (b) To draw an octahedron, begin with a parallelogram, representing a square viewed from the edge (I). Next, draw a line upward from the center (II), and finally, draw a line downward as if it comes out from behind the plane in the center (III).*

9.2 Molecular Shapes and the VSEPR Model

Learning Objective

Predict the shape of a molecule or ion using the VSEPR model.

Review

The basic postulate of the VSEPR theory is very simple—*electron groups (electron domains) in the valence shell of an atom stay as far apart as possible because they repel each other.* As we describe in this section, this simple concept allows us to predict how the electron domains will arrange themselves around the central atom in a molecule or polyatomic ion. Before we look at this, let's examine the two types of electron domains we encounter in molecules.

One type consists of *bonding domains*, which contain electron pairs involved in bonds between two atoms. A bonding domain can consist of one electron pair (a single bond), two electron pairs (a double bond), or three electron pairs (a triple bond). *Nonbonding domains* make up the other type. A nonbonding domain consists of one unshared electron pair (also called a lone pair). Study Example 9.1 in the text to be sure you know how to identify these two types of electron domains. Here's another example.

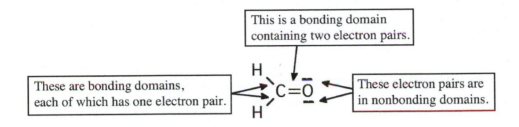

This is a bonding domain containing two electron pairs.

These are bonding domains, each of which has one electron pair.

These electron pairs are in nonbonding domains.

If all the electron pairs are in bonding domains, the arrangement that minimizes repulsions also defines the shape of the molecule or ion. When some electron pairs in the valence shell of a central atom are unshared (i.e., when they are *lone pairs*), the description we use for arrangement of the bonded atoms isn't the same as for the arrangement of electron domains. In ammonia, for example, the theory predicts that the electron domains are arranged tetrahedrally, but there are only three hydrogen atoms to attach to the nitrogen. When we look at how the nitrogen and three hydrogens are arranged, we see a pyramid with the nitrogen at the top. Therefore, even though the electron domains are distributed tetrahedrally, we describe the ammonia molecule as pyramid-shaped. (In fact, we say that NH_3 has a *trigonal* pyramidal shape because the pyramid has a three-sided base.)

The point of this is that when we apply the VSEPR theory to a molecule or ion, there are *two* shapes that concern us. One is the arrangement of the *electron domains* in the valence shell of the central atom. The other is the arrangement of the *bonded atoms* around the central atom. Remember, however, ***the shape of a molecule is described according to the arrangement of its atoms, not the predicted orientations of the electron domains.*** Although we use the theory to tell us where the electron pairs are, we use the orientations of the *atoms* when describing the molecular structure.

You should be able to predict the shape of the molecule or ion if you are able to draw its Lewis structure following the simple rules in Figure 8.12 in the text. The shape will be one of those shown in Figures 9.4, 9.5, 9.7, and 9.10. The procedure is as follows:

1. Draw the Lewis structure for the molecule or ion.

2. Count the number of electron groups (domains) around the central atom. (Remember, a bonding domain can consist of a single, double, or triple bond.)

3. Determine the arrangement of electron groups around the central atom:

two groups	linear	five groups	trigonal bipyramidal
three groups	planar triangular	six groups	octahedral
four groups	tetrahedral		

4. Attach the necessary number of bonded atoms and then determine which shape from Figures 9.4, 9.5, 9.7, and 9.10 it corresponds to. (Try to draw the appropriate structure so you can visualize the shape.) Be particularly careful if the fundamental geometry is trigonal bipyramidal. Lone pairs always go in the equatorial plane.

Study Examples 9.1 through 9.4 in the text.

Self-Test

2. Draw Lewis structures and, without referring to Figures 9.4, 9.5, 9.7, and 9.10, predict the geometry of each of the following:

(a) SO_4^{2-} (b) SF_2

(c) CS_2 (d) BrF_6^+

(e) PH_3 (f) SeF_4

(g) HCO_2^- (formate ion)   (h) BrF_2^-

9.3   Molecular Structure and Dipole Moments

Learning Objective

Explain how the geometry of a molecule affects its polarity.

Review

In this section we see that even when the bonds in a molecule are polar bonds, the effects of the individual bond dipoles can cancel if the molecule is symmetrical. If there are no lone pairs in the valence shell of an atom M in a molecule MX_n, and if all the atoms X that surround M are the same, then the molecule will have one of the symmetrical structures discussed in Section 9.2 and will be nonpolar.

Usually, molecules in which the central atom has lone pairs in its valence shell will be polar. The effects of their bond dipoles don't cancel. Two exceptions are linear molecules in which the central atom has three lone pairs, and square planar molecules in which the central atom has two lone pairs. These exceptions are illustrated in Section 9.3 in the text.

Self-Test

3. Which of the following molecules will be polar?

    (a) SCl_2 (b) SF_6 (c) ClF_3 (d) SO_3 _____

4. The molecule CH_2Cl_2 has a tetrahedral shape but is a polar molecule. Why?

9.4   Valence Bond Theory

Learning Objective

Describe how the valence bond theory views bond formation.

Review

The two main bonding theories based on the principles of wave mechanics are valence bond theory (VB theory) and molecular orbital theory (MO theory). The VB theory retains the image of individual atoms coming together to form bonds. The MO theory, in its simplest form, considers the molecule after the nuclei are in their proper positions and examines the electron energy levels that extend over *all* the nuclei. Both theories ultimately give the same results when they are refined by eliminating simplifying assumptions.

Valence bond theory

According to VB theory, a pair of electrons can be shared between two atoms when orbitals (one from each atom) overlap. By *overlap* we mean that two orbitals from different atoms simultaneously share some region in space. This overlap provides the means by which the electrons spread themselves over both nuclei.

Two other main principles in VB theory are:

1. Only two electrons, with their spins paired, can be shared between two overlapping orbitals.

2. The strength of a bond is proportional to the amount of overlap of the orbitals.

Atoms tend to form bonds that are as strong as possible because this lowers the energy of the atoms the most. This means that when several atoms surround some central atom, they tend to position themselves to give the maximum possible overlap of the orbitals that are used for bonding. According to VB theory, this is what determines molecular shapes and bond angles.

Self-Test (Use separate sheets of paper to answer these questions.)

5. Describe the formation of the chlorine–chlorine bond in Cl_2 using the principles of VB theory.

6. Hydrogen telluride, H_2Te, has a H—Te—H bond angle of 89.5°. Explain the bonding in H_2Te in terms of the overlap of orbitals.

7. Arsine, AsH_3, has H—AS—H bond angles equal to 92°. Explain the bonding in AsH_3.

9.5 Hybrid Orbitals and Molecular Geometry

Learning Objective

Explain how hybridization refines the valence bond theory of bonding.

Review

Because many molecules have bond angles that can't be explained by the directional properties of simple atomic orbitals, it is necessary to consider how these simple orbitals can "mix" or blend when bonds are formed. Mixing simple atomic orbitals gives *hybrid orbitals* with directional properties that differ from the basic atomic orbitals. In general, hybrid orbitals form stronger bonds than simple atomic orbitals because they give better overlap with orbitals of other atoms. Be sure you learn the directional properties of the hybrids described in text Figure 9.19. These are summarized in the following table.

Hybrid Type	Orbitals Mixed	Orientations of Orbitals
sp	$s + p$	linear
sp^2	$s + p + p$	planar triangular
sp^3	$s + p + p + p$	tetrahedral
sp^3d	$s + p + p + p + d$	trigonal bipyramidal
sp^3d^2	$s + p + p + p + d + d$	octahedral

When unshared electron pairs exist in the valence shell of an atom, they are also found in hybrid orbitals. Study the explanations of the bonding for water and ammonia in Section 9.5, as well as Examples 9.7 to 9.8 of the text. Notice that the central atom needs one hybrid orbital for each attached atom *plus* one hybrid orbital for each lone pair.

Free rotation of groups of atoms around a single bond is possible because such rotation doesn't affect the overlap of the orbitals that form the bond, and therefore it does not affect appreciably the strength of the bond. Free rotation about bonds makes possible large numbers of different conformations. (In Section 9.6, you will see that quite a different situation exists with respect to rotation about a double bond.)

VSEPR theory helps us select the kind of hybrid orbitals an atom uses to form its bonds

We can use the VSEPR theory to figure out which kind of hybrid orbitals are used by an atom when it forms bonds. The VSEPR theory predicts the orientations of the electron groups around an atom, which we then use to deduce the kinds of hybrid orbitals that are used.

Example 9.1   Using VSEPR theory to deduce hybridization

Which kind of hybrid orbitals are used by iodine in the ICl_2^- ion?

Analysis:

We can anticipate the kind of hybrid orbitals the central atom will use if we know the geometry of the molecule. Therefore, the first step is to construct the Lewis structure. Next, we apply the VSEPR theory. Once we know the geometry around the central atom, we can select the hybrid orbitals used.

Assembling the Tools:

We will use the tool for drawing Lewis structures in Figure 8.12 and then the tools in this section to decide on the shape. Finally we apply VSEPR theory to decide on the hybrid orbitals.

Solution:

First, let's draw the Lewis structure for the ion.

There are five electron pairs (domains) around iodine, which the VSEPR theory predicts should be arranged in a trigonal bipyramid. The hybrid that has this orientation is sp^3d, so we conclude that in this ion the iodine uses two sp^3d hybrids to form the bonds and the other three to house the three lone pairs.

Is the Answer Reasonable?

Here are some things to check: Be sure you've accounted for all the valence electrons in the Lewis structure. If so, you've likely obtained the right Lewis structure. Next, count again the number of domains around the central atom. This number equals the number of hybrid orbitals needed. In this ion, we have 5 domains around the iodine atom, so we will need 5 hybrid orbitals. The only set that consists of 5 orbitals is sp^3d. Our answer seems reasonable.

Coordinate covalent bonds and hybrid orbitals

According to valence bond theory, a coordinate covalent bond is formed by the overlap of a filled orbital of one atom with an empty orbital of another atom. In the discussion describing the formation of BF_4^- from BF_3 and F^-, notice that sufficient hybrid orbitals are made available to hold all of the electrons in the bonds.

Self-Test

8.   Why do atoms often use hybrid orbitals to form bonds instead of unhybridized atomic orbitals?

---

9.   Use the VSEPR theory to predict which kinds of hybrid orbitals the central atom uses to form its bonds in each of the following molecules. To do this, use a separate sheet of paper to construct the Lewis structure for each.

   (a) $AsCl_5$ _____

   (b) $SeCl_2$ _____

   (c) $AlCl_3$ _____

   (d) NF_3 _____

10.   Use orbital diagrams to give explanations using hybrid orbitals of the bonding and geometry of the following:

   (a) $GeCl_4$

   (b) $SeCl_4$

   (c) SCl_2

   (d) IF_5

11.   Explain, in terms of valence bond theory, the reaction for the formation of an ammonium ion from an ammonia molecule and a hydrogen ion.

12.   Which type of hybrid orbitals are used by silicon in the $SiCl_6^{2-}$ ion? Give an orbital diagram for silicon that illustrates the bonding in this ion.

9.6   Hybrid Orbitals and Multiple Bonds

Learning Objective

Describe the nature of multiple bonds using orbital diagrams and hybridization.

Review

In this section we see that two kinds of bonds can be formed by the overlap of orbitals:

σ bonds   in which the electron density is concentrated along an imaginary line joining the nuclei.

π bonds   in which the electron density is divided into two regions that lie on opposite sides of an imaginary line joining the nuclei.

In every molecule that you will study, a single bond is a σ bond. A double bond consists of *one* σ and *one* π bond; a triple bond consists of *one* σ bond and *two* π bonds. Thus, every bond between two atoms is composed of at least a σ bond. In complex molecules, the molecular geometry is determined by this σ bond framework, and the kind of hybrid orbitals that an atom uses is determined by the number of σ bonds it forms and the number of unshared pairs of electrons it has on the central atom. The following example illustrates how we apply these concepts.

Example 9.2   Determining the kinds of hybrids that atoms use in molecules

What kinds of hybrid orbitals do the carbon and nitrogen atoms use in the molecule

$$
\begin{array}{ccccccc}
 & H & H & H & H & & \\
 & | & | & | & | & & \\
H- & C & =C & -C & -N & -H \\
 & & & | & \cdot\cdot & \\
 & & & H & &
\end{array}
$$

Analysis:

For a complex molecule that has more than one "central atom" we can treat each individually using the tools that were just learned. Since hydrogen is bonded only to one other atom, we will ignore them and focus on the carbon and nitrogen atoms. We will need to count the number of bonding domains around each atom and then the number of nonbonding lone electron pairs. Once that information is known, we can assign a structure.

Assembling the Tools:

We can use the tool for drawing Lewis structures in Figure 8.12, to determine the number of bonding and nonbonding domains around each carbon and nitrogen atom. Then the tools in this section are used to decide on the shape. Finally we apply VSEPR theory to decide on the hybrid orbitals.

Solution:

The carbon on the left and the one in the center each form three σ bonds and neither has any unshared electrons. Therefore, each of these carbons needs three hybrid orbitals, so *sp*² hybrids are used. The carbon at the right forms four σ bonds and therefore uses *sp*³ hybrids. The nitrogen forms three σ bonds, which requires three hybrid orbitals, and it needs a fourth hybrid to house the lone pair; the total is four, so *sp*³ hybrids are used by the nitrogen.

$$H-\overset{\overset{\displaystyle H}{|}}{C}=\overset{\overset{\displaystyle H}{|}}{C}-\overset{\overset{\displaystyle H}{|}}{\underset{\underset{\displaystyle H}{|}}{C}}-\overset{\overset{\displaystyle H}{|}}{\underset{}{N}}-H$$

sp²   *sp³*

Is the Answer Reasonable?

All we can do here is recheck our work to be sure we haven't made any mistakes.

An important feature of double bonds is the restricted rotation around the axis of the bond. This is because such rotation misaligns the unhybridized *p* orbitals that form the π bond. As a result, rotation around a double bond axis involves bond breaking, which is very difficult.

Self-Test

13. Which kinds of hybrid orbitals are used by the carbon and oxygen atoms in acetic acid, which has the following structure?

$$H-\overset{\overset{\displaystyle H}{|}}{\underset{\underset{\displaystyle H}{|}}{C}}-\overset{\overset{\displaystyle :O:}{\|}}{C}-\overset{}{\underset{}{O}}-H$$

    _____

    _____

    _____

    _____

14. Explain why it should be possible to isolate three compounds with the formula $C_2H_2Cl_2$ and in which there are carbon–carbon double bonds.

    _____

    _____

    _____

    _____

15. Why doesn't it matter whether or not rotation about a triple bond is restricted?

    _____

    _____

9.7   Molecular Orbital Theory Basics

Learning Objective

Use molecular orbitals to explain the bonding of simple diatomic molecules

Review

According to molecular orbital theory (MO theory), when two orbitals overlap, their electron waves interact by constructive and destructive interference to give bonding and antibonding molecular orbitals. A *bonding orbital* is one that concentrates electron density between nuclei and leads to a lowering of the energy when occupied by electrons. An *antibonding orbital* concentrates electron density in regions that do not lie between nuclei. When occupied by electrons, an antibonding MO leads to a raising of the energy. Remember that compared to the atomic orbitals from which they are formed, bonding MOs are lower in energy and antibonding MOs are higher in energy.

The electronic structure of a molecule is obtained by feeding the appropriate number of electrons into the molecule's set of molecular orbitals. The rules that apply to the filling of molecular orbitals are the same as those for atomic orbitals.

1.   An electron enters the lowest energy MO available.

2.   No more than two electrons with spins paired can occupy any given MO.

3.   When filling an energy level consisting of two or more MOs of equal energy, electrons are spread over the orbitals as much as possible with their spins in the same direction.

You should study the shapes of the σ and σ^* orbitals formed by the overlap of *s* orbitals as well as the shapes of the σ, σ^*, π, and π^* orbitals that arise from the overlap of *p* orbitals. Also study the order of filling of the MOs given by the energy level diagram in Figure 9.41 and Table 9.1.

Self-Test

16.   On a separate piece of paper, construct the MO energy level diagrams and give the molecular orbital electron populations for (a) Li_2^- and (b) F_2^-.

17.   Which should be more stable, C_2^+ or C_2^-? Explain.

_____

_____

18.   Which has the more stable bond, NO or NO^+? (See Practice Exercise 9.23 in the text.)

_____

9.8   Delocalized Molecular Orbitals

Learning Objective

Compare and contrast resonance structures to delocalized molecular orbitals.

Review

Molecular orbital theory avoids resonance by allowing for the simultaneous overlap of more than two orbitals in such a way as to produce a large π-type cloud that extends over three or more atoms. This kind of an extended molecular orbital is said to be *delocalized* because the bond is not localized between just two atoms. One of the most important molecules having delocalized MOs is benzene.

Molecules in which there are delocalized bonds are more stable than they would be if the bonds were localized. The extra stability produced by delocalization is called the delocalization energy.

Self-Test

19. Draw resonance structures for the nitrate ion and show how the bonding would be represented using a delocalized MO.

20. How is the benzene molecule represented to show the delocalized molecular orbitals of the ring?

9.9 Bonding in Solids

Learning Objective

Use band theory to explain bonding in solids and physical properties.

Review

Solids all have measurable electrical properties. One of these is the ability to conduct electricity. Some solids, particularly the metals, conduct electricity quite well. Other materials, such as nonmetals and many solid organic compounds such as glass and Teflon, are considered to be insulators. A third class of materials in between these two are called semiconductors. Not surprisingly, many of these are in the group of elements we have classified as metalloids.

The band theory attempts to describe how these three types of material act. For an electron to move from one atom to another, there must be an empty orbital that the electron can move to, and the energy needed to move from one orbital to another should be very small for a conducting material and very large for a nonconducting material.

In a metal such as sodium the valence shell has a $3s^1$ electron. Since the $3s$ orbital can hold two electrons, an electron can move freely from one sodium atom to another. In magnesium, both of the s orbitals are occupied, but these are p orbitals with almost the same energy and electrons can use those empty p orbitals to move. In this example we have valence electrons that are said to occupy closely spaced energy levels in the valence band. The empty orbitals that the electrons move to are called the conduction band. For the Na and Mg, the valence band and conduction band have overlapping energy levels and it is very easy to conduct electricity. Insulators have their nearest empty orbitals, needed for conduction, at an energy level that is so high that conduction just does not occur. Semiconductors have their empty orbitals of the conduction band closer to the valence band and, with just a little energy, electrons can be made to flow.

An important feature of this section is the discussion of the solar cell that takes advantage of the properties of semiconductors. Light emitting diodes are another use for the semiconductor materials.

Self-Test

21. Chlorine, like sodium, has one empty p orbital. Why is chlorine an insulator?

_____

22. Would you expect the ionization energy of an element to be related to the element's ability to conduct electricity?

_____

9.10 Bonding of the Allotropes of the Elements

Learning Objective

Use hybridization and valence bond theory to explain allotropes of nonmetallic elements.

Review

This section emphasizes an important point, that is, large atoms generally do not form π bonds. The reason for this is that effective sidewise overlap of p orbitals is limited to the Period 2 elements. The existence of double and triple bonds in carbon compounds adds to the ability to have a wide variety of organic compounds that are needed for life. Oxygen and nitrogen also form multiple bonds in many compounds using π bonds.

The second emphasis of this section is the discussion of allotropes, which are different physical forms of a given element. Obvious examples are the molecular oxygen, O_2, and ozone, O_3, allotropes of oxygen. Carbon has graphite, diamond, and C_{60}, buckminsterfullerene, as its allotropes. The allotropes of sulfur and phosphorus are also mentioned.

Self-Test

23. Which allotrope of phosphorus spontaneously ignites in air? Which allotrope is used to make matches?

_____

24. Some have suggested that alternative life-forms might be silicon based because silicon falls just below carbon in the periodic table. What considerations, discussed in this section, suggest that this is not likely.

_____

Answers to Self-Test Questions

1. Check your answers by referring to the drawings in Figure 9.4 in the text.
2.

(a) $\left[\begin{array}{c} \ddot{O} \\ | \\ \ddot{O}-S-\ddot{O} \\ | \\ \ddot{O} \end{array} \right]^{2-}$ tetrahedral

(b) $\ddot{F}-S-\ddot{F}$ nonlinear

(c) $\ddot{S}=C=\ddot{S}$ linear

(d) $\left[\begin{array}{c} F \quad\quad F \\ F-I-F \\ F \quad\quad F \end{array} \right]^{+}$ octahedron

(e) $H-\overset{\cdot\cdot}{P}-H$ trigonal pyramidal

(f) $\begin{array}{c} F \quad F \\ F-I \\ \quad F \end{array}$ distorted tetrahedron

(g) $\left[\begin{array}{c} H \\ O \\ \| \\ H-C \\ \diagdown O \end{array} \right]^{-}$ planar triangular

(h) $\ddot{F}-\ddot{Br}-\ddot{F}$ linear

3. SCl_2 and ClF_3 are polar. (Draw Lewis structures; note lone pairs.)
4. The C—Cl and C—H bonds differ in polarity, so the bond dipoles can't cancel to give a nonpolar molecule.
5. Overlap of the half-filled p orbitals of the chlorine atoms produces the bond. See Figure 9.18 for F_2.
6. Hydrogen $1s$ orbitals overlap two half-filled p orbitals of tellurium. Since p orbitals are at 90°, the bond angle is very close to 90°.

Te (in H_2Te)   (••) (••)(•×)(•×)   • = Te electron
                   × = H electron
                  $5s$ $5p$

7. Hydrogen 1*s* orbitals overlap 3 half-filled *p* orbitals of arsenic.

    As (in AsH₃)   • = As electron
                                          x = H electron
                    4*s*   4*p*

8. Hybrid orbitals overlap better with orbitals on neighboring atoms and form stronger bonds.
9. (a) sp^3d (b) sp^3 (c) sp^2 (d) sp^3

10. (a) Ge (in GeCl₄)      x = Cl electron
                           sp^3

    molecule is tetrahedral

    (b) Se (in SeCl₄)
                          sp^3d *d*

         x= Cl electron

    molecule has a distorted tetrahedral shape

    (c) S (in SCl₂)   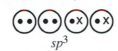   x= Cl electron
                        sp^3

    nonlinear or bent molecular shape

11.   ⎡ This orbital overlaps with the
                                       ⎣ vacant orbital of the H⁺ ion

    N (in NH₄⁺)      x = H electron
                   sp^3

    (d) I (in IF₅)   ⬤ x x x x x ⬤⬤⬤
                       sp^3d^2 *d*
         x= F electron

    square pyramidal molecular shape

12. *sp³d²* hybrids

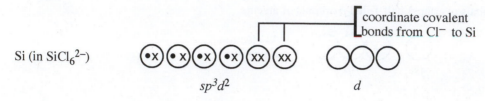

Si (in SiCl₆²⁻)

sp³d²   *d*

X = Cl electron

13.

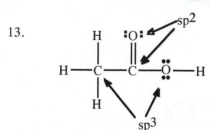

14. Restricted rotation around the double bond permits structures II and III (below) to be isolated as well as structure I.

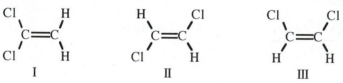

I   II   III

15. The atomic arrangement is linear.

16.

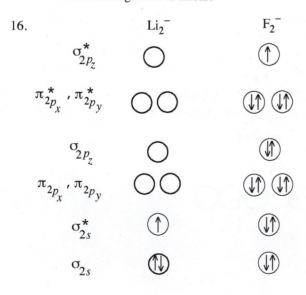

17. C_2^+, bond order $= \dfrac{5-2}{2} = \dfrac{3}{2}$

C_2^-, bond order $= \dfrac{7-2}{2} = \dfrac{5}{2}$

Therefore C_2^- is more stable than C_2^+.

18. NO, bond order is $\dfrac{8-3}{2} = \dfrac{5}{2}$

NO^+, bond order is $\dfrac{8-2}{2} = \dfrac{6}{2}$

NO^+ has a more stable bond than NO.

19.

$$\left[\ \overset{:\ddot{O}:}{\underset{:\ddot{O}\ \ \ \ \ddot{O}:}{N}}\ \right]^- \longleftrightarrow \left[\ \overset{:\ddot{O}:}{\underset{:\ddot{O}\ \ \ \ddot{O}:}{N}}\ \right]^- \longleftrightarrow \left[\ \overset{:\ddot{O}:}{\underset{:\ddot{O}\ \ \ \ddot{O}:}{N}}\ \right]^- \qquad \left[\ \overset{O}{\underset{O\ \ \ \ \ O}{N}}\ \right]^-$$

20.

21. Chlorine exists as Cl_2 so all of its electrons are paired and there are no empty orbitals in the valence band. The next higher orbital, the $4s$, requires too much energy to reach for easy conduction of electrons.

22. Yes, to some degree, the ability to dislodge an electron from an atom will be related to the element's conductivity.

23. White phosphorous, red phosphorus

24. Silicon will not form multiple (double and triple) bonds as carbon does. Also, silicon apparently does not form long chains or rings as found in carbon compounds.

Tools for problem solving

In this chapter you learned to apply the following concepts as tools in solving problems. Study each one carefully so that you know what each is used for. When faced with solving a problem, recall what each tool does and consider whether it will be helpful in finding a solution. You might want to tear these pages out to use along with solving problems in this chapter.

Basic molecular shapes (Section 9.1)

The five basic geometries are linear, planar triangular, tetrahedral, trigonal bipyramid, and octahedral.

VSEPR model (Section 9.2)

Electron groups repel each other and arrange themselves in the valence shell of an atom to yield minimum repulsions, which is what determines the shape of the molecule.

Using VSEPR to predict molecular shape and polarity (Section 9.2)

To obtain a molecular structure, follow these steps: (1) draw the Lewis structure, (2) count electron domains, (3) select the basic geometry, (4) add atoms to bonding domains, and (5) ignore nonbonding domains to describe the shape.

Molecular shape and polarity (Section 9.2)

We can use molecular shape to determine whether a molecule will be polar or nonpolar.

Valence bond theory criteria for bond formation (Section 9.4)

A bond requires overlap of two orbitals sharing two electrons with paired spins. Both orbitals can be half–filled, or one can be filled and the other empty. We use these criteria to establish which orbitals atoms use when bonds are formed.

Hybrid orbitals formed by *s* and *p* atomic orbitals (Section 9.5)

Two electron domains are formed by sp hybrid orbitals.

Three electron domains are formed by sp^2 hybrid orbitals.

Four electron domains are formed by sp^3 hybrid orbitals.

Five electron domains are formed by sp^3d hybrid orbitals.

Six electron domains are formed by sp^3d^2 hybrid orbitals.

The VSEPR model and hybrid orbitals involving *s* and *p* electrons (Section 9.5)

Lewis structures permit us to use the VSEPR model to predict molecular shape, which then allows us to select the correct hybrid orbitals for the valence bond description of bonding. We represent the central atom by M, the atoms attached to the central atom by X, and lone pairs by E. We can then signify the number of bonding and nonbonding domains around M as a formula MX_nE_m, where n is the number of bonding domains and m is the number of nonbonding domains. The following structures are obtained for two, three, and four domains.

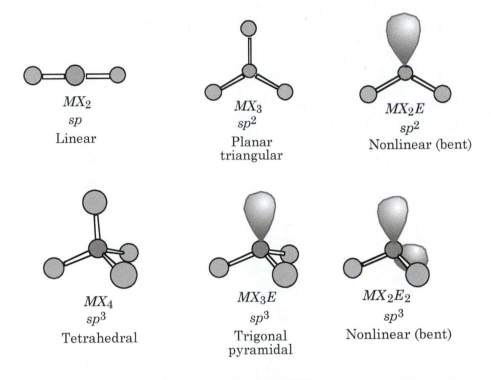

MX_2
sp
Linear

MX_3
sp^2
Planar
triangular

MX_2E
sp^2
Nonlinear (bent)

MX_4
sp^3
Tetrahedral

MX_3E
sp^3
Trigonal
pyramidal

MX_2E_2
sp^3
Nonlinear (bent)

The VSEPR model and hybrid orbitals involving *d* electrons (Section 9.5)

The following VSEPR structures are obtained for five and six domains using the generalized formula MX_nE_m.

MX_5	MX_4E	MX_3E_2	MX_2E_3
sp^3d	sp^3d	sp^3d	sp^3d
Trigonal bipyramidal	See-saw or distorted tetrahedral	T-shaped	Linear

MX_6	MX_5E	MX_4E_2
sp^3d^2	sp^3d^2	sp^3d^2
Octahedral	Square pyramidal	Square planar

Sigma (σ) and pi (π) bonds and hybridization (Section 9.6)

The Lewis structure for a polyatomic molecule lets us apply these criteria to determine how many σ and π bonds are between atoms and the kind of hybrid orbitals each atom uses. Remember that the shape of the molecule is determined by the framework of σ bonds, with π bonds used in double and triple bonds.

Calculating bond order in MO theory (Section 9. 7)

$$\text{Bond order} = \frac{(\text{number of bonding } e^-) - (\text{number of antibonding } e^-)}{2 \text{ electrons/bond}}$$

Filling molecular orbitals (Section 9.7)

Electrons populate MOs following the same rules that apply to atomic orbitals in an atom.

Tendency toward multiple bond formation (Section 9.10)

Atoms of Period 2 have a stronger tendency to form multiple bonds with each other than do those in Periods 3 and below.

Summary of Useful Information

The various structures obtained by applying the VSEPR theory can be summarized as shown in the table below. In the formulas in the table, the symbol M stands for the central atom, X stands for an atom bonded to the central atom, and E stands for a lone pair of electrons in the valence shell of the central atom. Thus, MX_2E_3 represents a molecule in which the central atom is bonded to two other atoms and has three lone pairs in its valence shell. To use the information in the table, construct the Lewis structure for the substance. Then write its formula using the symbols as defined above. Find the formula in the table and read the name for the molecular shape alongside.

Formula	Structure	Formula	Structure
MX_2	linear	MX_5	trigonal bipyramidal
MX_3	planar triangular	MX_4E	distorted tetrahedral
MX_2E	nonlinear (bent)	MX_3E_2	T-shaped
MX_4	tetrahedral	MX_2E_3	linear
MX_3E	trigonal pyramidal	MX_6	octahedral
MX_2E_2	nonlinear (bent)	MX_5E	square pyramidal
		MX_4E_2	square planar

Correlation between VSEPR theory and orbital hybridization.

Number of sets of electrons in valence shell of central atom	Geometric arrangement of electron pairs	Hybrid orbitals that give this geometry
2	linear	sp
3	planar triangular	sp^2
4	tetrahedral	sp^3
5	trigonal bipyramidal	sp^3d
6	octahedral	sp^3d^2

Notes:

Chapter 10

Properties of Gases

This chapter and the next examine the physical properties of the states of matter. Gases are the most easily understood, and many of the properties of gases are quite familiar to everyone. We will study the four variables that control gas behavior—pressure, volume, temperature, and amount (which we will measure in moles). These variables are interrelated, because one cannot be changed without changing one or more of the others. The relationships among them are expressed in the *gas laws*, which we describe in this chapter. We will also see how these laws can be explained in terms of a single theory, the kinetic theory of gases.

10.1   A Molecular Look at Gases

Learning Objective

Describe the properties of gases at the macroscopic and molecular levels the differences between real and ideal gases.

Review

The purpose of this section is to review familiar properties of gases and show that a very simple molecular model of this state of matter is able to explain gas behavior. Let's review the list.

Gases can be compressed.

Gases exert a pressure that depends on the amount of gas in a container.

Gas pressure increases with increasing temperature.

Gases have low density, meaning a given volume doesn't contain much matter.

Gases expand to fill any container they're in.

Gases freely mix with each other.

All of these properties can be explained by a model that describes a gas as a collection of widely spaced molecules in rapid motion. Their collisions with the walls of the container produce the pressure, which increases with temperature because molecules go faster and hit harder as their temperatures increase.

Self-Test

1.   What four physical properties of gases are interrelated?

_____

2. If we compare gases at different temperatures, would the speed at which they mix with each other increase, decrease, or stay the same as the temperature is raised? Explain your answer in terms of the molecular model of a gas.

_____

_____

10.2 Measurement of Pressure

Learning Objective

Explain the measurement of pressure using barometers and manometers.

Review

Pressure, as you learned in Chapter 6, is the ratio of force to area. The earth exerts a gravitational force that pulls on the surrounding envelope of air—the atmosphere—and causes the air to be most concentrated near the earth's surface where it exerts a pressure that we call the *atmospheric pressure*. The simplest device to measure it is the Torricelli barometer.

At sea level, the atmospheric pressure holds a column of mercury in a barometer at an average height of about 760 mm. The pressure unit called the *standard atmosphere (atm)* was originally defined as the pressure exerted by a column of mercury 760 mm high at a temperature of 0 °C. With the introduction of the SI, the *atmosphere* was redefined in terms of the *pascal*, the SI unit of pressure (1 atm = 101,325 Pa = 101.325 kPa). The pressure unit *bar* is also defined in terms of the pascal (1 bar = 100 kPa). A smaller unit of pressure called the *torr* is defined as 760 torr = 1 atm. For most purposes, we can use the relationship 1 torr = 1 mm Hg, meaning a pressure of 1 torr will support a column of mercury 1 mm high. This makes measuring pressures in torr relatively simple by using mercury as the fluid in barometers and manometers (the topic to be discussed next).

You will frequently find it necessary to convert between torr and atm, so it is important that you remember the relationship

$$1 \text{ atm} = 760 \text{ torr} = 760 \text{ mm Hg}$$

To check yourself in calculations, remember that *the pressure expressed in torr is always numerically larger than if it is expressed in atm*. This will help you avoid errors. For example, if you are converting 0.700 atm to torr and use the conversion incorrectly, you would obtain an answer of 9.21×10^{-4} torr. But the number of torr should be much larger than the number of atm, so the answer must be wrong and we can check the calculation. (The correct answer is 532 torr.)

Your teacher may also want you to learn the relationships to convert between the atmosphere and pascal units, and between units of the pascal and bar. If so, you will need to memorize the following:

$$1 \text{ atm} = 101,325 \text{ Pa}$$

$$1 \text{ bar} = 100,000 \text{ Pa}$$

Pressures of trapped gases are measured using manometers

Manometers, either the open-end or the closed-end type, are employed to measure the pressure of a confined gas. The difference in the heights of the two liquid levels in a U-shaped tube is proportional to the difference between the pressures exerted by the gases on opposite sides. If mercury is used in the manometer, the height difference is measured in mm Hg, which gives the pressure difference in torr. (Remember, 1 mm Hg = 1 torr.) Study the discussions of open– and closed–end manometers in Section 10.2 in the text. The value we calculate for h_{Hg}, expressed in millimeters, is the difference in torr between the pressures exerted by the gases connected to the two arms of the manometer. Study Example 10.1 in the text to see how this is done and then try the example below.

Example 10.1 Examining the difference between a closed-end and open-end manometer

There are two types of manometer, the open-end and the closed-end manometers. What is the use of one manometer compared to the other? Would you classify a barometer as a closed-end manometer or as an open-end manometer?

Analysis:

The only difference between the two types of manometer is whether or not one end is open or closed. In both of the manometers we use mercury as the liquid (although other liquids can be used) and the second end is open to the system being measured.

Solution:

The solution to this question is found when we identify the pressure on the closed and open ends of the manometer that is not connected to the gas we are measuring. In the open-end manometer this pressure is atmospheric pressure (around 760 torr). In the closed-end manometer the pressure in the closed end is the very small, almost zero, pressure of mercury. In both cases the pressure is

$$P_{\text{open-end}} \;=\; P_{\text{atmosphere}} \;\pm\; \Delta h_{Hg} = \text{the difference in height of two columns of mercury}$$

Since one column is about 760 torr, and small differences are usually observed, the pressures we measure with an open-end manometer are relatively high.

$$P_{\text{closed-end}} \;\approx\; 0.0 \;+\; \Delta h_{Hg} = \Delta h_{Hg} = \text{the difference in height of two columns of mercury}$$

Since one column is about 0.0 torr, and small differences are usually observed, the pressures we measure with a closed-end manometer are relatively small.

Look at the barometer in Figure 10.3 in the text. The clue to answering this question is to note that the top of the column is sealed. Therefore a barometer is a closed-end manometer. In order to make measurements of atmospheric pressure, the tube must be longer than the highest expected pressure (at least 760 mm, perhaps a bit more).

Is the Answer Reasonable?

We check our arguments with the discussion in the text and see that they agree.

Self-Test

3.   A pressure of 732 torr has what value in atm?   _____

4.   On a TV weather report, the pressure was reported to be 29.2 in., meaning "inches of mercury."

   What does this correspond to in torr?   _____In atm?_____

5. A pressure of 0.844 atm has what value in torr? _____

6. A pressure of 746 torr has what value in kilopascals (kPa)? _____

7. The pressure in a storm center was reported to be 967 mb. Express this pressure in

    (a) atm _____ (b) torr _____

8. In using an open-end manometer containing mercury, what measurements are required in determining the pressure of a confined gas?

    _____

9. In using a closed-end manometer containing mercury, what measurements are required in determining the pressure of confined gas?

    _____

10. On a day when the atmospheric pressure was 784 torr, the manometer reading on a gas sample was found to be 82 torr. What is the gas pressure in torr if the manometer is of the closed-end design?

    _____ The open-end design? _____

10.3   Gas Laws

Learning Objective

Describe and use the gas laws of Dalton, Charles, Gay-Lussac, and the combined gas law.

Review

In this section we examine some of the gas laws. We begin with the relationships among *P, V,* and *T* for a fixed (constant) amount of gas. These deal with situations in which we change one or two of the variables and observe (or calculate) how one of the others changes. Historically, the scientists who studied gases held one variable constant and observed the relationship between the other two. This led to three separate gas laws. We will also examine one of the laws that deals with situations in which the amount of gas changes.

Pressure–volume law (Boyle's law). Gases generally obey the rule that their volumes are inversely proportional to their pressures, provided that the temperature and the amount of gas, *n*, are kept constant.

$$V \propto 1/P \text{ (constant } T \text{ and } n) \quad \text{or} \quad PV = \text{constant}$$

The hypothetical gas that would obey this relationship (as well as the others that we will discuss) exactly is called an *ideal gas.*

Temperature–volume law (Charles' law). Provided we express the temperature of a gas in kelvins, the volume of a fixed amount of gas is directly proportional to the temperature if the pressure is kept constant.

$$V \propto T \quad \text{(constant } P \text{ and } n) \quad \text{or} \quad \frac{V}{T} = \text{constant}$$

Notice, once again, that for this relationship to hold, *temperatures must be expressed in kelvins.*

Pressure–temperature law (Gay-Lussac's law). Provided that the volume is held constant, the pressure of a fixed quantity of gas is directly proportional to the Kelvin temperature.

$$P \propto T \quad \text{(constant } V \text{ and } n) \quad \text{or} \quad \frac{P}{T} = \text{constant}$$

As with Charles' law, calculations using this law require Kelvin temperatures.

The combined gas law.

A single equation can be written that expresses all three of these laws.

$$\frac{PV}{T} = \text{constant}$$

Usually, we use these gas laws in calculations where one or two of the variables change and we wish to know the effect on another. If we use subscripts of 1 and 2 to identify the P, V, and T before and after the change, we can express the combined gas law as given by the equation below.

$$\frac{P_1 V_1}{T_1} = \frac{P_2 V_2}{T_2} \qquad \text{(Equation 10.1)}$$

When $P_1 = P_2$, the equation reduces to that of Charles' law (the temperature–volume law). When $T_1 = T_2$, the combined gas law equation expresses Boyle's law (the pressure–volume law). When $V_1 = V_2$, the combined gas law equation reduces to Gay-Lussac's law (the pressure–temperature law).

Solving the combined gas law for one of its variables

Examples 10.3 and 10.4 in the text illustrate how to apply the combined gas law in calculations. In using this equation, it is necessary to solve it for the unknown quantity. This is where some students have difficulty, especially if their algebra skills are rusty. Here is a tip that may help you avoid mistakes.

Suppose we wished to solve the equation for P_2. The first step is to rewrite the equation as shown below.

$$P = P \times \frac{V}{V} \times \frac{T}{T}$$

The volume and temperature ratios will ensure that volume and temperature units cancel and the answer will be in pressure units.

Next, we have to decide where to put the 1's and 2's. To follow the reasoning here, first look at the left side of Equation 10.1.

Notice that the P_1 and V_1 are *both* above the fraction bar and T_1 is below the fraction bar. No matter how we rearrange things, if P_1 is on one side of the fraction bar (either in the numerator or denominator), then V_1 will be on the same side.[1] Furthermore, whichever side of the fraction bar we find P_1 and V_1 on, T_1 will be on the *opposite* side. Therefore, let's insert the subscripts 1 in the equation we're solving for P_2. In doing this, we have to realize that P_1 is considered to be in the numerator, as though it's in the fraction $P_1/1$.

---

[1] This applies as long as P_1 and V_1 are on the same side of the equals sign.

$$\boxed{P_1 \text{ and } V_1 \text{ are both in the numerator.}}$$

$$P = P_1 \times \left(\frac{V_1}{V}\right) \times \left(\frac{T}{T_1}\right)$$

$$\boxed{T_1 \text{ is in the denominator.}}$$

Once the 1's are in place, the 2's follow. Here's the equation solved for P_2.

$$P_2 = P_1 \times \frac{V_1}{V_2} \times \frac{T_2}{T_1}$$

Even if your algebra skills are good, the approach just described serves as a good check to see if you've solved the equation correctly.

Example 10.2   Using the combined gas law equation

Consider a sample of 2.15 g of argon, one of the noble gases, occupying a volume of 695 mL at 22.4 °C under a pressure of 782 torr. It is desired to change this amount of sample to a final state that is at a temperature of 39.6 °C and a pressure of 693 torr.

(a) Are enough data provided to enable another scientist to duplicate the sample in its initial state? If not, what additional data would be needed?

(b) For this specific quantity of sample to be in its final state, what other variable must change? To what value?

Analysis:

We are asked two specific questions about the state of a gas and also what will happen if we change one of the states; in this case both the temperature and the pressure are changed. We need an equation that can do this all at once.

Assembling the Tools:

The tool we need is the combined gas law equation. This equation also helps us answer both (a) and (b).

Solution:

(a)   We do not need any extra information, in fact, the mass of argon is extra data that is not needed.

(b)   To use Equation 10.1 above, we identify the variables as follows:

$V_1 = 695$ mL $P_1 = 782$ torr $T_1 = 22.4$ °C + 273.15 °C $= 295.6$ K
$V_2 = ?$ $P_2 = 693$ torr $T_2 = 39.6$ °C + 273.15 °C $= 312.8$ K

We can enter the data in the correct places.

$$\frac{782 \text{ torr} \times 695 \text{ mL}}{295.6 \text{ K}} = \frac{782 \text{ torr} \times V_2}{312.8 \text{ K}}$$

Then we rearrange the equation algebraically to read

$$V_2 = 695 \text{ mL} \times \frac{693 \text{ torr}}{782 \text{ torr}} \times \frac{312.8 \text{ K}}{295.6 \text{ K}}$$

When we do the math and cancel the units the answer to this problem is 652 mL.

Is the Answer Reasonable?
The temperatures and pressures do not change very much, and the values indicate that temperature will increase the volume and the pressure change decreases the volume. The result is that we expect the answer to be close to the volume we started with, and it is.

Self-Test

11. The pressure on 425 mL of a gas is changed at constant temperature from 722 torr to 786 torr. What is the new volume?

12. To change the volume of a gas at 955 torr from 233 mL to 500 mL at constant temperature, what new pressure must be provided?

13. What do we call the hypothetical gas that would obey the pressure–volume law exactly?

14. At constant temperature and amount, the volume of an ideal gas is inversely proportional to

    (a) $\frac{1}{P}$ (b) P (c) P^2 (d) $\sqrt{P}$

15. What will be the new volume if 854 mL of oxygen is cooled from 48.2 °C to 18.7 °C at constant pressure?

16. A sample of 750 mL of nitrogen gas is heated from 20 °C to 100 °C. In order to keep its pressure constant, to what new volume should the gas be allowed to change?

17. A sample of 656 mL of oxygen gas at 873 torr is cooled from 55.0°C to 28.0 °C in a sealed container. What gas variable changes and to what new value?

18. What will be the final volume of a 515 mL sample of helium if its pressure is changed from 773 torr to 653 torr and its temperature is changed from 32 °C to 57 °C?

19. In order for the pressure of a fixed mass of gas at 13.2 °C to change from 2 atm to 3 atm and its volume to go from 4 L to 5 L, what must the final temperature, in °C, become?

    _____

20. What is the final pressure in torr exerted by 475 mL of oxygen at 655 torr and 27.2 °C if its temperature goes to 54.3 °C and its volume changes to 407 mL?

    _____

10.4   Stoichiometry Using Gas Volumes

Learning Objectives

Perform stoichiometric calculations using the gas laws and Avogadro's principle.

Review

Avogadro's principle deals with varying amounts of gas

Avogadro's principle describes how changing the amount of gas affects the other gas variables. If we keep T and P constant, the volume increases when we increase the amount of gas. (This is what you observe when you blow up a balloon; adding more air makes the balloon bigger.)

$$V \propto n \quad \text{(at constant } T \text{ and } P\text{)}$$

Similarly, if we keep T and V constant, the pressure increases when we increase the amount of gas. (This is what happens when you pump air into a tire. The tire doesn't get larger, but the pressure inside it increases.)

$$P \propto n \quad \text{(at constant } T \text{ and } V\text{)}$$

Avogadro's principle leads to the concept of a molar volume—the volume occupied by one mole of a gas. For comparisons, conditions of standard temperature and pressure (STP) are chosen. STP corresponds to 0 °C (273 K) and 1 atm. Under these conditions, 1 mol of an ideal gas occupies a volume of 22.4 L.

When gases react, their volumes are in the same ratio as their coefficients if the volumes are compared at the same T and P. (This is sometimes called the *law of combining volumes*.) Under these conditions, gas stoichiometry is greatly simplified as illustrated below.

For reactions involving gases, Avogadro's principle provides a new way to state stoichiometric equivalencies. For example, from the coefficients in the equation

$$N_2(g) + 3H_2(g) \longrightarrow 2NH_3(g)$$

we know that in terms of *moles* 1 mol $N_2 \Leftrightarrow 3$ mol H_2. We can now rephrase this in terms of gas *volumes* provided that the volumes are compared at identical pressures and temperatures.

$$1 \text{ vol. } N_2 \Leftrightarrow 3 \text{ vol. } H_2$$

When we measure gaseous reactants and products at the same temperature and pressure, gas stoichiometry problems become very simple. Study Examples 10.3 and 10.4 in the text. Notice how we deal with problems in which the T and P are *not* the same for all the gases.

Example 10.3 Stoichiometry calculations using gas volumes

Carburetion is the process of mixing the correct volumes of gases in a chemical reaction. This is the job of a carbureator in automobiles. (Modern cars often use fuel injection to precisely mix air and gasoline.) Home barbecue grills also have a carburetor that must be changed or adjusted depending on whether you are burning propane, C_3H_8, or methane or natural gas, CH_4. How many liters of oxygen are needed to react with 1.00 liter of propane and 1.00 liter of methane? (Assume all gases are at the same temperature and pressure.)

Analysis:

We just learned that the volumes of gases, at the same temperature and pressure, in a chemical equation will react in the same proportions as the stoichiometric coefficients. Therefore, we need a balanced chemical equation for each gas and then we can apply our stoichiometry conversions.

Assembling the Tools:

The tools we need are Avogadro's principle and our techniques for writing and balancing chemical equations.

Solution:

The equation for methane: $CH_4 + 2O_2 \longrightarrow CO_2 + H_2O$

The equation for propane: $C_3H_8 + 5O_2 \longrightarrow 3CO_2 + 4H_2O$

These equations tell us that two volumes of oxygen react with each volume of methane while it takes five volumes of oxygen to react with each volume of propane. We can now set up and solve the equations.

$$1.00 \text{ L CH}_4 \times \frac{2 \text{ L O}_2}{1 \text{ L CH}_4} = 2.00 \text{ L O}_2$$

$$1.00 \text{ L C}_3\text{H}_8 \times \frac{5 \text{ L O}_2}{1 \text{ L C}_3\text{H}_8} = 5.00 \text{ L O}_2$$

It appears that it takes two and a half times more oxygen to burn a liter of propane as compared to methane.

Is the Answer Reasonable?

The math is simple so you should just cancel the units to be sure that the problem was set up properly. The fact that propane has three times as many carbon atoms and uses 2.5 times the oxygen also seems reasonable.

Self-Test

21. Use the combined gas law to calculate the molar volume of a gas at 25 °C and 2 atm.

_____

22. What is the standard molar volume (including its units)?

_____

23. What is Avogadro's principle? _____
_____

24. What do we call the reference conditions for gas temperature and pressure, and what are their values?

_____

25. A sample of 18.4 L of a gas was trapped at STP. How many moles of the gas was this?

_____

26. In the following reaction, all substances are gases.

$$2NO + O_2 \longrightarrow 2NO_2$$

Assuming that the gases are measured at the same values of P and T, how many liters each of NO and O_2 are needed to prepare 2.0 L of NO_2?

_____

27. Nitrogen and hydrogen can be made to combine to give ammonia according to the following equation:

$$N_2 + 3H_2 \longrightarrow 2NH_3$$

All the substances are gases. If three liters of nitrogen are used, what volume of hydrogen is needed and what volume of ammonia is made (assuming all volumes are measured under the same conditions of temperature and pressure)?

_____

28. One step in the industrial synthesis of sulfuric acid is the following reaction:

$$2SO_2(g) + O_2(g) \longrightarrow 2SO_3(g)$$

If 640 L of O_2, initially at 740 torr and 20.0 °C, is used to make SO_3, how many moles and how many grams of SO_2 are needed?

_____

10.5   Ideal Gas Law

Learning Objective

Explain and apply the ideal gas law, and explain how it incorporates the other gas laws.

Review

The ratio PV/T is proportional to the number of moles of gas, n. The proportionality constant is given the symbol R. Inserting R into the proportionality and clearing fractions gives the usual form of the *ideal gas law*,

$$PV = nRT \qquad \text{(Equation 10.2)}$$

The ideal gas law gives us the relationship among all four physical variables of a gas—pressure, volume, temperature, and number of moles. The *universal gas constant*, R, has different numerical values, depending on the units used for P and V. The value we will use in calculations is 0.0821 L atm/mol K. Remember that when you use this value of R in the ideal gas law equation, the unit of pressure must be atm and the unit of volume must be liter (the temperature, of course, must be in kelvins). The ideal gas law equation can be used in the following kinds of calculations.

1. To find the *volume* that a given amount of gas will occupy at some value of pressure and temperature.

2. To find the *pressure* that a given amount of gas will exert if it is kept at a given temperature in some particular volume.

3. To find the *temperature* a given amount of a gas must have if it is confined in a specified volume and is to exert a certain pressure.

4. To use the measured values of P, V, and T to find the number of moles (n) of gas in the sample. If we then divide the mass of the gas sample in grams by the number of moles, we get the *molar mass*, which is numerically the same as the molecular mass.

5. To calculate the density of a gas at a particular temperature and pressure, and to use gas density to calculate the molar mass of a gas.

Be sure to study the worked examples in this section of the text and answer the practice exercises. Then work on the exercises below.

Example 10.4 Stoichiometric calculations using gases

Sodium hydroxide granules can remove carbon dioxide from an airstream by its reaction according to the following equation.

$$NaOH(s) + CO_2(g) \longrightarrow NaHCO_3(s)$$

The air of a room measuring 5.00 m × 10.00 m × 3.50 m containing CO_2 at a concentration of 1.00×10^{-6} mol/L is passed through a bed of NaOH granules. What is the minimum number of grams of NaOH needed to remove all of the CO_2 from this quantity of air?

Analysis:

Don't let all the business about the room's dimensions throw you. With a 1:1 mole ratio of NaOH to CO_2, the stoichiometry is simple; it's getting the number of moles of CO_2 that requires work. Once we find the number of moles of CO_2 we automatically know the number of moles of NaOH. Then it's a moles-to-grams conversion to find the number of grams of NaOH. We're told that the concentration of CO_2 in the room is 1.00×10^{-6} mol L^{-1}. So we need the volume of the room in liters to start with. That is easily converted to moles of CO_2 and then we can do a simple stoichiometric conversion to obtain the mass of NaOH using the chemical reaction given above.

Assembling the Tools:

The tools we need are the process and conversion factors needed for stoichiometry calculations.

Solution:

First we calculate the volume of the room as

$$5.00 \text{ m} \times 10.00 \text{ m} \times 3.50 \text{ m} = 175 \text{ m}^3$$

To convert 175 m^3 to liters is just a unit-conversion problem like those in Chapter 3.

$$1 \text{ m}^3 = 1 \times 10^3 \text{ L}$$

So 175 m^3 = 175 × 10^3 L. (Or 175 cubic meters contain 175 thousand liters.) Now we can calculate the number of moles of CO_2 using the concentration and volume to calculate moles.

$$(\text{vol. in L}) \times \frac{1.00 \times 10^{-6} \text{ mol } CO_2}{\text{L}} = \text{mol } CO_2$$

$$(175 \times 10^3 \text{ L}) \times \frac{1.00 \times 10^{-6} \text{ mol } CO_2}{\text{L}} = 0.175 \text{ mol } CO_2 = 0.175 \text{ mol NaOH}$$

Because the formula mass of NaOH is 40.00, we can convert the moles of NaOH to g NaOH.

$$0.175 \text{ mol NaOH} \times \frac{40.00 \text{ g NaOH}}{1 \text{ mol NaOH}} = 7.00 \text{ g NaOH}$$

We find that the CO_2 in the room air requires a minimum of 7.00 g NaOH to be removed by the chemical reaction given.

Is the Answer Reasonable?

We review the process of calculating the volume of the room, converting to mol of CO_2, then converting to mol NaOH, and then g NaOH. All seems in order and we are confident the results are okay.

Example 10.5 Calculating the molar mass and molecular formula of a gas

A gaseous compound of sulfur and fluorine with an empirical formula of SF has a density of 4.102 g L^{-1} at 25.0 °C and 750 torr. What is the molecular mass and the molecular formula of this compound?

Analysis:

Whenever you are stuck about what to do, go back to the basic definitions and let them lead you. In this problem, remember that a molecular mass is always the ratio of the grams of a substance to the moles of the substance—the "grams per mole." The ideal gas law, will help us because we know R and the variables within the equation tell us what to look for (P, V, T, and n).

Assembling the Tools:

The tools needed include the concept of the molar mass, and the ideal gas law.

Solution:

We're told that 1 L of the gas has a mass of 4.102 g, so what we have to do next is find out how many moles this mass equals while being able to occupy 1 L at 750 torr and 25.0 °C. Volume, V, is written as 1.000 L since the density's units can be taken as an exact number. Before we calculate n, we have to convert 750 torr into atm (0.987 atm) and 25.0 °C into kelvins (298 K) in order to use $R = 0.0821$ L atm mol^{-1} K^{-1}. Using these figures, the number of moles n is found by rearranging the ideal gas law, $PV = nRT$, into

$$n = \frac{PV}{RT} = \frac{0.987 \text{ atm} \times 1.00 \text{ L}}{0.0821 \text{ L atm mol}^{-1}\text{K}^{-1} \times 298 \text{ K}} = 4.03 \times 10^{-2} \text{ mol}$$

The ratio of grams to moles is

$$\frac{4.102 \text{ g}}{4.03 \times 10^2 \text{ mol}} = 102 \text{ g mol}^{-1}$$

The molar mass, therefore, is 102. The formula mass for the empirical formula, SF, calculates to be 51.07. This is half the molecular mass (102/51.07), so each subscript in SF has to be multiplied by 2. The molecular formula of the compound is S_2F_2.

Is the Answer Reasonable?

The fact that empirical formula mass is exactly half of the molar mass gives us confidence. We can also check the calculations in more detail. First we check the rearranged gas law and see it is correct. We also check that we used the correct units and you should cancel them above. Finally we can estimate that the numerator is close to 1.0 and the denominator is close to 30 and 1/30 = 0.033. Our estimated value is close to our answer of 0.04. We also see that mass divided by moles is about 100, close to the 102 we got.

Self-Test

29. Write the equation for the ideal gas law.   _____

30. Write the value of R, including the units.   _____

31. What volume will 1.867 g of O_2 occupy if its temperature is 14.3 °C and its pressure is 683 torr?

    _____

32. If 0.233 g of N_2 is confined at 22.4 °C in a volume of 175 mL, what pressure will it have?

    _____

33. If a 0.165 gram sample of CO_2 is to have a pressure of 3.54 atm when kept in a container with a volume of 35.0 mL, what temperature in degrees Celsius must it have?

    _____

34. A 0.390 gram sample of a gas at 25 °C occupied a volume of 350 mL at a pressure of 740 torr. Calculate its molecular mass.

    _____

35. What is the density of argon gas at 35 °C and 673 torr?

    _____

36. Arrange the following gases in order of increasing density (lowest to highest) if the densities are compared at the same T and P.

    N_2, CH_4, C_4H_{10}, H_2, CO_2 _____

10.6   Dalton's Law of Partial Pressures

Learning Objective

Describe Dalton's law of partial pressures, and apply it to collecting a gas over water.

Review

Gases in a mixture exert their own pressures, called *partial pressures*, independently of the other gases present. The total pressure of the mixture is a simple function of the partial pressures, P_A, P_B, P_C, and so forth.

$$P_{total} = P_A + P_B + P_C + \ldots$$

Called *Dalton's law of partial pressures*, this law is useful in doing calculations involving gases collected over water. When collected in this way, the gas is really a mixture. It contains the desired gas mixed with water vapor. Often we want to calculate the pressure of the collected gas without the water vapor ($P_{dry\ gas}$). All we need to do is subtract the partial pressure of the water vapor from the total pressure of the gas mixture. (The total pressure is all we can actually measure; partial pressures are calculated.)

$$P_{dry\ gas} = P_{total} - P_{water}$$

The partial pressure of water vapor, when it is in the presence of liquid water, is called the vapor pressure of the liquid water. All liquids, including water, have their own *vapor pressures* that depend only on the temperature of the liquid that's in contact with the vapor. For any given liquid the value of its vapor pressure increases with increasing temperature of the liquid. The vapor pressure of water at various temperatures is found in Table 10.2 in the text.

The partial pressure, P_A, of a gas A in a mixture is a fraction of the total pressure, P_{total}. For a given V and T, pressure is proportional to n. Therefore,

$$\frac{P_A}{P_{total}} = \frac{n_A}{n_{total}} = X_A$$

where X_A is called the mole fraction of A. The mole fraction is the ratio of the number of moles of a substance (such as A) to the total number of moles of all the substances present.

$$X_A = \frac{n_A}{n_A + n_B + n_C + \ldots} = \frac{n_A}{n_{total}}$$

Multiplying a mole fraction by 100 gives the *mole percent*.

$$mol\ \%\ A = X_A \times 100\%$$

The relationship between the partial pressure of a gas and the total pressure is often written as

$$P_A = X_A P_{total}$$

Example 10.6 Calculating the volume of dry gas after collection over water

A 500 mL sample of nitrogen is collected over water at a temperature of 20 °C. The atmospheric pressure at the time of the experiment is 746 torr. What would be the volume of the nitrogen at this same temperature if it were free of water and at a pressure of 760 torr?

Analysis:

The problem asks us to determine the volume of nitrogen at a different pressure while the temperature is held constant. The other task is to correct the pressure for the vapor pressure of water, The simplest method is to correct for the vapor pressure of water, then determine the volume at the desired pressure of dry nitrogen.

Assembling the Tools:

The tools needed include the combined gas laws and Dalton's law of partial pressures.

Solution:

The initial state of the gas is that it is wet with a total pressure of 746 torr in a volume of 500 mL at a temperature of 20 °C. The vapor pressure of water is 17.54 torr at this temperature (from Table 10.2 in the text). We calculate the pressure of dry nitrogen as

$$P_{total} = 746 \text{ torr} = P_{nitrogen} + 17.54 \text{ torr}$$

$$P_{nitrogen} = 728 \text{ torr}$$

Once we find that the pressure of the dry nitrogen is 728 torr, we look at the combined gas law equation (or just Boyle's law). Since the temperature is constant at 20 °C, we can cancel it and save the step of converting to kelvins. Canceling T_1 and T_2 from the combined gas law equation, we have

$$P_1 V_2 = P_2 V_2$$

$$(728 \text{ torr})(500 \text{ mL}) = (760 \text{ torr})(V_2)$$

$$V_2 = 479 \text{ mL of dry nitrogen}$$

Is the Answer Reasonable?

Since all variables except the pressure are held constant we can focus on the fact that the two pressures are close to the same value and therefore our answer should be close to that too. Also, since the pressure increased slightly from 728 torr to 760 torr we expect the volume to be less than 500 mL, and it is.

Self-Test

37. Suppose that the total pressure on a mixture of helium and argon is 676 torr and the partial pressure of the helium is 422 torr.

    (a) What is the partial pressure of the argon? _____

    (b) What is the mole fraction of argon in the mixture? _____

38. Suppose that all of the argon were removed from the sample in the previous question but the volume of the container was left the same and the temperature was not changed. What would a pressure gauge read for the remaining gas?

_____

39. The average composition of the air we exhale in terms of the partial pressures of the gases is as follows: N_2, 569 torr; O_2, 116 torr; CO_2, 28 torr; and $H_2O(g)$, 47 torr. Calculate the composition of the air in mole percents.

_____

Graham's law of effusion

Diffusion is not the same as *effusion*. Effusion is the movement of a gas into a vacuum through an extremely small hole in the wall of the gas container, the intent being to minimize collisions of the effusing gas molecules with other gas molecules or with themselves. Effusion is an uncommon phenomenon because it has to be set up experimentally. Diffusion, a common phenomenon, is simply the spreading out and intermingling of the molecules of one gas with another.

According to *Graham's law of effusion*, the rate of effusion of a gas is inversely proportional to the square root of the density of a gas, or to the square root of the molecular mass of the gas, provided that the pressure and temperature of the gas are kept constant. Equation 10.7 is the easiest equation to use in calculations.

Self-Test

40. State the law of gas effusion.

_____

_____

41. At room temperature and 760 torr the density of helium is 0.000160 g/mL. Under these conditions the density of hydrogen is 0.0000818 g/mL. How much more rapidly will hydrogen effuse than helium through the same pinhole?

    (a) 1.95 times as rapidly   (c)   1.22 times as rapidly

    (b) 1.11 times as rapidly   (d)   1.40 times as rapidly   _____

42. Arrange the following gases in order of increasing (smallest to largest) rates of effusion.

    N_2, CH_4, C_4H_{10}, H_2, CO_2

_____

43. Look back at your answer to Self-Test Question 40. What relationship is there between the density of a gas and its rate of effusion?

_____

10.7   The Kinetic–Molecular Theory of Gases

Learning Objectives

State the main postulates of the kinetic theory of gases, and show how it explains the gas laws on a molecular level.

Review

The gas laws describe *observations*; they *explain* nothing. They would be true even if atoms and molecules were yet to be discovered. The kinetic theory of gases, on the other hand, answers the question, "What *must* be true about a gas at the submicroscopic level to explain why the gas laws exist?" This section takes each of the gas laws in turn and in a qualitative way, uses the kinetic theory and its model of an ideal gas to explain them.

The *kinetic theory of gases* proposes a model for the structure of a gas in the form of a set of postulates. They are given in Section 10.7 in the text. They are simple and should be learned. In summary, the theory describes a gas as a collection of extremely tiny particles in random motion. Collision of the particles explains *how* gases have pressures and how the average kinetic energy of the particles is proportional to the temperature. The higher the average kinetic energy of gaseous particles, the higher the pressure and the higher the temperature.

The theory accounts for the gas laws without much difficulty:

Pressure–temperature law ($P \propto T$, at constant n and V). If raising the temperature makes the average kinetic energy of the gas particles increase, they hit the walls harder, giving a higher pressure. Also called Gay-Lussac's law.

Temperature–volume law ($V \propto T$, at constant n and P). If gas particles are given higher average kinetic energies and velocities by giving the gas a higher temperature, the pressure tends to increase. To keep the pressure constant, the collisions with the walls have to be spread out over a larger area, which is accomplished by letting the particles take up a larger space or volume. Also called Charles' law.

Pressure–volume law ($P \propto V$, at constant n and T). If the volume is reduced, the container's walls are hit more frequently by gas particles, thereby increasing the pressure. Also called Boyle's law.

Effusion law (effusion rate $\propto 1/\sqrt{\text{gas density}}$, at constant pressure and temperature). A gas having a low density has less massive particles than a gas with a high density. At the same temperature, the lighter particles move with a greater average root mean square speed than the heavier particles. This means that particles of the less dense gas get from one place to another more quickly than those of the more dense gas, so the gas with the lower density must effuse more rapidly.

Law of partial pressures ($P_{\text{total}} = P_A + P_B + P_C + \dots$). According to the model of an ideal gas, its particles do not repel or attract neighboring particles. This causes the particles to act independently, so those of each gas contribute separately toward the total pressure. Therefore, their partial pressures add up to the total pressure, not some value less than or greater than the total.

Avogadro's principle ($V \propto n$, at constant P and T; or $P \propto n$, at constant V and T). At the same T, regardless of the gas, gas molecules have the same average kinetic energy and they exert the same average force per hit when they strike a unit area of the walls. The frequency of these hits per unit area is proportional to the molar concentration of the gas (moles per unit of volume), so the total force per unit area—the pressure—is proportional to the moles of the gas and not the molecular mass of the gas. Hence $P \propto n$, which is one way of stating the law.

Gas compressibility. Gases, but not liquids and solids, are relatively easily compressed because the volume occupied by a gas is mostly empty space.

Absolute zero. When all of the particles in a gas stop moving, the gas particles must have zero average kinetic energy, regardless of their mass. Since gas temperature is proportional to this average kinetic energy, a state of zero kinetic energy must be the coldest condition possible—absolute zero (0 K or –273.15 °C).

Self-Test

44. What are the postulates of the kinetic theory of gases?

_____

_____

_____

45. The Kelvin temperature of a gas is proportional to what property of the gas particles?

_____

46. What is the fundamental difference between the gas laws—taken as a group— and the postulates of the kinetic theory?

_____

_____

47. In your own words explain each of the gas laws in terms of the model of the ideal gas.

_____

_____

48. Which of the following statements are false?

(a) Compressing the gas molecules into a smaller volume causes them to become smaller._____

(b) Raising the temperature makes the gas molecules strike the walls of a container with greater average force._____

(c) The theory predicts that at absolute zero, molecules should be motionless._____

(d) Increasing the volume of a gas, while keeping the temperature constant, causes the molecules to slow down, so they don't hit the walls as hard and the pressure is less._____

(e) The pressure of a gas is related to the number of collisions per second with a given area of the walls as well as to the force with which the molecules collide. _____

(f) Raising the temperature causes a gas to expand because raising the temperature makes the gas molecules larger. _____

10.8   Real Gases

Learning Objectives

Explain the physical significance of the corrections involved in the van der Waals equation for the differences between real and ideal gases.

Review

Real gases do not obey the ideal gas law perfectly. There are two reasons for this that can be traced to inaccuracies in the original kinetic theory explanation for ideal gas behavior. First, the molecules of a real gas have finite volumes (the original kinetic theory described gas particles as points without volume). Second, real molecules have weak attractions toward each other (a fact ignored by the theory).

The finite volumes of the gas molecules cause the ideal gas law to be disobeyed when gases are squeezed into very small volumes at high pressure. The attractions between the molecules cause paths of the molecules to swerve as the molecules pass near each other. As a result, real molecules travel longer distances between collisions, so they don't collide with the walls as often and produce a lower pressure. This phenomenon becomes most important when the molecules are slowed down by cooling them to low temperatures. Therefore, high P and low T are conditions under which there are the greatest deviations from ideal gas behavior.

To adapt the ideal gas law to fit real gases, van der Waals applied corrections to the real P and V. He subtracted a term, nb (where n = moles and b is a factor that depends on the size of the gas molecules), from the measured volume of the gas. Because the actual pressure is smaller than the pressure an ideal gas would exert, a correction was added to the real pressure. He added the term n^2a/V^2 (where n = moles and a is a correction factor that is proportional to the strengths of the attractive forces between the molecules) to the measured pressure of the gas. The final result is called the van der Waals equation of state.

$$\left(P + \frac{n^2a}{V^2}\right)(V - nb) = nRT$$

Be sure you know the significance of the van der Waals factors a and b.

Self-Test

49.   What are the two principal reasons why real gases deviate from ideal gas behavior?

_____

50.   Why is the measured volume of a real gas larger than the volume that would be occupied by an equivalent number of moles of an ideal gas?

_____

_____

51. Why is the measured pressure of a real gas smaller than the pressure that would be exerted by an equivalent number of moles of an ideal gas?

_____

_____

52. Krypton has a larger van der Waals *a* constant than neon. Which of the two gases has the greater attractive forces between their molecules?

_____

53. Which gas would be expected to have the larger value of the van der Waals constant *b*, HF or HCl? Why?

_____

Answers to Self-Test Questions

1. Pressure, volume, temperature, and mass (or moles)
2. Increase. Molecules move faster and can mix faster at the higher temperature.
3. 0.963 atm
4. 742 torr, 0.976 atm
5. 641 torr
6. 99.5 kPa
7. (a) 0.954 atm, (b) 725 torr
8. Atmospheric pressure and the difference in heights of the liquid level in the manometer.
9. The difference in heights of the liquid level in the manometer.
10. 82 torr, 866 torr
11. 3.90×10^2 mL
12. 445 torr
13. An ideal gas
14. b
15. 776 mL
16. 955 mL
    Pressure changes to 801 torr.
    5.60×10^2 mL
    2
    21. °C
    22.
    23. H
        pre STP
    24. Stan of gases have equal numbers of moles when compared under identical conditions of
        rature.
        and pressure, 273 K and 760 torr

25. 0.821 mol
26. 2.0 L of NO and 1.0 L of O_2
27. 9 L of H_2 and 6 L of NH_3
28. 56.5 mol of SO_2; 3.62×10^3 g of SO_2
29. $PV = nRT$
30. $R = 0.0821$ L atm/mol K
31. 1.53 L
32. $P = 1.15$ atm
33. 402 K or 129 °C
34. 28.0 g/mol
35. 0.713 g L^{-1}
36. H_2, CH_4, N_2, CO_2, C_4H_{10}
37. (a) 254 torr (b) 0.376
38. 422 torr (This question gets at the fundamental meaning of partial pressure.)
39. 74.9 mol % N_2, 15.3 mol % O_2, 3.7 mol % CO_2, 6.2 mol % H_2O
40. The rate of effusion of a gas is inversely proportional to the square root of the gas density.
41. d
42. C_4H_{10}, CO_2, N_2, CH_4, H_2
43. At a given *T* and *P*, rate of effusion is inversely proportional to gas density.
44. Compare your answer with the three postulates of the kinetic theory given in Section 10.7 of the text.
45. To the average kinetic energy of the gas particles.
46. The gas laws describe observations. The kinetic theory theorizes about the structure of a gas that is consistent with the observations.
47. Compare what you write with the explanations given in Section 10.7.
48. (a), (d), (f)
49. Individual gas particles have real volumes; individual particles in a gas sample do have small attractions for each other.
50. Because the molecules of a real gas have finite volumes and therefore take up some space.
51. Molecular attractions cause real gas molecules to travel farther between collisions with the walls, thereby reducing the number of collisions per second.
52. Kr
53. HCl, because HCl is a larger molecule than HF.

Tools for problem solving

In this chapter you learned to apply the following concepts as tools in solving problems. Study each one carefully so that you know what each is used for. When faced with solving a problem, recall what each tool does and consider whether it will be helpful in finding a solution.

You might want to tear these pages out to use along with solving problems in this chapter.

Combined gas law (Section 10.3)

$$\frac{P_1 V_1}{T_1} = \frac{P_2 V_2}{T_2}$$

Ideal gas law (Section 10.5)

$$PV = nRT$$

Determination of molar mass (Section 10.5)

$$\text{molar mass} = \frac{g}{V}\frac{RT}{P} = d\frac{RT}{P}$$

Dalton's law of partial pressures (Section 10.6),

$$P_{\text{total}} = P_A + P_B + P_C + \cdots$$

When a gas is collected over water,

$$P_{\text{total}} = P_{\text{water}} + P_{\text{gas}}$$

Mole fraction and mole fraction related to partial pressure (Section 10.6)

$$X_A = \frac{n_A}{n_{total}} = \frac{P_A}{P_{total}}$$

Graham's law of effusion (Section 10.6)

$$\frac{\text{effusion rate }(A)}{\text{effusion rate }(B)} = \sqrt{\frac{d_B}{d_A}} = \sqrt{\frac{M_B}{M_A}}$$

---

van der Waals equation of state for real gases (Section 10.8)

$$\left(P + \frac{n^2 a}{V^2}\right)(V - nb) = nRT$$

Summary of Important Equations

Combined gas law equation:

$$\frac{P_1 V_1}{T_1} = \frac{P_2 V_2}{T_2}$$

Ideal gas law equation:

$$PV = nRT$$

Dalton's law of partial pressures:

$$P_{total} = P_A + P_B + P_C + \ldots$$

Mole fraction, X_A, of component A in a mixture:

$$X_A = \frac{n_A}{\text{total number moles of all components}}$$

Mole fraction of a gas, X_A, and its partial pressure, P_A:

$$P_A = X_A P_{total}$$

Graham's Law equation:

$$\frac{\text{effusion rate } (A)}{\text{effusion rate } (B)} = \sqrt{\frac{d_B}{d_A}} = \sqrt{\frac{M_B}{M_A}}$$

Notes:

Chapter 11

Intermolecular Attractions and the Properties of Liquids and Solids

In Chapter 10 you learned about the physical properties of gases, and you saw that gas behavior is essentially independent of the chemical composition of the gas molecules. In this chapter we turn our attention to the other two states of matter, where attractions between molecules play a dominant role, and where physical properties are strongly affected by the chemical makeup and structure of the molecules.

11.1   Intermolecular Forces

Learning Objective

Describe the major intermolecular forces including dipole attractions, hydrogen bonding, and London forces.

Review

Gases compared to liquids and solids

Intermolecular attractions are attractions that exist between molecules. They are most significant when molecules are close together and are practically insignificant when molecules are far apart. In gases, the molecules hardly feel the intermolecular attractions at all because the particles are so widely spaced. Differences in the strengths of these attractions caused by differences in chemical makeup are so small that all gases behave in nearly the same way. This is why we are able to have gas laws and the concept of an ideal gas. But in liquids and solids, where the molecules are practically touching, the intermolecular attractions are very strong and differences in their strengths are significant and are reflected in differences in physical properties.

Self–Test

1.   Under what conditions do the properties of real gases deviate *most* from the predicted properties of an ideal gas?

Intermolecular forces

The attractions between neighboring molecules (*intermolecular attractions*) are always much weaker than the chemical bonds (*intramolecular attractions*) within molecules. Although the strengths of bonds determine the chemical properties of substances, it is the strengths of intermolecular attractions that determine many of the physical properties. The principal kinds of intermolecular attractions are *dipole–dipole attractions, hydrogen bonds, London forces, ion–dipole attractions, and ion–induced attractions.*

Dipole–dipole attractions occur between polar molecules and are about 1% as strong as normal covalent bonds.

Hydrogen bonding is an especially strong type of dipole–dipole attraction (about 5 times the strength of the usual dipole–dipole attraction). It occurs between molecules in which hydrogen is covalently attached to nitrogen, oxygen, or fluorine. Therefore, molecules that contain O—H and N—H bonds experience hydrogen bonding. (This is something to look for in the structure of a molecule when you wish to know if hydrogen bonding will be a factor.) HF is the only fluorine–containing molecule that hydrogen bonds.

London forces (also called *dispersion forces*) result from attractions between *instantaneous dipoles* and *induced dipoles* in neighboring molecules. London forces flicker on and off and their average strength is usually less than most dipole–dipole attractions. Remember that London forces are present in *all* substances.

Large molecules and atoms have larger, more easily distorted electron clouds than small molecules and atoms. As a result, the strengths of London forces increase as the sizes of the atoms in a molecule become larger. When comparing molecules that contain atoms of about the same size, the longer the chain of atoms, the greater is the *total* strength of the London forces of attraction. Molecular shape also affects the strengths of London forces; even with the same number of atoms, small compact shapes yield weaker London forces than long chain–like shapes.

When there are many atoms in a molecule, the total attractions, produced by London forces often outweigh other forces, such as dipole–dipole attractions.

Ion–dipole attractions are simply the attractions between ions and polar molecules, like the attraction of a sodium ion for the partially negative end of a permanently polarized water molecule.

Ion–induced dipole attractions result when an ion induces a dipole in a nearby molecule and thus sets up an attraction between the two.

Self–Test

2. How do the strengths of London forces, dipole–dipole attractions and hydrogen bonds compare?

    _____

3. What kinds of attractive forces (both intermolecular forces and chemical bonds) are present between the particles in each of the following substances?

    (a) argon   _____

    (b) potassium sulfate, K_2SO_4 _____

    (c) methyl alcohol, CH_3OH _____

    (d) ammonia, NH_3 _____

    (e) pentane, $CH_3CH_2CH_2CH_2CH_3$ _____

4. For each pair of substances below, choose the one that has stronger intermolecular attractions.

   (a) PF_3 or PCl_3 _____   (d)   HF or HCl   _____

   (b) CH_4 or SiH_4 _____   (e) Cl_2 or Br_2 _____

   (c) CH_4 or CH_3Cl _____   (f) C_2H_6 or C_6H_{14} _____

5. Arrange the following in order of increasing (smallest to largest) strengths of intermolecular attractions.

   (A)

$$\underset{\displaystyle CH_3-\underset{\displaystyle |}{\overset{\displaystyle OH}{\overset{\displaystyle |}{CH}}}-CH_3}{}$$

   (B)

$$CH_3-CH_2-CH_3$$

   (C)

$$CH_3-\overset{\displaystyle O}{\overset{\displaystyle \|}{C}}-CH_3$$

_____

6. In the preceding question:

   (a)   Which experience hydrogen bonding?   _____

   (b)   Which experience moderate dipole–dipole forces?   _____

   (c)   Which experience London forces?   _____

7. Explain in your own words what types of interactions are grouped under the term *London forces*.

_____

_____

_____

11.2   Intermolecular Forces and Physical Properties

Learning Objective

Explain how the molecular attractive forces affect macroscopic properties.

Review

The physical properties of liquids and solids depend on both the tightness of packing of their particles and on the strengths of the intermolecular attractions. However, some properties depend more on one of these factors than the other.

Properties that depend mostly on tightness of packing

The closeness of packing is the reason that liquids and solids resist compression when pressure is applied to them, and it is also primarily responsible for the slow rates of diffusion in these states of matter.

Liquids and solids are nearly *incompressible* because there is almost no empty space into which to squeeze the molecules when pressure is applied. *Diffusion* is slow in liquids because to move from one place to another the particles must work their way around and past so many other near neighbors. In human terms, it's like attempting to cross a crowded room; you must squeeze past so many other people that movement is slow. Diffusion in solids is nearly nonexistent; the particles are not free to move about the way they can in liquids.

Properties that depend mostly on intermolecular attractions

Properties mentioned in this section that are controlled mostly by the strengths of intermolecular attractions are retention of shape and volume, surface tension, wetting of a surface by a liquid, viscosity, and evaporation and sublimation.

Shape and volume. Attractive forces hold the particles close together in liquids and solids, so their volumes don't change when transferred from one container to another. In solids, the attractive forces are even greater and the particles cannot easily move away from their equilibrium positions. This rigidity allows solids to keep their shapes when transferred from one container to another.

Surface tension. To understand *surface tension*, as well as many of the other physical properties of liquids and solids, it is necessary to understand the relationship between attractive forces and potential energy, so let's review these important concepts. First, *whenever something feels an attractive force, it has potential energy*. A large reservoir of water high in the mountains has potential energy because the water feels the earth's gravitational attraction. Allowing the water to flow to a lower altitude releases some of this potential energy, which we can use to generate electricity or for some other energy–consuming activity. If there were no attractive forces between the earth and the water, it would not flow downhill and it would not release energy that we could use. Second, *the greater the sum total of the attractive forces felt by something, the lower will be its potential energy*. Consider the water, for example. You probably know that the earth's gravitational attraction decreases with increasing distance from the earth's center. When the water flows to a lower altitude, its comes closer to the center of the earth, so it feels a *greater* gravitational attraction at the same time that its potential energy decreases.

Now let's consider the phenomenon known as *surface tension*. This property of liquids arises because molecules at the surface feel fewer attractions than the molecules within the liquid. This is because the molecules at the surface are only partially surrounded by other molecules, whereas those inside the liquid are completely surrounded. This causes the molecules within the liquid to experience greater *total* attractive forces than molecules at the surface, and as a result, the molecules in the interior of the liquid have lower potential energies than those at the surface. The *surface tension* is a quantity that's related to the energy difference between a molecule at the surface and a molecule inside the liquid. We can also say that this is the amount of energy needed to bring a molecule from the interior to the surface. When the intermolecular attractions are large, the surface tension is large. The tendency of a liquid to achieve a minimum potential energy causes the liquid to minimize its surface area and thereby minimize the number of higher–energy surface molecules.

Wetting of a surface. For a liquid to engage in the *wetting* of a surface, the attractive forces between the liquid and the surface must be about as strong as between molecules of the liquid. Substances with low surface tensions easily wet solid surfaces because the attractions in the liquid are weak. Water has a large surface tension and can wet a glass surface because the surface contains oxygen atoms to which water molecules can hydrogen bond. Water doesn't wet waxy or greasy surfaces because the water molecules are only weakly attracted to hydrocarbon molecules by London forces. The London forces are much weaker than the hydrogen bonding between water molecules in the liquid, so molecules of water tend to cling together and don't spread out over the greasy surface.

Viscosity. *Viscosity* is the internal "friction" within a liquid and makes it difficult for a liquid to flow. Viscosity increases with increasing intermolecular attractions and increasing molecular size. Viscosity decreases with temperature.

Evaporation and sublimation. *Evaporation* of a liquid or *sublimation* of a solid occurs when molecules leave their surfaces and enter the vapor space that surrounds them. Remember that it is always the molecules with very large kinetic energies that escape, so the average kinetic energy of molecules left behind is lowered, and this means that the temperature is also lowered. (In other words, evaporation leads to a cooling effect, and so is endothermic.)

Factors affecting the rate of evaporation

Two factors control the rate of evaporation. One is the temperature and the other is the strengths of the intermolecular attractions. Increasing the temperature increases the rate of evaporation. Study Figure 11.18 in the text to be sure you understand why this is so. Substances with weak intermolecular attractive forces evaporate more rapidly than those that have strong intermolecular attractive forces. Study Figure 11.19.

Example 11.1   Comparing evaporation rates

For the following pairs of substances, choose the one that will have the greater rate of evaporation at a given temperature.

    (a) $CH_3CH_2CH_3$ or $CH_3CH_2CH_2CH_2CH_3$

    (b)

Analysis:
We just learned that intermolecular forces are inversely related to evaporation rates. If the relative attractive forces can be determined, we can then deduce which substance will have the greater evaporation rate.

Assembling the Tools:
We just learned about the intermolecular forces that attract molecules together and will use those as our tools.

Solution:
(a) We see that these two compounds are not polar and they have no capability to hydrogen bond. They do have London attractive forces. The longer carbon chain (pentane) has more sites at which one molecule can have instantaneous dipoles attracting a neighboring molecule. The total London attractive forces will be stronger for pentane as compared to the shorter propane. Since propane has the weaker attractive forces, it will evaporate at a greater rate than pentane.

(b) These molecules are generally the same size and should have roughly equivalent London forces of attraction. The compound on the left (butanol or butyl alcohol) has an —OH group that allows hydrogen bonding to occur. The compound on the right (diethyl ether) is polar around the sp^3 oxygen atom that gives diethyl ether a bent structure. The hydrogen bonding is expected to be stronger than dipole–dipole attractions. From that we conclude that diethyl ether has a greater evaporation rate because it has smaller attractive forces.

Is the Answer Reasonable?
We review our reasoning and see no errors. For part (a) we may have experience with propane as a gas for heating or cooking, which would agree with our conclusion that it evaporates more readily than pentane.

Self–Test

8. Which of the properties discussed in this section are primarily the result of the tightness of packing of molecules in liquids and solids?

   _____

9. In ethylene glycol (antifreeze), the intermolecular attractive forces are much greater than between molecules in gasoline. Molecules that evaporate from ethylene glycol, therefore, must have much more kinetic energy than most of the molecules that evaporate from gasoline. Yet, if you spill these two liquids on yourself, the gasoline gives a much greater cooling effect than the ethylene glycol. Why? (Think!)

   _____

   _____

10. Arrange the molecules in Question 5 in order of increasing rate of evaporation (slowest to fastest).

    _____

11. Which of the molecules below is expected to have the greater viscosity? Why?

$$HO—CH_2—CH_2—OH \qquad\qquad CH_3—CH_2—OH$$
   ethylene glycol   ethanol

   _____

   _____

11.3 Changes of State and Dynamic Equilibria

Learning Objective

Use the concept of dynamic equilibrium to understand changes of state.

Review

Changes of state include:

solid $\longrightarrow$ liquid	liquid $\longrightarrow$ vapor	vapor $\longrightarrow$ solid
solid $\longrightarrow$ vapor	liquid $\longrightarrow$ solid	vapor $\longrightarrow$ liquid

Changes of state can take place under conditions of dynamic equilibrium. Equilibrium is one of the most important concepts for you to grasp in this chapter. In a *dynamic equilibrium,* two opposing events are taking place at equal rates, and the dynamic events associated with a *phase change,* or change of state, such as those involving evaporation and *condensation,* provide simple illustrations. For example, when a liquid evaporates into an enclosed space, the concentration of molecules in the vapor rises until the rate at which molecules condense becomes equal to the rate at which they evaporate. Once this happens, the *number* of molecules in the vapor remains constant with time; in the time it takes for a hundred molecules to evaporate into the vapor, another hundred molecules condense and leave the vapor. Thus a dynamic equilibrium is a condition in which an unending sequence of balanced events leads to a constant composition within the equilibrium system. Melting and freezing also involve an equilibrium, and the *melting point* of a solid is simply the temperature at which dynamic equilibrium exists between the solid and liquid phases. Sublimation also involves an equilibrium, one between the solid and vapor phases of a substance.

Self–Test

12. In some high–altitude cities, such as Denver, Colorado, the temperature may stay well below the freezing point of water but snow disappears gradually without ever melting. What happens to it?

    _____

13. What is another term that has the same meaning as "phase change"?

    _____

14. What changes are constantly taking place during the equilibrium involved in sublimation?

    _____

15. Dynamic equilibrium was discussed earlier in the text when we dealt with a chemical process in aqueous solution. What was that process?

    _____

11.4   Vapor Pressures of Liquids and Solids

Learning Objective

Explain what is meant by vapor pressure and how it reveals the relative strengths of intermolecular attractions.

Review

When a liquid evaporates into an enclosed space, the vapor exerts a pressure called the *vapor pressure*. When the rates of evaporation and condensation become equal, the pressure exerted by the vapor is called the *equilibrium vapor pressure of the liquid*. Usually, when we use the term *vapor pressure*, we mean equilibrium vapor pressure. Its value is independent of the size of the container, the surface area of the liquid, or the amount of liquid present, as long as at least some liquid remains at equilibrium.

A similar situation exists for solids. The vapor that's in equilibrium with a solid also exerts a pressure called the *equilibrium vapor pressure of the solid*, which is also independent of the size of the container, the surface area of the solid, or the amount of solid that remains at equilibrium.

Vapor pressures differ for different substances, and when measured at the same temperature, can be used to compare the strengths of intermolecular attractions. When the intermolecular attractions are strong, vapor pressures are low.

Vapor pressure depends on temperature, as we saw in Chapter 10 when we discussed collecting gases over water. For any given liquid, the vapor pressure increases as the temperature rises. This is because at the higher temperature, a larger fraction of the molecules have sufficient energy to escape from the liquid's surface. The higher rate of evaporation at the higher temperature requires a greater concentration of molecules in the vapor to establish equilibrium, so the vapor pressure is higher. For the same reason, the equilibrium vapor pressure of a solid also rises with increasing temperature. Graphs depicting the change in vapor pressure with increasing temperature all have similar shapes as shown in Figure 11.23 in the text.

Self–Test

16. The following are the vapor pressures of some common solvents at a temperature of 20 °C. Arrange these substances in order of increasing strengths of their intermolecular attractive forces.

carbon disulfide, CS_2	294 torr
ethyl alcohol, C_2H_5OH	44 torr
methyl alcohol, CH_3OH	96 torr
acetone, $(CH_3)_2CO$	186 torr

17. On a separate sheet of paper, sketch a graph that shows how vapor pressure varies with temperature for a typical liquid.

11.5 Boiling Points of Liquids

Learning Objective

Understand the concepts of boiling points and normal boiling points.

Review

Boiling occurs when bubbles of vapor form below the surface of a liquid. The *boiling point* is the *temperature* at which a liquid boils, and it is the temperature at which the liquid's vapor pressure becomes equal to the atmospheric pressure exerted on the liquid's surface. The temperature at which the vapor pressure equals 1 atm is the liquid's *normal boiling point*. Substances with strong intermolecular attractions generally have high boiling points.

Example 11.2   Predicting relative boiling points

Arrange the following in order of increasing normal boiling point.

$$(CH_3)_2CHNH_2, \ (CH_3)_2CH_2, \ (CH_3)_2CO$$

Analysis:

The normal boiling point is defined as the temperature at which the vapor pressure of a pure liquid is equal to 760 torr. In previous problems we asked which compound had the highest (or lowest) vapor pressure at a given temperature. Now we need to predict the temperature (at least in a relative sense) at which each of these compounds will have a vapor pressure of 760 torr. We can get a clue by examining Figure 11.23. Here we can see that a substance with a lower vapor pressure at a given temperature will have a higher boiling point than a substance with a higher vapor pressure. We conclude that substances with higher vapor pressures have lower boiling points. Now the problem is easy since we have had experience determining relative vapor pressures.

Assembling the Tools:

We will use our tool that describes the effect of intermolecular attractive forces on boiling points. We also need several other tools such as drawing Lewis structures and determining polarity of molecules to assess the strengths of intermolecular attractions.

Solution:

The first compound, $(CH_3)_2CHNH_2$, contains the —NH_2 group that allows hydrogen bonding in addition to the ever present London forces. The $(CH_3)_2CH_2$ molecules are attracted to each other only by London forces. Finally, $(CH_3)_2CO$ is polar with a double–bonded oxygen atom. We then deduce that:

$(CH_3)_2CH_2$, is the least polar and has the highest vapor pressure and the lowest boiling point.

$(CH_3)_2CHNH_2$ forms hydrogen bonds and has the lowest vapor pressure and the highest boiling point.

$(CH_3)_2CO$ is polar and it should have a vapor pressure and boiling point between the other two compounds.

Are the Answers Reasonable?

We check our reasoning about the intermolecular attractive forces of the molecules, their relative vapor pressures, and finally the boiling points. We also see a direct relationship between the intermolecular attractive forces and normal boiling points.

Self–Test

18. Why does a pressure cooker cook foods more rapidly than when they are boiled in an open pot?

_____

_____

_____

19. Arrange the substances in Question 16 in increasing order according to their expected boiling points.

_____

20. What happens to the temperature if heat is added more rapidly to an already boiling liquid? Will an egg cook more quickly in a rapidly boiling pot or a pot that gently simmers? Explain. _____

_____

_____

21. Why is boiling point a useful physical property for the purposes of identifying liquids?

_____

22. Which would be expected to have a higher boiling point, argon or krypton? Explain your reasoning.

   _____

23. Which would be expected to have a higher boiling point, hydrogen sulfide or hydrogen chloride?

   _____

24. Which would be expected to have the higher vapor pressure at 20 °C, C_3H_8 or C_5H_{12}? Would your answer be the same at 40 °C? Explain.

   _____

25. Suppose a vacuum pump is attached to two separate containers, one holding ethanol and the other ethylene glycol. When the pump is turned on, the pressure above each liquid will drop. As the pressure is reduced, which liquid will boil first? Why?

$$HO-CH_2-CH_2-OH \qquad\qquad CH_3-CH_2-OH$$
$$\text{ethylene glycol} \qquad\qquad\qquad \text{ethanol}$$

   _____

   _____

11.6   Energy and Changes of State

Learning Objective

   Calculate energy changes involved in changes of state

Review

Figure 11.27 in the text describes the shapes of heating and cooling curves, which show how the temperature of a substance changes as heat is added or removed. The flat portions of the graphs correspond to phase changes that represent potential energy changes associated with solid ⇌ liquid and liquid ⇌ vapor phase changes. Learn the general shapes of such graphs and what kinds of energy changes take place along the various line segments.

   Every change of state is accompanied by an energy change. They are *enthalpy* changes because phase changes take place under conditions of constant pressure. Study the definitions of *molar heat of fusion, molar heat of vaporization,* and *molar heat of sublimation* given in this section. Melting of a solid, evaporation of a liquid, and sublimation are changes that absorb energy (they are endothermic), so their values of ΔH are positive. Phase changes in the opposite direction—freezing of a liquid and condensation of a vapor to either a liquid or to a solid—release energy and are exothermic. Their values of ΔH are equal in magnitude to but opposite in sign to the heats of fusion, vaporization, and sublimation.

   The size of the enthalpy change associated with a change of state depends on the strengths of the intermolecular attractions in the substance and also by how much these forces are altered during the change of state. The $\Delta H_{vaporization}$ is larger than ΔH_{fusion}, and the $\Delta H_{sublimation}$ is larger than $\Delta H_{vaporization}$.

Measuring heats of vaporization

In Section 11.9 in the text we learned how heats of vaporization can be measured by observing how the vapor pressure changes with temperature. The *Clausius–Clapeyron equation* relates the heat of vaporization to the variation of vapor pressure with temperature.

Self–Test

26. If you know the values of the molar heat of fusion and the molar heat of vaporization, you can calculate the value of the molar heat of sublimation. Why can this be done, and how can it be done?

_____

27. The molar heat of vaporization of benzene (C_6H_6) is 30.8 kJ/mol and water has a molar heat of vaporization of 40.7 kJ/mol. Which of these substances has the larger intermolecular attractive forces?

_____

28. Referring to the data in Question 27, which substance, water or benzene, should have the larger molar heat of sublimation?

_____

29. How many kilojoules are required to vaporize 10.0 g of liquid benzene? (See Question 27.)

_____

30. Arrange the following in increasing order of their expected molar heats of vaporization: HCl, NH_3, CH_4, HBr, He.

_____

31. According to Figure 11.26 in the text, what would the boiling point of water be if it were not for hydrogen bonding? Could life as we know it exist without hydrogen bonding?

_____

32. Which of the following compounds should have the higher heat of vaporization?

    (a) $CH_3CH_2CH_3$ or $CH_3CH_2CH_2CH_2CH_3$ _____

    (b)

    $$H-\overset{\overset{\displaystyle H}{|}}{\underset{\underset{\displaystyle H}{|}}{C}}-\overset{\overset{\displaystyle H}{|}}{\underset{\underset{\displaystyle H}{|}}{C}}-\overset{\overset{\displaystyle H}{|}}{\underset{\underset{\displaystyle H}{|}}{C}}-\overset{\overset{\displaystyle H}{|}}{\underset{\underset{\displaystyle H}{|}}{C}}-O-H \quad \text{or} \quad H-\overset{\overset{\displaystyle H}{|}}{\underset{\underset{\displaystyle H}{|}}{C}}-\overset{\overset{\displaystyle H}{|}}{\underset{\underset{\displaystyle H}{|}}{C}}-O-\overset{\overset{\displaystyle H}{|}}{\underset{\underset{\displaystyle H}{|}}{C}}-\overset{\overset{\displaystyle H}{|}}{\underset{\underset{\displaystyle H}{|}}{C}}-H$$

    _____

11.7   Phase diagrams

Learning Objective

Understand and effectively use phase diagrams.

Review

A typical *phase diagram* has three lines that serve as temperature–pressure boundaries for the three physical states of the substance—solid, liquid, and gas. Any point on one of these lines represents a temperature and pressure in which there is an equilibrium between *two* phases. The three equilibrium lines intersect at a common point called the *triple point*. Each substance has a unique triple point corresponding to the temperature and pressure at which solid, liquid, and gas are simultaneously in equilibrium with each other.

The liquid–vapor line terminates at the *critical point*, corresponding to the *critical temperature* and *critical pressure*. Below the critical temperature, a substance can be liquefied by the application of pressure. Above the critical temperature, no amount of pressure can create a distinct liquid phase. Above the critical temperature the substance is said to be a *supercritical fluid*.

Given a temperature and a pressure, we can find the point on the phase diagram that corresponds to those coordinates. If the point lies in the gas, liquid or solid region, the substance will consist of one phase. If the coordinates lie on one of the lines, the substance will be in dynamic equilibrium with two of the phases. If the coordinates correspond to the intersection of the three equilibrium lines, then those are the coordinates of the triple point.

If two pairs of temperature and pressure are specified, P_1, T_1 and P_2, T_2, these two points can be connected with a straight line. From the start, P_1, T_1, to the end, P_2, T_2, we can clearly see what, if any, phase changes the system experiences.

Self–Test

33. According to the phase diagram for water (Figure 11.28), determine what phase will exist at

    (a)   –20 °C and 330 torr   _____

    (b)   78 °C and 4.58 torr   _____

    (c)   85 °C and 780 torr   _____

    (d)   30 °C and 500 torr   _____

    (e)   95 °C and 285 torr   _____

34. What changes occur if water at –20 °C and 200 torr is gradually warmed at a constant pressure?

    _____

35. What changes will be observed if water is pressurized from 1.00 torr to 1000 torr at 0 °C?

    _____

36. (a) At what temperature does liquid and gaseous water exist in equilibrium at 1 atm? (b) What do we call this temperature?

    (a) _____ °C   (b) _____

11.8   Le Châtelier's Principle and Changes of State

Learning Objective

Use Le Châtelier's principle to understand changes of state.

Review

Le Châtelier's principle must be learned. We will use it frequently when studying both physical and chemical equilibria. The principle, for example, lets us generalize about physical changes as follows.

1. An increase in temperature shifts an equilibrium in the direction of an endothermic change.

2. An increase in pressure shifts an equilibrium in a direction that favors a decrease in volume.

Self–Test

37. State Le Châtelier's principle. _____

    _____

38. The equilibrium between a liquid and a vapor can be written as

    $$liquid \rightleftharpoons vapor$$

    (a) The change is endothermic in one direction and exothermic in the other. Rewrite the equation, placing heat as a reactant or product, whichever is appropriate.

    _____

    (b) If the temperature of a container in which this equilibrium is established is decreased while the volume is kept the same, what happens to the position of equilibrium and the relative amounts of liquid and vapor in the container?

    _____

    _____

11.9   Determining Heats of Vaporization

Learning Objective

Understand how heats of vaporization are determined experimentally.

Review

If we look back at the graph in Figure 11.23 in the text, we see that all of the vapor pressures increase with increasing temperature. We also note that the sequence of these curves is ordered and the compound with the lowest heat of vaporization is at the lower temperatures and the one with the highest heat of vaporization is at the highest temperature. Importantly, the shapes of these curves also are very similar, and to a mathematician they suggest an exponential curve. Indeed, there is one equation that links them all, the Clausius–Clapeyron equation (Equation 11.3 in the text). This equation suggests that if we plot the natural logarithm of the vapor pressure versus the reciprocal of the Kelvin temperature (ln P versus $1/T$ shown in Figure 11.32), we should get a straight line with a slope equal to $-\Delta H_{vap}/R$. Another method that can be used with just two data points is shown in Equation 11.4 in the text. The example below illustrates how this is used.

Example 11.3   Using the Clausius – Clapeyron equation

Butane, C_4H_{10}, is the liquid fuel in disposable cigarette lighters. The normal boiling point of butane is –0.5 °C and its heat of vaporization is 24.27 kJ/mol. What is the approximate pressure of butane gas over liquid butane in a cigarette lighter at room temperature (25.0 °C)? Express your answer in units of atm. (See Facets of Chemistry 11.1.)

Analysis:

The Clausius–Clapeyron equation accurately describes the relationship between vapor pressure and temperature. From the statement of the problem we know two temperatures, and will use the two–point version of this equation. The only unknown is the final pressure that the problem asks us to calculate.

Assembling the Tools:

The Clausius–Clapeyron equation is the tool to use in this case.

Solution:

We write the equation:

$$\ln \frac{P_1}{P_2} = \frac{\Delta H_{vap}}{R} \left(\frac{1}{T_2} - \frac{1}{T_1} \right)$$

Now let's summarize the data we have: $\Delta H_{vap} = 24.27$ kJ/mol $= 24.27 \times 10^3$ J/mol

$$T_1 = -0.5 + 273.2 = 272.7 \text{ K} \quad P_1 = 1.00 \text{ atm}$$

$$T_2 = 25.0 + 273.2 = 298.2 \text{ K} \quad P_2 = ?$$

Converting $\ln(P_1/P_2) = \ln P_1 - \ln P_2$, we then insert the numbers into the equation to get

$$\ln(1.00 \text{ atm}) - \ln(?) = \frac{24.27 \times 10^3 \text{ J mol}^{-1}}{0.0821 \text{ L atm mol}^{-1} \text{ K}^{-1}} \left(\frac{1}{298.2 \text{ K}} - \frac{1}{272.7 \text{ K}} \right)$$

We need to convert J to L atm so the units cancel. We do this by equating the two values of *R*.

$$R = 8.314 \text{ J mol}^{-1} \text{ K}^{-1} = 0.0821 \text{ L atm mol}^{-1} \text{ K}^{-1}$$

After canceling the mol^{-1} and the K^{-1} we create a conversion factor from this equality and get

$$(0.0821 \text{ L atm}/8.314 \text{ J})$$

We use this to convert the 24.27 x 10^3 J mol^{-1} to 239.7 L atm mol^{-1}. Using that value for the heat of vaporization and since the natural log, ln, of 1.00 is zero, we get

$$- \ln (?) = - 1.286$$

The antiln of 1.286 is 3.62 atm.

Is the Answer Reasonable?

We see that as the temperature increased, the pressure increased, which fits our concepts of the kinetic molecular theory. We also note that our units cancel properly, so the result seems okay.

Self–Test

39. A certain compound has a vapor pressure of 245 torr at 290 K and a vapor pressure of 354 torr at 315 K. What is the heat of vaporization of this compound?

40. A friend asserts that she can determine the heat of vaporization of any compound using only one measurement of the compound's vapor pressure at a given temperature. Is your friend correct? Defend your conclusion.

11.10 Structures of Crystalline Solids

Learning Objective

Describe how crystal structures are based on repeating molecular geometries.

Review

A crystalline solid is characterized by a highly organized, regular, repeating pattern of particles. To describe such structures, we use the concept of a lattice, which is a pattern of points that have the same repeat distances arranged at the same angles as the particles in the crystal. Study the discussion of two–dimensional lattices to be sure you understand the concept.

When describing crystals, we often use the term crystal lattice for the three–dimensional lattice that describes their structures. Especially important is the concept that a single kind of lattice can be used to characterize many different structures (just as many different wallpaper patterns can be created using the same basic two–dimensional lattice).

In a lattice, the smallest repeating unit is called the *unit cell*. The number of kinds of lattices (or unit cells) is very small, and all substances can be described in terms of this limited set. The most symmetrical lattices (and unit cells) are cubic. There are three kinds of cubic unit cells—*simple cubic, face–centered cubic,* and *body–centered cubic.* Figure 11.38 in the text illustrates how different substances can have the same kind of unit cell but with different cell dimensions.

Counting atoms in a unit cell

You should learn how to calculate the number of atoms in a unit cell. This requires that you realize that we have to assemble the parts actually enclosed within the cell into whole atoms, so we can count their number. Remember the following:

An atom entirely within a unit cell counts as 1 atom.

An atom at a corner contributes 1/8 of an atom to the unit cell.

An atom in a face contributes 1/2 of an atom to the unit cell.

An atom along an edge contributes 1/4 of an atom to the unit cell.

The rock salt structure, which is the structure of NaCl, is quite common among the alkali halides. You should learn how the atoms are arranged in the unit cell. Study Figures 11.40 and 11.43.

Closest–packed structures

There are two kinds of closest–packed structures—cubic closest packed (ccp) and hexagonal closest packed (hcp). Using letters A, B, and C to designate the relative orientations of layers of atoms, the ccp structure has the layer–stacking arrangement A–B–C–A–B–C ..., whereas the hcp structure has the layer–stacking arrangement A–B–A–B–A–B .... Study Figures 11.44 through 11.46.

Amorphous solids

Unlike crystals, *amorphous solids* lack long–range order and resemble liquids in their arrangements of particles. They are often called *supercooled liquids*. Examples include glass and many plastics.

Self–Test

41.   Why do crystals have such regular surface features?

_____

42.   What is the difference between a lattice and a structure based on the lattice?

_____

_____

43. Below is an illustration of atoms arranged in a two–dimensional lattice. On the drawing, sketch the outline of a unit cell for the lattice.

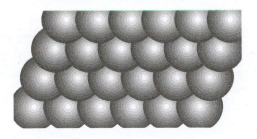

44. How many atoms are there per unit cell in the structure in the preceding problem?

_____

45. The compound ZnS forms a face–centered cubic structure as shown in Figure 11.42. All four Zn^{2+} ions are located entirely within the unit cell. How many S^{2-} ions are there per unit cell?

_____

46. The unit cell for ZnS is shown in Figure 11.42. Could the compound $AlCl_3$ crystallize with this same arrangement of ions in a cubic unit cell? Explain your answer.

_____

_____

47. On a separate sheet of paper, make sketches of simple cubic, face–centered cubic and body–centered cubic unit cells. Compare your drawings to Figures 11.36, through 11.39.

48. How do the unit cells of KCl and NaCl compare? _____

_____

49. Which kind of unit cell is depicted in each of the following?

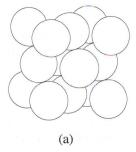   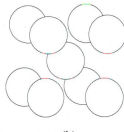   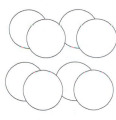

(a)   (b)   (c)

(a) _____   (b) _____   (c) _____

50. The radius of a K^+ ion is 133 pm and the radius of a Cl^- ion is 181 pm. The salt KCl has the same kind of unit cell that NaCl has (Figure 11.40). Calculate the length of an edge of this unit cell in units of picometers.

_____

51. Below is a sketch of part of the first layer in a closest–packed structure. If the second layer is started by placing an atom over point B and the third layer is started by placing an atom over point A, is the result a ccp or hcp structure?

_____

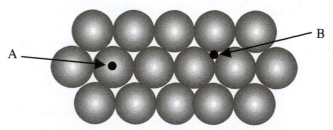

52. Why aren't the structures of amorphous solids described by lattices?

_____

_____

11.11   Determining the Structure of Solids

Learning Objective

Illustrate the basics of X–ray crystallography.

Review

When a crystal is bathed in X rays, it produces a *diffraction pattern* that can be recorded on film or by other detection devices. From the angles at which the diffracted beams emerge from the crystal and the wavelength of the X rays, the distances between planes of atoms can be computed using the *Bragg equation*,

$$n\lambda = 2d\sin\theta$$

where n is an integer, λ is the wavelength of the X rays, d is the distance between planes of atoms, and θ is the angle at which the X rays are diffracted. By a complex procedure, scientists can use these interplanar distances to figure out the structure of the crystal. As it turns out, X rays have the appropriate wavelengths, which are neither too long nor too short, to successfully create a diffraction pattern.

X–ray diffraction data yields information about distances between atoms within a crystal, from which the sizes of unit cells and even atomic sizes can be measured. Study Example 11.7.

Self–Test

53. Why do diffracted X–ray beams occur in only certain directions when a crystal is bathed in X rays?

    _____

    _____

54. Draw three additional sets of parallel lines through the points in Figure 11.50. Do all the sets of lines have the same spacing?

    _____

55. In the Bragg equation (Equation 11.5):

    (a) What does the symbol θ stand for?   _____

    (b) What does the symbol λ stand for?   _____

    (c) What does the symbol d stand for?   _____

11.12   Crystal Types and Physical Properties

Learning Objective

Understand that intermolecular forces determine the properties of crystal types.

Review

In this section we assign crystalline solids to four crystalline types: *ionic crystals, molecular crystals, covalent crystals,* and *metallic crystals*. For each type you should learn the kinds of particles found at the lattice sites and the kinds of attractive forces that exist between them. You should be able to relate these forces to the physical properties of the solid. Study Table 11.7 and Example 11.8.

Self–Test

56. What is the electron–sea model of a metallic crystal?

    _____

57. One of the elements was once named columbium, and that name is still used by some people. It is shiny, soft, ductile, and melts at 2468 °C. It also conducts electricity. What type of crystal does columbium form?

    _____

58. Antimony pentachloride forms white crystals that do not conduct electricity and that melt at 2.8 °C to give a nonconducting liquid. What kind of solid are $SbCl_5$ crystals?

    _____

59. Strontium fluoride forms brittle, nonconducting crystals that melt at 1450 °C and give an electrically conducting liquid. The solid itself is not an electrical conductor. What type of solid is SrF_2?

Answers to Self–Test Questions

1. High pressure and low temperature
2. Hydrogen bonding is stronger than ordinary dipole–dipole attractions, which are usually stronger than London forces.
3. (a) London forces; (b) ionic bonding, covalent bonding, London forces; (c) covalent bonding, hydrogen bonding, London forces; (d) covalent bonding, dipole–dipole forces, London forces; (e) covalent bonding, London forces
4. (a) PF_3, (b) SiH_4, (c) CH_3Cl, (d) HF, (e) Br_2, (f) C_6H_{14}
5. (B) < (C) < (A)
6. (a) only (A), (b) only (C), (c) all of them
7. London forces include instantaneous dipole attractions, induced dipole–induced dipole attractions, induced dipole–ion attractions.
8. Retention of volume, incompressibility, rate of diffusion
9. At room temperature gasoline has many molecules that have enough energy to escape the liquid and it evaporates very quickly. This removes heat very quickly. While the ethylene glycol molecules carry away more energy per molecule that evaporate, there are many fewer molecules evaporating each second and the total energy loss is much less than the gasoline molecules.
10. (A) < (C) < (B)
11. Ethylene glycol. Because of the two —OH groups, it will have stronger intermolecular attractions than ethanol.
12. It sublimes.   13.   Change of state
14. Evaporation from the solid; condensation to the solid phase
15. Ionization of weak acids and weak bases
16. CS_2 < $(CH_3)_2CO$ < CH_3OH < C_2H_5OH
17. See Figure 11.23
18. The higher pressure causes the water to boil at a higher temperature.
19. CS_2 < $(CH_3)_2CO$ < CH_3OH < C_2H_5OH
20. Temperature remains constant. Adding heat just makes the liquid boil more rapidly. An egg will cook in the same time whether the water is boiling or gently simmering. Rapid boiling, however, wastes energy by vaporizing more water molecules than needed.
21. It is easily measured.
22. Krypton; krypton's electron cloud is more easily polarized resulting in a larger London force of attraction.
23. HCl (more polar)
24. C_3H_8 (weaker total London forces). Yes, the relative forces would be the same.
25. Ethanol will boil first. It has a higher vapor pressure than ethylene glycol, so as the pressure is reduced, the external pressure will match the vapor pressure of ethanol before that of ethylene glycol.
26. Enthalpy is a state function, so $\Delta H_{sublimation} \cong \Delta H_{fusion} + \Delta H_{vaporization}$.

27. Water   28. Water   29. 3.95 kJ

30. $He < CH_4 < HBr < HCl < NH_3$

31. Water would boil at about –80 °C. Life depending on liquid water couldn't exist because it would
be   too cold and most biochemical reactions would be too slow to support life. If the current earth
     temperatures existed, there would be virtually no liquid water. Additionally, replication of life
     depends on hydrogen bonding in DNA.

32. (a) $CH_3CH_2CH_2CH_2CH_3$, (b) $CH_3CH_2CH_2CH_2OH$

33. (a) solid, (b) gas, (c) liquid, (d) liquid, (e) gas

34. solid $\rightarrow$ liquid $\rightarrow$ gas

35. It begins as a solid. At 0.01 °C, all three phases would appear. Above 0.01 °C it would all be gas.

36. (a) 100 °C, (b) the boiling point

37. Check your answer in Section 11.7 of the text.

38. (a) heat + liquid vapor, (b) The position of equilibrium shifts to the left. There will be more
    liquid and less vapor.

39. +11.2 kJ

40. If the friend measures the boiling point, it gives the second set of data where $P = 760$ torr and the
    temperature is the normal boiling point.

41. Because they have such ordered internal structures.

42. A lattice is a symmetrical array of points; a structure based on a lattice has chemical units associated
    with the lattice points.

43.

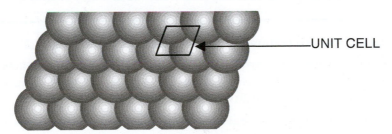

UNIT CELL

44. One.

45. Four.

46. No. $AlCl_3$ would require three times as many anion sites as cation sites, but the ratio on the ZnS
    structure is one to one.

47. See Figures 11.36 through 11.39.

48. Both are fcc, but the edge length is greater for KCl than for NaCl because K^+ is larger than Na^+.

49. (a) face–centered cubic (fcc), (b) body–centered cubic (bcc), (c) simple cubic

50. Edge length = 2(133 pm) + 2(181 pm) = 628 pm

51. hcp

52. Because amorphous solids lack long–range order.

53. Constructive and destructive interference

54. More sets can be drawn. Their spacings differ.

55. (a) The angle at which X rays are reflected from a set of planes of atoms in a crystal. (b) The
    wavelength of the X rays. (c) The distance between planes of atoms that are giving the reflection.

56. Positive ions of the metal at lattice sites surrounded by a sea of mobile electrons.

57. Metallic (the element is niobium, Nb)

58. Molecular
59. Ionic

Tools for problem solving

In this chapter you learned to apply the following concepts as tools in solving problems. Study each one carefully so that you know what each is used for. When faced with solving a problem, recall what each tool does and consider whether it will be helpful in finding a solution.

You might want to tear these pages out to use along with solving problems in this chapter.

Intermolecular forces: dipole–dipole, hydrogen bonds, and London forces (Section 11.1)

Dipole–dipole attractions, hydrogen bonding, and London forces are the three intermolecular forces that are responsible for the intermolecular attractions between molecules in the liquid and solid states. Ion–dipole and ion–induced dipole attractions are the interactions of ions with polar and nonpolar substances.

Estimating intermolecular attractive forces (Section 11.1)

From molecular structure, we can determine whether a molecule is polar or not and whether it has N—H or O—H bonds. This lets us predict and compare the strengths of intermolecular attractions. You should be able to identify when dipole–dipole, London, and hydrogen bonding occurs. See the summary in Table 11.3.

Factors that affect rates of evaporation: temperature and intermolecular forces (Section 11.2)

They allow us to compare the relative rates of evaporation based on temperature and the strengths of intermolecular forces in substances. They also allow us to compare the strengths of intermolecular forces based on the relative magnitudes of vapor pressures at a given temperature.

Factors that affect vapor pressure (Section 11.4)

The two factors that affect the vapor pressure of a substance are the temperature of the system and the intermolecular forces.

Boiling points and intermolecular forces (Section 11.5)

Liquids boil when their vapor pressure equals the atmospheric pressure. The normal boiling point of a liquid is the boiling point of the liquid at 1 atm. Since the boiling point is related to vapor pressure, it too is a measure of intermolecular forces.

Enthalpy changes during phase changes (Section 11.6)

Each phase change has an associated enthalpy; fusion, vaporization, and sublimation are endothermic, and condensation, deposition, and crystallization are exothermic. Two of these changes include the $\Delta H_{vaporization}$ and $\Delta H_{sublimation}$, allow us to compare the strengths of intermolecular forces in substances.

Heat of fusion (Section 11.6)

The amount of heat required to melt a given amount of a substance is given by

$$q = n \times \Delta H_{fusion}$$

Similar equations are used for the other phase changes.

Phase diagrams (Section 11.7)

We use a phase diagram to identify temperatures and pressures at which equilibrium can exist between phases of a substance, and to identify conditions under which only a single phase can exist. The triple point and critical point can also be determined.

Le Châtelier's principle (Section 11.8)

Le Châtelier's principle enables us to predict the direction in which the position of equilibrium is shifted when a dynamic equilibrium is upset by a disturbance.

Clausius–Clapeyron equation (Section 11.9)

The relationship, Equation 11.3, between vapor pressure, heats of vaporization, and temperature is

$$\ln P = \frac{-\Delta H_{vap}}{RT} + C$$

This may be written as a "two–point" relationship as in Equation 11.4,

$$\ln \frac{P_1}{P_2} = \frac{\Delta H_{vap}}{R}\left(\frac{1}{T_2} - \frac{1}{T_1}\right)$$

Cubic unit cells (Section 11.10)

The three basic cubic lattice structures are simple cubic, face–centered cubic, and body–centered cubic. By knowing the arrangements of atoms in these unit cells, we can use the dimensions of the unit cell to calculate atomic radii and other properties.

Counting atoms in unit cells (Section 11.10)

The number of atoms in the unit cell can be counted by adding up the parts of the atoms that are contained in the unit cell: ⅛ for an atom in a corner, ¼ for an atom on an edge, ½ for an atom in a face, and 1 for an atom wholly in the unit cell.

Bragg equation (Section 11.11)

$$n\lambda = 2d \sin \theta$$

Properties of crystal types (Section 11.12)

By examining certain physical properties of a solid (hardness, melting point, electrical conductivity in the solid and liquid states), we can often predict the nature of the particles that occupy lattice sites in the solid and the kinds of attractive forces between them.

Summary of Important Information

Intermolecular attractions and physical properties

As the strengths of intermolecular attractions *increase*, there is an *increase* in

surface tension

viscosity

molar heat of vaporization

molar heat of sublimation

normal boiling point

critical temperature

and a *decrease* in

rate of evaporation

vapor pressure

Chapter 12

Mixtures at the Molecular Level: Properties of Solutions

The emphasis in this chapter is on the *physical* properties common to all solutions, not on their chemical properties. The latter depend on the chemicals themselves.

12.1 Intermolecular Forces and the Formation of Solutions

Learning Objective

Describe how intermolecular forces affect the solution process for solids, liquids, and gases.

Review

When two substances are in contact, the random motions of their molecules tend to cause them to mix. In the absence of forces to prevent this, mixing occurs and the substances dissolve in one another. We can also view the mixed and unmixed states in terms of their relative probabilities once the two are in contact, as we do here for two gases. The mixed state has a vastly higher probability than the unmixed one, so the system changes spontaneously from a state of low probability to one of high probability. This illustrates a general principle of nature: *a system, left to itself, will tend toward the most probable state*.

For the mixing of gases, there are virtually no attractive forces to hinder the natural tendency toward mixing, so gases always mix freely. However, for the dissolving of liquids or solids in liquid solvents, even though probability is still a significant factor, intermolecular attractions also play a very important role. Analysis of the solution process for various solute–solvent combinations leads to the "*like dissolves like rule*", where "like" refers to similarities in polarity. This rule tells us that *substances with similar strengths of intermolecular attractions tend to be soluble in each other*. On the other hand, when the intermolecular attractions are much different in the solute and solvent, the substances are mutually insoluble.

For solutions of molecular substances, the intermolecular forces to be considered are London forces, dipole–dipole attractions, and hydrogen bonding. For an ionic compound dissolving in a liquid, ion–dipole attractions are important.

The ability of water to *solvate* (*hydrate*) ions largely explains why ionic compounds dissolve better in water than in any other solvent.

Self–Test

1. What are the driving forces behind the formation of the following solutions at room temperature and pressure? (Go beyond the like dissolves like rule and examine the solution process in detail.)

    (a)   Ethyl alcohol in water   _____

    _____

    (b)   Potassium bromide in water _____

    _____

    (c)   The fragrance of a flower bouquet spreading in a room's air

    _____

2. Wax consists of nonpolar molecules. In terms of intermolecular attractions, explain why wax does not dissolve in water.

    _____

    _____

3. Why should the hydration of the ions of NaCl help this compound to dissolve in water?

    _____

4. Liquids X and Y consist of nonpolar molecules. Will X tend to dissolve in Y?   _____

    Explain. _____

    _____

5. NaNO$_3$, an ionic solid, does not dissolve in gasoline. Explain.

    _____

6. Which of the following is probably more soluble in benzene, C$_6$H$_6$? Why?

    (a)   CH$_3$ —CH$_2$—CH$_2$—O—CH$_3$ (b)   CH$_2$ —CH —CH$_2$

              OH   OH   OH

    _____

    _____

12.2 Heats of Solution

Learning Objective

Detail the energy changes that take place when two substances form a solution.

Review

It costs energy to separate solute particles from each other and to separate solvent molecules from each other, because forces of attraction have to be overcome. But when new forces of attraction become established between solute and solvent molecules, energy is liberated. The *heat of solution* is the net enthalpy change, and enthalpy diagrams are helpful to show these relationships.

For salts, we have information about *lattice energies* and *hydration energies* from independent sources, and the net values of these energies are roughly of the same magnitude as measured heats of solution.

Two liquids form an *ideal solution* when the net heat of solution is zero, indicating that there is no change in forces of attraction between molecules in the separated components when compared to the forces between them in the solution.

For gases, no energy is needed to separate the molecules. Energy must be expended to open pockets in the solvent to accept the gas molecules; an exception is water, which already has empty spaces in its structure. Energy is released when the gas molecules enter the pockets. When gases dissolve in organic solvents, the process is often endothermic because more energy must be expended opening pockets than is released by gas molecules entering the solution. When gases dissolve in water, the process is usually exothermic because pockets already exist in water to accept the solute, so only the exothermic contribution occurs.

Self–Test

7. What is another technical name for "heat of solution?"

    _____

8. How does the lattice energy affect the heat of solution of a solid solute?

    _____

9. Consider a sample of NaCl and a separate beaker of water. The potential energy of this system

    _____ when the NaCl is converted to its gaseous ions. Why?
    (increases or decreases)

    _____

    The separated ions now enter the water. Each becomes surrounded by water molecules, which is a

    phenomenon called _____.

    The potential energy of the system _____ as a result of this.
                                        (increases or decreases)

    Why? _____

10. For a particular salt, the true lattice energy is –436 kJ/mol and the hydration energy is –548 kJ/mol. If the heat of solution for the salt were a function just of these two energies, what is its value?

   _____

   The formation of the solution is therefore exothermic or endothermic?

   _____

11. When one liquid is dissolved in another, we can envision a 3–step process. Which step or steps is(are) endothermic and why?

   _____

   _____

   Which step or steps is(are) exothermic and why? _____

   _____

   _____

12. What is true about two liquids and their solution if the solution is properly described as ideal?

   _____

   _____

13. Briefly explain why gases dissolve in water exothermically.

   _____

   _____

12.3   Solubility as a Function of Temperature

Learning Objective

Explain how temperature affects the solubility of a solute in a solvent.

Review

We can include the heat of solution as a product (when it is negative) or as a reactant (when it is positive) in the equilibrium expression for a saturated solution. When we do this, it is easy to apply Le Châtelier's principle to predict how the equilibrium will shift if the solution is heated or cooled. As a rule, gases become less soluble in water as the temperature increases.

Self–Test

14. Suppose a saturated solution involves the following equilibrium.

$$\text{solute}_{\text{undissolved}} \rightleftharpoons \text{solute}_{\text{dissolved}} + \text{heat}$$

(a)   Is the formation of the solution endothermic or exothermic?

_____

(b)   If the saturated solution is heated, will more solute come *out* of solution or go *into* solution?

_____

15. Write the equilibrium expression of a gas dissolving in water to make a saturated solution, and include the heat of solution either as a product or as a reactant.

_____

16. Write the equilibrium expression for a saturated solution of a typical gas in an organic solvent, and include the heat of solution as a product or reactant.

_____

12.4 Henry's Law

Learning Objective

Using Henry's law, perform calculations involving the pressure of a gas in solution.

Review

The relevant equilibrium is

$$\text{gas} + \text{solvent} \rightleftharpoons \text{solution}$$

As the gas dissolves, a large decrease in the volume of this system occurs, so an increase in pressure—a volume–reducing action—will shift the equilibrium to the right.

Henry's law (pressure–solubility law) describes the direct proportionality between gas solubility and pressure that applies for many gases.

$$C_{\text{gas}} = k_{\text{H}}\, P_{\text{gas}}$$

where C_{gas} is the concentration of gas in the solution, P_{gas} is the partial pressure of the gas over the solution, and k_{H} is the proportionality constant. Equation 12.5 gives a useful alternate expression of the law. Study Example 12.1.

Some important gases, such as CO_2, SO_2, and NH_3, are helped into aqueous solution by strong intermolecular forces or by reacting to some extent with the solvent.

Self–Test

17. At 20 °C, the solubility of oxygen in water is 43.0 mg/L at a pressure of 762 torr. Calculate its solubility in these units at a pressure of 162 torr.

_____

18. Perform the following calculations and answer the question regarding gas solubility.

(a) Concentrated hydrochloric acid is a 12.0 M solution of HCl in water. How many liters of HCl gas (at 16 °C and 1.25 atm) dissolve in 1.00 L of water to give this solution?

_____

(b) At 16 °C and 1.25 atm, the solubility of N_2 in water is 0.0194 g L^{-1}. How many liters of N_2 at this temperature and pressure dissolve in 1.00 L of water?

_____

(c) What chemical fact accounts for the difference in the amounts of these two gases that dissolve in 1.00 L of water?

_____

_____

12.5   Concentration Units

Learning Objective

Define solution concentrations: percent concentration, molal concentration, molar concentration, mole fraction, and mole percent, and convert between the different units.

Review

The molarity of a given solution varies slightly with temperature because liquids expand when heated. As the temperature rises, the solute is distributed in a larger volume, and the molarity decreases. Although molarity is very useful for problems dealing with stoichiometry, when we are concerned with the physical properties of a solution, we need to have concentration expressed in a way that is invariant with temperature changes.

Mole fractions (Section 10.6) as well as the new concentration expressions introduced in this section, mass fractions (or mass percents) and molalities, do not change as the temperature of a solution changes. This is because they give us, directly or indirectly, information about the ratio of particles of solute to particles of solvent, and that ratio doesn't change when a solution's temperature varies.

All expressions of concentration are ratios, with quantities specified in the numerator and denominator. For example, in Chapter 4 you learned that molarity is a ratio of moles of solute to liters of solution. To be able to work with the various ways of expressing concentration, it is essential that you know precisely how they are defined. This means you must know the units of the numerator and those of the denominator. Those who do not learn these units invariably flounder when trying to work problems involving concentrations.

Let's take a look at each of the concentration expressions presented in this chapter to see how we can extract from them information about the solution composition.

Mole Fraction

In Chapter 10 you learned that the units implied in a *mole fraction*, *X*, are

$$\frac{\text{number of moles of one component}}{\text{total number of moles of all components}}$$

It is also useful to remember that the sum of the mole fractions of all the components equals one. Therefore, if a solution of sugar in water is described as having 0.10 mole fraction of sugar, the mole fraction of water must be 0.90, because 0.10 + 0.90 = 1.00. If we wished to use this information to construct conversion factors, the following two are available.

$$\frac{0.10 \text{ mol sugar}}{0.90 \text{ mol water}} \qquad \frac{0.90 \text{ mol water}}{0.10 \text{ mol sugar}}$$

Notice that we've used the mole fraction to derive the relative amounts of the two components.

A mole percent, of course, is simply a mole fraction multiplied by 100%.

Percentage Concentration

The most commonly used *percentage concentration* is *mass percent* (also called weight percent or weight/weight percent). Mass percentage is defined as follows

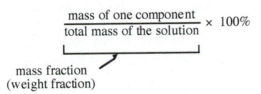

$$\underbrace{\frac{\text{mass of one component}}{\text{total mass of the solution}}}_{\substack{\text{mass fraction} \\ \text{(weight fraction)}}} \times \ 100\%$$

It's the mass fraction (also called weight fraction) multiplied by 100%. Mass is usually expressed in grams, but it can be expressed in whatever mass units you wish. The only condition is that the units must be the same in both numerator and denominator.

Quantities similar to mass percentage are parts per million (ppm) and parts per billion (ppb).

$$\text{ppm} = \text{mass fraction} \times 10^6$$

$$\text{ppb} = \text{mass fraction} \times 10^9$$

A convenient rule of thumb to remember is that 1 ppm is equivalent to 1 mg of solute dissolved in one liter of an aqueous solution. Similarly, one ppb is equivalent to 1 μg per liter of aqueous solution.

When we are given a mass percentage, it is not difficult to translate it to a ratio of masses. We just imagine having 100 g of solution. For example, suppose a solution were labeled 12.0% (w/w) C_2H_5OH in water. If we had 100 g of this solution, it would contain 12.0 g of C_2H_5OH and we could express the concentration as

$$\frac{12.0 \text{ g } C_2H_5OH}{100 \text{ g solution}}$$

We can use this to make two conversion factors that could be used in calculations.

$$\frac{12.0 \text{ g } C_2H_5OH}{100 \text{ g of solution}} \qquad \frac{100 \text{ g of solution}}{12.0 \text{ g } C_2H_5OH}$$

Other percentage concentrations are mass/volume (or weight/volume, w/v) and volume/volume (v/v). In the lab, if mass percentage is not specified, you can assume that this is what is meant. Review the discussion in Section 12.5 in the text.

Molality

The units for *molality* are moles of solute per kilogram of solvent. Thus, if an aqueous solution were labeled 0.25 m CH_3OH, we can represent the concentration as follows

$$0.25 \ m \ CH_3OH \ = \ \frac{0.25 \text{ mol } CH_3OH}{1.00 \text{ kg } H_2O}$$

Often, it's desirable to substitute 1000 g of solvent for 1 kilogram, so we can express the molality as

$$0.25 \ m \ CH_3OH \ = \ \frac{0.25 \text{ mol } CH_3OH}{1000 \text{ g } H_2O}$$

When we use molality in calculations, we can derive two conversion factors. For example,

$$\frac{0.25 \text{ mol } CH_3OH}{1000 \text{ g of solvent}} \qquad \frac{1000 \text{ g of solvent}}{\text{mol } CH_3OH}$$

Be sure you know the difference between *molarity*, **M**, and *molality*, **m**. They sound a lot alike but mean quite different things. Notice that the numerators are the same for both units, but the denominators are quite different.

Concentration Conversions

In converting among concentration units, remember that when you want to convert from a temperature, dependent concentration unit to a temperature independent unit, or *vice versa*, you also have to know the density of the solution. Density provides the link between a solution's mass and its volume, because density is mass per unit volume.

If conversions such as the one studied in Examples 12.4 and 12.5 of the text give you a headache, you're probably a member of a large club. You will find that organizing facts, both given and sought, into tables provides the best help. Let's see how this would work by redoing Example 12.4. The problem asked, "What is the molality of 10.0% (w/w) aqueous NaCl?" The *Analysis* of this problem noted that it asks us to go from one set of units to another, that is, from

$$\frac{10.0 \text{ grams of NaCl}}{100.0 \text{ grams of NaCl soln}} \quad \text{to} \quad \frac{\text{mole of NaCl}}{1 \text{ kg of water}}$$

A way to organize the approach to any of these kinds of problems is to set up a small table like the one below. First, we identify the information that's needed to compute the target concentration. We'll put such data in boxes. Next, we will extract information from the numerator and denominator of the given concentration and place it into the table. Here's the empty table where we've identified the locations of both the given data and the data required to find the answer.

Substance	Mass	Moles
NaCl	_____ g	_____ mol
H_2O	_____ g = _____ kg	
Total	_____ g	

The answer will be found by applying the defining equation for molality to our specific situation.

$$\text{Molality} = \frac{\boxed{\text{mole NaCl}}}{\boxed{\text{kg } H_2O}}$$

Now, let's begin to fill in the table. First, we take apart the given concentration to give us two table entries, mass of NaCl and mass of H_2O (shaded entries in the table). Then we calculate the other table entries needed to obtain the remaining units for the desired concentration. The table can now be completed. The mass of water is the difference between the total mass and the mass of NaCl; a grams–to–moles calculation is performed to calculate the number of moles of NaCl. (Notice that we haven't bothered to fill in places in the table that are not relevant to the problem.)

Substance	Mass	Moles
NaCl	10.0 g	0.171 mol
H_2O	90.0 g = 0.0900 kg	
Total	100.0 g	

The molality, m, is now found by taking the ratio

$$\text{molality} = \frac{0.171 \text{ mol NaCl}}{0.0900 \text{ kg } H_2O} = 1.90 \; m$$

Let's next work an example that shows how the use of a table can help us carry out another conversion from one concentration expression to another.

Example 12.1   Converting mole fractions into percents by mass

So–called 100 proof alcohol is a solution of ethyl alcohol in water. It has a mole fraction of C_2H_5OH of 0.235. Calculate the mass percent of ethyl alcohol in 100 proof ethyl alcohol. The molecular masses are: H_2O, 18.02; ethyl alcohol, C_2H_5OH, 46.07.

Analysis:
We are asked to carry out the following conversion from the given mole fraction of C_2H_5OH to its mass percent:

$$\frac{0.235 \text{ mol } C_2H_5OH}{\text{total of 1 mol}} \quad \text{to} \quad \frac{\text{grams } C_2H_5OH}{\text{total grams soln}} \times 100\% \quad \text{(the mass percent of } C_2H_5OH)$$

Notice that we are converting one temperature–independent concentration unit into another. We will not need the density to solve this problem. To begin we assume a total amount of solution equal to 1 mol. Then the mole fraction becomes equal to the number of moles of C_2H_5OH. From there we carry out the needed conversions.

Assembling the Tools:

We need the tools defining the mole fraction and mass percent. Also needed are our conversion tools from moles to grams. The concentration table will be useful for keeping track of data.

Solution:

We will enter the data from the mole fraction ratio into the table (shaded areas). The data needed to calculate the answer is identified in the table. By difference, we can calculate the moles of water in the solution. For the corresponding mass–percent concentration, we need to convert moles of C_2H_5OH to grams. For the total mass of the solution, we must also compute the mass of water that corresponds to 0.765 mol of water. Then we can add the masses of alcohol and water to find the total mass of the solution.

The number of grams of the components are found as follows.

For C_2H_5OH:

$$0.235 \text{ mol } C_2H_5OH \times \frac{46.07 \text{ g } C_2H_5OH}{1 \text{ mol } C_2H_5OH} = 10.8 \text{ g } C_2H_5OH$$

For H_2O:

$$0.765 \text{ mol } H_2O \times \frac{18.02 \text{ g } H_2O}{1 \text{ mol } H_2O} = 13.8 \text{ g } H_2O$$

The completed table is as follows, with the given data shaded and the data we need to calculate the answer in boxes.

Substance	Mass	Moles
C_2H_5OH	10.8g	**0.235 mol**
H_2O	13.8 g	0.765 mol
Total mass	24.6g	**1.000 mol**

The desired mass percent is found by:

$$\text{Mass percent of } C_2H_5OH = \frac{\text{grams } C_2H_5OH}{\text{Total grams}} \times 100\%$$

$$\frac{10.8 \text{ g}}{24.6 \text{ g}} \times 100\% \qquad\qquad =$$

$$\qquad\qquad =$$

43.9% C_2H_5OH

Is the Answer Reasonable?

We can check to be sure we have the right units for numerator and denominator for both the given and desired concentrations. If these are correct, the rest should be okay as well. You could also do some approximate arithmetic for the moles to grams calculations and the mass percent calculation. Do this on your own, and you should be satisfied we've obtained the correct answer.

Example 12.2   Finding mole fractions starting from molality

Ammonium nitrate is sometimes a component in nitrogen fertilizers. Suppose that an aqueous solution of NH_4NO_3 has a molal concentration of 2.480 m. What are the mole fractions of NH_4NO_3 and water in this solution?

Analysis:

The requested conversion is:

$$\frac{2.480 \text{ mol } NH_4NO_3}{1000 \text{ g } H_2O} \quad \text{to} \quad \frac{\text{mol } NH_4NO_3}{\text{total moles}}$$

All we have to calculate is the *total number of moles*. We're given 2.480 mol of NH_4NO_3. So we calculate the number of moles of H_2O in 1000 g and add the two quantities together. Again we notice that the starting and ending units are temperature independent and we do not need the density for this conversion.

Assembling the Tools:

We need the tools defining the mole fraction and mass percent. Also needed are our conversion tools from moles to grams. The concentration table will be useful for keeping track of data.

Solution:

We need to convert 1000 g of water to moles of water.

$$1000 \text{ g } H_2O \times \frac{1 \text{ mol } H_2O}{18.02 \text{ g } H_2O} = 55.49 \text{ mol } H_2O$$

The completed table follows.

Substance	Mass	Moles
NH_4NO_3		**2.480 mol**
H_2O	**1000 g**	55.49 mol
Total		57.97 mol

The mole fraction of water is simply

$$X_{H_2O}$$

$$= \frac{55.49}{57.97} = 0.9572 \text{ or } 95.72 \text{ mol\%}$$

The mole fraction of NH_4NO_3 is

$$X_{NH_4NO_3} = \frac{2.480}{57.97} = 0.04278 \text{ or } 4.278 \text{ mol}\%$$

Is the Answer Reasonable?

The sum of the fractions should add to 1.00 and the percents to 100% taking into account differences for rounding errors. Our answer looks good on that basis.

Example 12.3 Calculating the molarity and molality of a solution

What are (a) the molality and (b) the molarity of a solution of NaBr with a concentration of 5.00%? The density of the solution is 1.04 g/mL.

Analysis:

The conversion of 5.00% solution (we assume that wt/wt% is implied) to molality is similar to the previous two examples. The conversion to molarity involves the density to convert the mass of the solution to volume. We start by setting up a concentration table and then using the definitions of molarity and molality to answer the question.

Assembling the Tools:

We need the tools defining the mole fraction, mass percent, and molarity. Also needed are our conversion tools from moles to grams and volume to mass (density). The concentration table will be useful for keeping track of data.

Solution:

We set up a concentration table,

Substance	Mass	Moles
NaBr (molar mass = 102.89)	5.00 g	0.0486
H$_2$O	95.0 g = 0.095 kg	5.271

(a) From the table we calculate the molality as

$$molality = m = \frac{mol_{solute}}{kg_{solvent}} = \frac{0.0486 \text{ mol NaBr}}{0.095 \text{ kg}} = 0.512 \ m$$

(b) To determine the molality we take one liter (1000 mL) of the solution and determine its mass

$$1000 \text{ mL solution} \times 1.04 \text{ g mL}^{-1} = 1040 \text{ g solution}$$

This solution is 5% solute, NaBr. The mass of solute is

$$1040 \text{ g solution} \times 0.05 = 52 \text{ g NaBr}$$

Using the molar mass we calculate that we have 0.505 moles of NaBr. Now we have the moles of NaBr that are in one liter of solution, which is the molarity:

$$Molarity = M = 0.505 \text{ mol NaBr}/1.00 \text{ L} = 0.505 \ M \text{ NaBr}$$

Are the Answers Reasonable?

We check our definitions of molality and molarity, and they are correct. Checking the mathematics reveals no errors either. Finally, this is a fairly dilute aqueous solution and we expect the molarity and molality to have similar numerical values, and they do.

Self–Test

19. A solution for disinfecting clinical thermometers was made by dissolving together 235 g of ethyl alcohol and 53.0 g of water. Calculate the percentage by mass of each component.

20. A sample with a mass of 56.3 g of 15.7% sucrose solution was taken for an experiment. How many grams of sucrose were present?

21. A bottle bears the label: "1.364 molal sodium chloride." What are the two conversion factors made possible by this information?

22. If the bottle in the preceding question actually holds only 0.200 mol NaCl, what mass of *water* is present as the solvent?

23. What mass of solute in grams is needed to make a 0.355 molal glucose solution if you intended to use 250 g of water as the solvent? Use 180 as the molecular mass of glucose.

24. How would you prepare 500 g of a 5.500% sugar solution?

25. For an experiment you need 8.42 g of H_2SO_4, and it is available as 10.0% (w/w) H_2SO_4. How many grams of this solution should you take?

26. In a two–component solution of benzene (C_6H_6) in carbon tetrachloride (CCl_4), the mole fraction of benzene is 0.367. What is the mass percent of benzene in this solution?

27. A solution of $FeCl_3$ has a concentration of 3.85 *m*. What are the mole fractions and mole percents of $FeCl_3$ and H_2O in this solution?

28.   What is the molality of a 23.8% (w/w) LiCl solution?

_____

29.   What are the mole fractions and the mole percents of the components in 23.8% (w/w) LiCl (the same solution in Question 28)?

_____

30.   In a two–component solution of sucrose, table sugar, in water, the mole fraction of sugar is 0.0150. Calculate the percentage by mass of the sugar. Use 342 as the molecular mass of sugar.

_____

31.   A solution of NaCl has a concentration of 2.32 *m*. What are the mole fractions and mole percents of NaCl in this solution?

_____

32.   What is the percentage concentration of a sodium carbonate solution with a concentration of 1.04 mol/L? The density of the solution is 1.105 g/mL? (Be sure to remember that both units in 1.105 g/mL refer to the *solution*. The value of the density tells us that there are 1.105 g Na_2CO_3 *solution* per milliliter of Na_2CO_3 *solution*, or 1 mL Na_2CO_3 solution per 1.105 g Na_2CO_3 solution.)

_____

12.6   Colligative Properties

Learning Objective

Describe the colligative properties, Raoult's law, freezing point depression, boiling point depression, and osmosis of a solution and use them in calculations for solutions with molecular and ionic solutes.

Review

Raoult's Law

The vapor pressure of a liquid is lowered when a nonvolatile solute (a solute that cannot evaporate) is dissolved in it because the solute particles interfere with the escape of molecules of the liquid into the vapor state but not their return to the liquid solution. For a molecular solute, a simple relationship exists between the vapor pressure of the solution, $P_{solution}$, the vapor pressure of the pure solvent, $P°_{solvent}$ and the mole fraction of the *solvent* (not of the solute). This is the vapor–pressure concentration law or *Raoult's law*.

$$P_{solution} = X_{solvent} \times P°_{solvent}$$

(If the solute breaks up into ions, we have to modify this, but that is a subject for Section 12.9.)

In a solution of two volatile liquids, each undergoes evaporation and thereby contributes a partial pressure toward the total vapor pressure. Each component also has its evaporation hindered by the other component, and so is subject to Raoult's law. As a result, we use a variation of the Raoult's law equation to calculate the *partial pressure* of the vapor of each component in the solution. Thus, for component *A* we have

$$P_A = P_A^o \times X_A$$

where P_A is the partial vapor pressure of component A; P_A^o is the vapor pressure of A when it is a pure liquid; and X_A is the mole fraction of A in the solution.

The total vapor pressure exerted by a mixture of volatile substances is the sum of the individual partial vapor pressures. This is simply an application of Dalton's law of partial pressures.

Only *ideal solutions* obey Raoult's law exactly. In many real two–component solutions of volatile liquids we see larger pressures than predicted by Raoult's law. This happens when intermolecular forces in the solution are less than they are in the separated components. Such solutions form endothermically because it costs more energy to overcome forces of intermolecular attraction in the separate components than is recovered when new forces of attraction operate in the solution.

Experimental pressures lower than predicted by Raoult's law are due to intermolecular forces in the solution that are greater than they are in the separated components. Such solutions form exothermically because less energy is needed to overcome attractive forces in the separate components than is recovered when attractive forces start to operate in the newly forming solution.

Vapor pressure lowering by a solute is only one example of a *colligative property*, a property relating to the relative *numbers* of particles of the mixture's components, not on their chemical identities.

Self–Test

33. What is the vapor pressure of a 1.00 molal sugar solution at 25 °C? Sugar, $C_{12}H_{22}O_{11}$, is a nonvolatile, nonionizing solute and the vapor pressure of water at 25 °C is 23.8 torr.

   _____

34. At 20 °C the vapor pressure of pure toluene is 21.1 torr and of pure cyclohexane is 66.9 torr. What is the vapor pressure of a solution made of 25.0 g of cyclohexane and 25.0 g of toluene? (The formula for cyclohexane is C_6H_{12} and toluene is C_7H_8.)

_____

Boiling point elevation and freezing point depression

Two other colligative properties of solutions, *boiling point elevation* and *freezing point depression*, are chiefly exploited to obtain molecular masses.

To overcome the lowering of the vapor pressure by a nonvolatile solute, we have to raise the solution's temperature a small amount to get it to boil. To make the solution freeze, we must decrease the solution's temperature below the solvent's normal freezing point. The value of the change in temperature, Δt, is directly proportional to the solution's molal concentration. The proportionality constant is different for each solvent and is different for boiling and freezing. Depending on the actual physical change, the constant is called the *molal boiling point elevation constant, K_b or the *molal freezing point depression constant, K_f*. If we measure Δt, and can find K_b or K_f for the solvent in a table, then we can calculate m, our symbol here for molal concentration:

$$\Delta t = K_b m$$

$$\Delta t = K_f m$$

When we know m by this procedure, we can calculate the moles of solute present in the mass of solute used for the solution. When we know both mass (in grams) and moles, it's simply a matter of taking the ratio of grams to moles to find the molar mass. To review the procedure:

Step 1. Select a solvent, and dissolve a measured mass of solute in a measured mass of the solvent.

Step 2. Measure the boiling point of this solution or its freezing point. (Unusually, precise thermometers have to be used because Δt is nearly always small.)

Step 3. Use Δt and K_b or K_f to calculate m.

Step 4. Multiply m by the number of kilograms of solvent actually used to find the moles of solute. Notice how the units work out to give "mol solute":

$$\underbrace{\frac{\text{mol solute}}{\text{kg solvent}}}_{\text{molality}} \times \text{ kg solvent} = \text{mol solute}$$

Step 5. Divide the number of grams of solute by the number of moles of solute to find the grams per mole, the molar mass. This numerically equals the molecular mass.

Re–study Example 12.9 in the text to see how these steps were used.

Example 12.4 Determining the molar mass of a compound

Vapona is the ingredient in flea collars and pesticide "strips." A sample of 0.347 g of Vapona was melted together with 35.0 g of camphor, and this mixture was cooled to give a solid. The solid was pulverized and its melting point was found to be 35.99 °C. Using the identical thermometer, pure camphor was found to melt at 37.68 °C. For camphor, the freezing point depression constant is $K_f = 37.7\ \dfrac{°C}{m}$. What is the molar mass of Vapona?

Analysis:
We will be using the tool that tells us the relationship between the freezing point depression and the molality of a nondissociating solute such as Vapona. Once we know the molality, we use its definition to determine the moles of Vapona in the solution.

Assembling the Tools:
We need the tools defining the molality and its relationship to the freezing point of camphor. Also needed is the definition of molar mass.

Solution:
The freezing point depression equation is shown above. Our freezing point depression is 37.68 °C – 35.99 °C = 1.69 °C. The molality of our solution is

$$\Delta t / K_f\ =\ 1.69\ °C/\,(37.7\ °C/m) = 0.0448\ \text{molal}$$

The molality is defined as moles of solute dissolved in one kg of solvent. From this we can calculate the moles of solute as

m x kg solvent = 0.0448 molal x 0.035 kg camphor = 0.00157 moles of solute

Last, the definition of the mole is the mass of sample divided by the molar mass, so the molar mass is

0.347 g Vapona/0.00157 mol Vapona = 221 g/mol Vapona

The molar mass is 221 g/mol for this compound.

Is the Answer Reasonable?

We check our reasoning from calculating the molarity to moles to the molar mass and see no errors. The mass is a reasonable number for this method. In addition we can find no math errors when the calculations are repeated.

Self–Test

35. A solution of 8.32 g of PABA, once widely used as a sunscreen agent, in 150 g of chloroform boiled at 62.62 °C. At the same pressure and with the same thermometer, pure chloroform boiled at 61.15 °C. What is the molecular mass of PABA? For chloroform,

$$K_b = 3.63 \frac{°C}{m}$$

Osmosis

Whether the movement of a fluid through a semipermeable membrane is *osmosis* or *dialysis* depends on the kind of membrane separating the solutions or dispersions of unequal concentration. As this membrane becomes less and less permeable, we approach osmosis as the limiting case of dialysis in general.

With the help of Equation 12.17 in the text ($\Pi = MRT$) we can use data on temperature (T) and *osmotic pressure* (Π) to find moles per liter (M). From the molarity thus obtained and the volume of solution prepared for the experiment, we can obtain the number of moles of solute that were used. Then we take the ratio of grams of solute to moles of solute to obtain the molar mass, just as in the preceding section.

$$\frac{\text{grams solute}}{\text{moles solute}} = \text{molar mass}$$

The technique of using osmotic pressure to determine a molecular mass is particularly useful with high–molecular–mass substances, because even though a given mass gives a very low concentration in units of molarity, the osmotic pressure is still sizable enough for an accurate measurement.

Example 12.5 Estimating molar mass of large molecules

What is the molar mass of a substance that gives an osmotic pressure of 0.952 torr when 11.3 g of the compound is dissolved in 500 mL of water at 4 °C?   What would be the freezing point depression of the same solution?

Analysis:

The osmotic pressure equation is the same as the ideal gas law (Π rather than P). Since we know how to obtain the molar mass from the ideal gas law we will use the same process here. We have to be careful that the units match those of R.

Assembling the Tools:

We need the osmotic pressure equation along with the definition of molar mass to solve this problem. Unit conversions may also be needed.

Solution:

The necessary equation is $\Pi V = nRT$. We convert 0.952 torr to 0.00125 atm. Then 500 mL is converted to 0.500 L while 4 °C becomes 277 K. We calculate the number of moles as

$$n = \frac{\Pi V}{RT} = \frac{(0.00125 \text{ atm})(0.500 \text{ L})}{(0.0821 \text{ L atm mol}^{-1} \text{ K}^{-1})(277 \text{ K})} = 2.74 \times 10^{-5} \text{ mol}$$

The definition of a mole is $n = g/\text{molar mass}$. Rearranging this definition to calculate the molar mass we get

$$\text{Molar mass} = g/n = 11.3 \text{ g}/2.74 \times 10^{-5} \text{ mol} = 4.12 \times 10^{5} \text{ g/mol}$$

At this low concentration the molarity and molality are essentially the same. From the first part of this example we have 2.74×10^{-5} mol in 0.5 liters or 5.48×10^{-5} mol/L. At this low concentration the molarity and molality are almost equal in aqueous solution so we can use the molarity in the freezing depression equation. We get

$$\Delta t = K_f m = 1.86 \text{ °C } m^{-1})(5.48 \times 10^{-5} \text{ } m) = 1.02 \times 10^{-4} \text{ °C}$$

Notice that this temperature change is virtually unreadable on ordinary laboratory thermometers.

Are the Answers Reasonable?

For the first part we obtained a large molar mass as expected. We can also check to be sure that all units cancel properly. For the second part we check the units and estimate the answer. These all agree with our answer and we have confidence that they are correct.

Self-Test

36. In some sciences, a body fluid might be described as having a "high osmotic pressure." How should we translate this—as meaning a high or a low concentration?

    _____

37. When raisins or prunes are placed in warm water they soon swell up and the water becomes slightly sweet to the taste. What phenomenon is occurring, dialysis or osmosis?

    _____

38. If a steak on the grill is heavily salted as it grills, barbecue experts say that the "salt draws the juices." Explain how that happens.

_____

_____

39. An aqueous solution of a protein with a concentration of 1.30 g/L at 25 °C had an osmotic pressure of 0.0160 torr. What was its molecular mass?

_____

van't Hoff factor

If you know that a solute breaks up into ions when it dissolves, then, as a rough estimate of the effect of this breakup on colligative properties, multiply the molality (or molarity) by the number of ions each formula unit gives as it dissociates. This works best for very dilute solutions.

As solutions of electrolytes become increasingly concentrated, their ions behave less and less as fully independent particles. To compare the degrees of dissociation at different concentrations, the *van't Hoff factor* is determined. It is the ratio of the degrees of freezing point depression actually observed for the solution to the freezing point calculated by assuming that the electrolyte does not dissociate at all.

One way to become comfortable about this concept is to study Table 12.5 in the text. Look at the first row of data for NaCl. The last column tells us that if NaCl were 100% dissociated in solution, its van't Hoff factor would be 2.00. This is because two ions form from each NaCl unit. And in a very dilute solution (0.001 *m*), the van't Hoff factor is 1.97 or very close to 2.00. In this quite dilute solution, the electrolyte behaves as if it were almost 100% dissociated. In the more concentrated solution of 0.1 *m*, the van't Hoff factor is less, only 1.87. But even this is not far from 2.00, so the solute behaves as if it were still mostly dissociated. Notice in Table 12.5 that the salts that give the largest deviations are those made of ions with more than one charge. These are able to attract each other strongly, and so they are less able to act independently in a solution.

This section also describes how a *percent dissociation* can be estimated from freezing point depression data, but the method is not highly accurate.

The section also alerts you to the existence of some solutes that give weaker colligative properties than expected because of the *association* of solute particles, one with another. In particular, notice how molecules of benzoic acid associate by hydrogen bonding to give dimers (particles formed from two identical small molecules).

Self–Test

40. Calculate the freezing point of aqueous 0.250 *m* NaCl on the assumption that it is 100% dissociated.

_____

Now do this calculation on the assumption that it is not dissociated at all.

_____

41. The van't Hoff factor for a salt at a concentration of 0.01 m is 2.70. At a concentration of 0.05 m, its van't Hoff factor would be greater or less?

_____

Explain. _____

_____

12.7 Heterogeneous Mixtures

Learning Objective

Illustrate the characteristics of heterogeneous mixtures such as suspensions and colloids.

Review

To this point we have discussed solutions that are homogeneous mixtures of a solute in a solvent. For the most part, the solvent we have considered is water and our solutions are called aqueous solutions. We now turn our attention to heterogeneous mixtures. Heterogeneous mixtures are very important commercial consumer products ranging from salad dressings to shampoos to paints and many, many more products in the stores we frequent. Modern materials scientists are preparing useful heterogeneous mixtures using nanoparticles in creative ways.

Heterogeneous mixtures may be called dispersions. One type of dispersion is a suspension that consists of relatively large particles suspended in a solvent. These particles may settle quickly as a sand and water mixture will settle when you stop shaking it. Other suspensions take an extremely long time to separate and may even seem to be stable.

Colloids are dispersions where the particles have dimensions in the nanometer range. These particles can be kept suspended indefinitely due to Brownian motion of the solvent that keeps them from settling. Chemical means can also be used to keep colloids from separating. One method is to coat the tiny colloid particles with a charged molecule. Since each colloid particle will have the same charge (positive or negative) they will not tend to coalesce together. Such a colloidal dispersion can be "broken" by adding a substance that disrupts the coating or removes the charge. Acids, bases or salt solutions can break a colloidal dispersion. Dilute colloidal solutions may appear to be true solutions. However they exhibit the Tyndall effect as shown in Figure 12.24 in the text.

Self–Test

42. Identify the state (gas, liquid or solid) of the dispersed phase and the dispersing medium for the following.

    (a)   perfume in a bar of soap   _____   _____

    (b)   steam from a teapot   _____   _____

    (c)   moisturizing cream   _____   _____

    (d)   cream cheese   _____   _____

(e)   nail polish   _____   _____

(f)   peanut butter   _____   _____

43. How can you quickly test to see if a substance is a true solution or a colloidal mixture?

Answers to Self–Test Questions

1. (a) Both the tendency toward a more probable particle distribution and the intermolecular attractions between the molecules of these two polar liquids assist the formation of their solution.
(b) The hydration of the K^+ and Br^- ions is the major factor, but the tendency toward a more probable particle distribution is also important.
(c) When one gas intermixes in another, only the tendency toward a more probable particle distribution is operating.

2. Water molecules are attracted to each other far more strongly than they can be attracted to wax molecules, so the molecules of water remained separated in their own phase.

3. The very strong attractions between Na^+ and Cl^- ions within crystalline NaCl are sharply reduced when each kind of ion becomes surrounded and hydrated by water molecules.

4. Yes, *X* will dissolve in *Y*. There are no forces of attraction to inhibit the operation of the tendency toward a more probable particle distribution.

5. The strong forces of attraction between the Na^+ and NO_3^- ions in the crystalline salt can find no substitutes in new forces of attraction between these ions and the nonpolar molecules of gasoline.

6. (a), because it more closely resembles a hydrocarbon and will have intermolecular attractions that are similar to those in benzene.

7. Enthalpy of solution

8. The lattice energy is the energy released when separated ions or molecules attract each other and come together to form the crystalline lattice. When a solid dissolves, this energy must be overcome to separate the particles from each other, and is an endothermic contribution to the heat of solution. Therefore, the larger the lattice energy, the more likely the heat of solution will be positive.

9. Increases. Because it must receive outside energy to separate the ions. Hydration. Decreases. Because now energy is released as forces of attraction result in a lowering of the potential energy.

10. $\Delta H_{solution} = -112$ kJ/mol, exothermic. (It costs 436 kJ/mol to break up the crystal and 548 kJ/mol is returned, so the net is 112 kJ/mol released by the system.)

11. The endothermic steps are the separation of the molecules of the individual liquids from each other. The exothermic step is the intermingling of these separated molecules to form the solution.

12. The intermolecular forces of attraction in each of the separated liquids are identical to those in the solution and there is no heat of solution.

13. The forces of attraction between gas particles are zero or extremely small and little energy is needed to make room for the gas molecules in liquid water. Therefore, when a gas dissolves in water only the energy of solvation (which is always exothermic) is a factor. It essentially becomes the heat of solution.

14. (a) Exothermic, (b) Come out of solution

15. $Gas_{undissolved} \rightleftharpoons Gas_{dissolved} + heat$

16. $Gas_{undissolved} + heat \rightleftharpoons Gas_{dissolved}$

17. 9.14 mg/L

18. (a) 227 L HCl gas. (b) 0.0132 L N_2 gas. (c) HCl reacts almost completely with water to give H_3O^+ and Cl^-.
19. 81.6% ethyl alcohol and 18.4% water
20. 10.4 g
21. $\dfrac{1.364 \text{ mol NaCl}}{1000 \text{ g } H_2O}$ or $\dfrac{1000 \text{ g } H_2O}{1.364 \text{ mol NaCl}}$ (One could replace "1000 g H_2O" by 1 kg H_2O)
22. 147 g of H_2O
23. 16 g of glucose
24. Dissolve 27.5 g of sugar in water to make the final mass of the solution equal 500 g by adding water.
25. Weigh out 84.2 g of 10% H_2SO_4 solution.
26. 22.6% (w/w) C_6H_6
27. For $FeCl_3$, mole fraction = 0.0648; mole percent = 6.48%
    For H_2O, mole fraction = 0.9352; mole percent = 93.52%
28. 7.36 m LiCl
29. X_{LiCl} = 0.56 or 7.10 mol %, X_{water} = 0.929 or 92.9 mol %
30. 22.4% sugar
31. X_{NaCl} = 0.0203 or 2.03 mol %
32. 9.98% Na_2CO_3
33. P_{soln} = 23.4 torr
34. P_{total} = 45.1 torr
35. 137 g mol^{-1}
36. High concentration
37. Dialysis (Both sugar and water pass across the membrane, not just water, so it's dialysis, not osmosis.)
38. The salt forms a very concentrated solution on the surface of the meat and this draws water by dialysis from the less concentrated solution inside the meat and the meat cells.
39. 1.51×10^6 g mol^{-1}
40. −0.93 °C. −0.47 °C
41. Less. At the higher concentration, the ions would interact with each other more and so the solution would behave as if the solute were even less dissociated.
42. (a) liquid in a solid, (b) gas in a gas, (c) liquid in a liquid, (d) liquid in a solid, (e) solid in a liquid, (f) solid in a liquid
43. Shine the light from a laser pointer through the mixture and see if it exhibits the Tyndall effect (light scattering)

Tools for problem solving

In this chapter you learned to apply the following concepts as tools in solving problems. Study each one carefully so that you know what each is used for. When faced with solving a problem, recall what each tool does and consider whether it will be helpful in finding a solution.

You might want to tear these pages out to use along with solving problems in this chapter.

"Like dissolves like" rule (Section 12.1)

This rule uses the polarity and strengths of attractive forces, along with chemical composition and structure, to predict whether two substances can form a solution.

Henry's law (Section 12.4)

$$C_{gas} = k_H P_{gas} \quad \text{or} \quad \frac{C_1}{P_2} = \frac{C_2}{P_2}$$

Mass fraction; mass percent (Section 12.5)

$$\text{mass fraction} = \frac{\text{mass of solute}}{\text{mass of solution}}$$

$$\text{percentage by mass} = \%(w/w) = \frac{\text{mass of solute}}{\text{mass of solution}} \times 100\%$$

Molal concentration (Section 12.5)

$$\text{molality} = \frac{\text{moles of solute}}{\text{kg of solvent}}$$

Raoult's law (Section 12.6)

$$P_{solution} = X_{solvent}\, P^o_{solvent}$$

Raoult's law for two volatile solvents (Section 12.6)

$$P_{solution} = X_A\, P^o_A + X_B\, P^o_B$$

Freezing point depression and boiling point elevation   (Section 12.6)

Freezing point depression: $\Delta T_f = K_f m$

Boiling point elevation: $\Delta T_b = K_b m$

Osmotic pressure   (Section 12.6)

$\Pi V = nRT$ or $\Pi = MRT$

Notes:

Chapter 13

Chemical Kinetics

In this chapter we examine the factors that affect reaction rates and some ways of understanding how the factors work.

13.1   Factors that Affect the Rate of Chemical Change

Learning Objective

Understand and use the five conditions that affect how rapidly chemicals react.

Review

There are five factors that control the rate of a reaction.

1.   *The chemical nature of the reactants*: Some substances, because of their chemical bonds, just naturally react faster than others under the same conditions.

2.   *Ability of reactants to meet*: Many reactions are classified as *homogeneous reactions* because they are carried out in solution where the reactants can mingle on a molecular level. For *heterogeneous reactions*, at least two phases are present, and particle size is the controlling factor. For a given mass of reactant, the smaller the particle size, the larger the area of contact with the other reactants in other phases, and the faster the reaction is able to proceed.

3.   *Concentrations of the reactants*:   The rates of most reactions, both homogeneous and heterogeneous, increase with increasing reactant concentrations.

4.   *Temperature*: With very few exceptions, reactions proceed faster as their temperature is raised.

5.   *Catalysts*: The rates of many reactions are increased by the presence of substances called *catalysts*. A catalyst is a substance that is intimately involved in the reaction, but which is not used up during the reaction.

Example 13.1   Using the factors that affect reaction rates

If you wish to greatly increase the *rate* at which heat can be obtained from the combustion of coal, what might you do?

Analysis:

We need to determine what affects reaction rates and see if any applies to the given scenario.

Assembling the Tools:

Our tool is the five factors affecting reaction rates.

Solution:

We cannot do much about the chemical nature of the reactants and the first factor may not play an important role. The second consideration is the ability of the reactants to meet. Grinding the coal to dust particles will speed up combustion greatly. The third factor, increasing the concentration, works well for homogeneous reactions, but the heterogeneous burning of coal will have a limited effect unless we switch to air that has been enriched in oxygen. Raising the temperature will increase the reaction rate, but combustion reactions have a relatively high temperature already and little advantage may be gained at higher temperatures. Finally, adding a catalyst to the process can increase the reaction rate.

Are the Answers Reasonable?

Based on the information provided and the principles involved, a review of the answer shows that it is reasonable.

Self-Test

1. Why is it unlikely that the combustion of octane, C_8H_{18}, a component of gasoline, occurs by a single step as given by the equation

    $$2C_8H_{18}(g) + 25O_2(g) \longrightarrow 16CO_2(g) + 18H_2O(g)$$

    _____

2. Why would chemical manufacturers be interested in studying the factors that affect the rate of a reaction?

    _____

3. Elemental potassium reacts more rapidly with moisture than does sodium under the same conditions. Which of the factors discussed in this section is responsible for this?

    _____

4. Insects move more slowly in autumn than in the summer. Why?

    _____

5. What is one reason why coal miners are concerned about open flames during the mining operation?

    _____

6. Fire fighters are taught about the "fire triangle."

Eliminate any one of the three corners, and the fire is extinguished. Which factors discussed in this section affect the fire triangle?

_____

7. Why is pure oxygen more dangerous to work with than air?

_____

13.2   Measuring Reaction Rates

Learning Objective

Determine, from experimental data, the relative rates at which reactants disappear and products appear, and the rate of reaction, which is independent of the substance monitored.

Review

The term *kinetics* implies action or motion. For a chemical reaction, this motion is the speed at which the reaction takes place, the *rate of reaction,* and by that we mean the speed at which the reactants are consumed and the products formed.

The speeds of reactions range from very rapid to extremely slow. One of the benefits of studying reaction rates and the factors that control them is the insights we gain into the *mechanisms* of reactions—the individual chemical steps that produce the net overall change described by a balanced equation.

A *rate* is always a ratio in which a unit of time appears in the denominator. A reaction rate has units of molar concentration in the numerator, so the units of reaction rate are usually mol L^{-1} s^{-1} (mole per liter per second). The rate of a reaction generally changes (decreases) with time as the reactants are used up. The *instantaneous rate* can be measured at any particular instant by determining the slope of the concentration versus time curve as illustrated in Example 13.2 in the text.

The rate is usually measured by monitoring the substance whose concentration is most easily determined. Once we know the rate at which one substance is changing, we can calculate the rate for any other substance (study Example 13.1 in the text). This is because the relative rates of formation of the products and rates of disappearance of the reactants are related by the coefficients of the balanced equation. Notice that reaction rates are always given as positive quantities.

Example 13.2   Reaction rates of one substance can be used to calculate the rate of another

In the following reaction the rate of disappearance of O_2 is 6.34×10^{-5} mol O_2 L^{-1} s^{-1}. What is the rate of appearance of CO_2?

$$2C_8H_{18}(g) + 25O_2(g) \longrightarrow 16CO_2(g) + 18H_2O(g)$$

Analysis:

We can see that the units for the disappearance of O_2 are mol O_2 L^{-1} s^{-1}. The units for the appearance of CO_2 will be mol CO_2 L^{-1} s^{-1}. The only conversion needed is that mol O_2 needs to be converted to mol CO_2.

Assembling the Tools:

The tools we need are those for balancing equations and then determining mole-to-mole conversion factors.

Solution:

We perform the factor label conversion in one step based on the coefficients in the balanced equation.

$$6.34 \times 10^{-5} \text{ mol } O_2 \text{ } L^{-1} \text{ } s^{-1} \frac{16 \text{ mol } CO_2}{25 \text{ mol } O_2} = 4.06 \times 10^{-5} \text{ mol } CO_2 \text{ } L^{-1} \text{ } s^{-1}$$

The rate of appearance of CO_2 is, 4.06×10^{-5} mol CO_2 L^{-1} s^{-1}

Is the Answer Reasonable?

Cancel the units in the equation above to verify that the conversion was done properly. The ratio of 16/25 is less than one and our answer is less than the quantity we started with. These factors indicate the problem is solved correctly.

Self-Test

8.   In Figure 13.5 in the text, what is the instantaneous rate at which HI is reacting at $t = 200$ seconds?

9.   In Figure 13.5 in the text, what is the instantaneous rate at which H_2 is forming at $t = 200$ s?

10. In the combustion of octane,

$$2C_8H_{18}(g) + 25O_2(g) \longrightarrow 16CO_2(g) + 18H_2O(g)$$

what would be the rate of formation of CO_2 if the concentration of octane was changing at a rate of 0.25 mol L^{-1} s^{-1}?

_____

11. Referring to the preceding question, if the rate of formation of H_2O is 0.900 mol L^{-1} s^{-1}, what is the rate of reaction of C_8H_{18}?

_____

At what rate will O_2 be reacting under these conditions?

_____

12. If octane is burning at a constant rate of 0.25 mol s^{-1}, how many moles of it will burn in 15 minutes?

_____

13.3   Rate Laws

Learning Objective

Use experimental initial rate data to determine rate laws.

Review

Concentration and rate are related by the *rate law* for a reaction. For example, for the reaction

$$xA + yB \longrightarrow \text{products}$$

in which x and y are coefficients of reactants A and B, respectively, the rate law will be of the form

$$\text{rate} = k[A]^n[B]^m$$

Remember that square brackets, [ ], around a chemical formula stand for the molar concentration (with the units of mol L^{-1}) of the substance. The proportionality constant, k, is the *rate constant*. The exponents give the *order of the reaction* with respect to each reactant, and their sum is the *overall order of the reaction*. Once you have the rate law for a reaction, it can be used to calculate the rate for any given set of reactant concentrations for the temperature at which k was determined. Keep in mind that k varies with temperature, and therefore, so does the reaction rate.

It is very important to remember that the exponents (n and m in this example) are not necessarily equal to the coefficients x and y, and can only be known for sure if they are determined from experimentally measured data. These data have to show how the rate changes when the concentrations change. In analyzing data like those in Table 13.2 of the text, observe how the rate changes when the concentration of one of the reactants changes while the concentrations of the other reactants are held constant.

Study Table 13.3 in the text and note how it applies to Examples 13.5 and 13.6. After working Practice Exercises 13.11, 13.12 and 13.13, try the following Self-Test.

Example 13.3   Determining a rate law

Using the information in the table below determine the rate law and the rate constant (with units) for the reaction

$$2NO(g) + O_2(g) \longrightarrow 2NO_2(g)$$

Initial Concentration (M)		Initial Rate
O_2	NO	(mol L^{-1} s^{-1})
0.00173	0.00347	3.29×10^{-6}
0.00345	0.00347	6.55×10^{-6}
0.00173	0.0139	5.24×10^{-5}

Analysis:

We need to determine how a change in concentration affects the reaction rate. This can be translated into an appropriate exponent for the substance in the rate law. We start by writing the general rate law for these reactants as Rate = $k[O_2]^x[NO]^y$. We need to see how one reactant affects the rate while the other is held constant. We have just discussed that procedure.

Assembling the Tools:

Our tool is the general form of the rate law and the methods that can be used to deduce exponents for the rate law.

Solution:

To determine the exponents x and y we construct a ratio of two rate laws using data from two different rows in our table.

$$\frac{\text{Rate}_1}{\text{Rate}_2} = \frac{k[O_2]_1^x[NO]_1^y}{k[O_2]_2^x[NO]_2^y}$$

When data are entered into the above ratio, we will be able to identify the values of x and y. Using the procedure outlined above we enter the data from the first two lines of the table as (units have been omitted for clarity)

$$\frac{3.29 \times 10^{-6}}{6.55 \times 10^{-6}} = \frac{k[0.00173]^x[0.00347]^y}{k[0.00345]^x[0.00347]^y}$$

From this ratio we cancel the rate constants, k, and the $[0.00347]^y$ terms to get

$$\frac{3.29 \times 10^{-6}}{6.55 \times 10^{-6}} = \frac{[0.00173]^x}{[0.00345]^x}$$

Evaluating this ratio we get $0.4977 = (0.501)^x$. From this we conclude that x must be 1. (Note: both 0.4977 and 0.501 are very close to 0.5.)

To determine the value of y we substitute rows 1 and 3 of the table to get

$$\frac{3.29 \times 10^{-6}}{5.24 \times 10^{-5}} = \frac{k[0.00173]^x[0.00347]^y}{k[0.00173]^x[0.0139]^y}$$

Canceling the identical terms we get $0.0628 = (0.250)^y$. Trying various integers for y we find that $(0.250)^2 = 0.0625$ and conclude that $y = 2$ and the rate law is Rate $= k [SO_2]^1[O_2]^2$.

We can now calculate the rate constant by substituting the data from any row in the table into the rate law. Using the data from row 1 we get

$$3.29 \times 10^{-6} \text{ mol L}^{-1} \text{ s}^{-1} = k (0.00173 \text{ mol L}^{-1})^1 (0.00347 \text{ mol L}^{-1})^2$$

Solving for k we obtain $154 \text{ L}^2 \text{ mol}^{-2} \text{ s}^{-1}$.

Are the Answers Reasonable?

We can recheck our canceling and the mathematical calculations. Another check is to calculate the rate constant for several lines in the table. If the rate law is correct, all rate constants will be close to each other. For the rate constant the units seem correct for a third–order reaction and an estimate of the result agrees with the answer obtained.

Self-Test

13. What is the order of the reaction for the NO in Example 13.6 in the text?

   _____

14. In Practice Exercise 13.14, what is the order of the reaction with respect to A and B, and what is the overall order of the reaction?

   _____

15. What are the units for the rate constant in the rate equation developed in Question 14?

   _____

16. When the concentration of a particular reactant was increased by a factor of 10, the rate of the reaction was increased by a factor of 1000. What is the order of the reaction with respect to that reactant?

   _____

17. For a reaction, $2A + D \longrightarrow$ products, the following data were obtained:

Initial Concentration (M)		Initial Rate
A	B	(mol L^{-1} s^{-1})
0.213	0.126	1.64×10^{-2}
0.426	0.126	3.27×10^{-2}
0.213	0.504	2.62×10^{-1}

(a) What is the rate law for the reaction? _____

(b) What is the value of the rate constant? (With correct units, too) _____

13.4   Integrated Rate Laws

Learning Objective

Use the basic results of integrated rate laws to determine the order of a reaction and calculate the time dependence of concentration for zero-, first-, and second-order reactions.

Review

For a first-order reaction, the rate constant can be obtained from a graph of the *natural logarithm* of the reactant concentration, ln[A], versus time. The slope of the line equals the rate constant. Alternatively, Equations 13.5 and 13.6 relate concentration and time to the rate constant for a first–order reaction.

Working with natural logarithms

To use Equations 13.5 and 13.6, you have to work with natural logarithms and their antilogarithms. These are easily handled on your calculator. To obtain the natural logarithm of a number, enter the number and then press the key labeled "ln *x*." (On some calculators, the key may have a different label; check your instruction manual.) The value that appears is the natural logarithm of the number. For example, the natural logarithm of 25.6 is 3.243.

$$\ln 25.6 = 3.243$$

Use this to check to see that you're using your calculator correctly. Notice that *the number of digits following the decimal point equals the number of significant figures in the number whose logarithm you're determining.* That's the rule for logarithms and significant figures.

To take the antilogarithm (antiln), you use the key labeled e^x on your calculator. (On some calculators, you press an "inverse" key followed by the "ln x" key.) For example, the antiln of 41.69 is 1.3×10^{18}.

$$\text{Antiln}(41.69) = e^{41.69} = 1.3 \times 10^{18}$$

Notice that when we take the anti-natural logarithm, the result has only two significant figures — the number of digits after the decimal in 41.69.

Applying the integrated rate laws

Finding the concentration after a specified time is illustrated in Example 13.7. Notice that we calculate the logarithm of the concentration ratio from k and t. Taking the antilogarithm gives the numerical value for the ratio. Then we substitute the known concentration and solve for the concentration we want to find.

Finding the time required for the concentration to drop to some particular value is simple. Just substitute the initial and final concentrations into the concentration ratio, take the logarithm of the ratio, substitute the value of the rate constant, and then solve for t. Notice that the units of t are of the same kind as the units of the rate constant; if k has units of s^{-1}, then t will be in seconds; if k has units of hr^{-1}, then t will be in hours.

For second-order reactions, a graph of the reciprocal of the concentration versus time gives a straight line. The slope of this line equals k. Calculations for second-order reactions are done with Equation 13.10. This equation doesn't involve logarithms, so it is easier to manipulate algebraically. Study Example 13.10 to learn how to use it.

Half-lives

The *half-life* of a reaction, $t_{1/2}$, is the time required for half of a given reactant to disappear. For a first-order reaction, $t_{1/2}$ is independent of the initial reactant concentration. The value of $t_{1/2}$ is inversely proportional to the initial reactant concentration for a second-order reaction.

You should be able to use Equations 13.7 and 13.10, and you should be able to use the way $t_{1/2}$ varies with initial concentration to determine whether a reaction is first- or second-order.

Example 13.4   Finding initial concentrations using integrated rate laws

The rate constant for a certain first-order reaction is 3.64×10^{-4} s^{-1}. If the concentration of the reactant Q in a reaction vessel is measured to be 8.54×10^{-6} mol L^{-1} after three hours and fourteen minutes, what was the concentration of Q when the reaction started?

Analysis:

Our task is to calculate what the starting concentration was when the concentration after 3 hours and 14 minutes of reaction was measured. This requires the integrated rate law. We review the information provided and see that if we convert the time to seconds we have all the information needed to obtain the answer.

Assembling the Tools:

The tool for solving this type of problem is the integrated rate equation for a first-order process. One form of the equation is

$$\ln[Q]_0 - \ln[Q]_t = kt$$

Solution:

We will first convert the time to 194 minutes and then to 1.164×10^4 seconds. Entering the data into the equation we get

$$\ln[Q]_0 - \ln(8.54 \times 10^{-6} \text{ mol L}^{-1}) = (3.64 \times 10^{-4} \text{ s}^{-1})(1.164 \times 10^4 \text{ s})$$

$$\ln[Q]_0 = 4.24 - 11.67 = -7.43$$

$$[Q]_0 = 5.9 \times 10^{-4} \text{ mol L}^{-1}$$

Is the Answer Reasonable?

Most important is that the initial concentration of Q is larger than the value at three hours and fourteen minutes after the reaction starts. We can recheck our setup and the calculations for more assurance.

Self-Test

18. The decomposition of SO_2Cl_2 is a first-order reaction. In an experiment, it was found that the SO_2Cl_2 concentration was 0.034 M after 145 min. The value of k at the temperature at which the experiment was performed is 1.7×10^{-5} s^{-1}. How many *moles* of the reactant were originally present in the apparatus, which has a volume of 350 mL?

19. A certain reactant disappears by a first-order reaction that has a rate constant $k = 4.7 \times 10^{-3}$ s^{-1}. If the initial concentration of the reactant is 0.700 mol/L, how long will it take for the concentration to drop to 0.050 mol/L?

20. For the reactant described in the preceding question, what will the concentration of the reactant be after 64.0 min if its initial concentration is 1.30 mol L^{-1}?

21. A certain reaction follows the stoichiometry

$$2D \longrightarrow \text{products}$$

It has the rate law Rate = $k[D]^2$. The rate constant for the reaction equals 6.3×10^{-3} L mol^{-1} s^{-1}.

(a) How many seconds will it take for the concentration of D to drop from 0.864 mol/L to 0.0768 mol/L?

(b) If the initial concentration of D is 0.255 mol/L, what will its concentration be after 2.54 min?

22. In a certain reaction, the half-life of a particular radioactive isotope is 30 minutes. The initial concentration of the isotope is 2.0 μmol/L, and the reaction is first-order.

    (a) What will be the concentration of the reactant after 2.5 hours?

    _____

    (b) How long will it take for the concentration to be reduced to 0.0312 μmol/L?

    _____

23. A certain first-order reaction has $t_{1/2} = 8273$ years. What is the rate constant for the reaction with units of s^{-1}?

    _____

24. The decomposition of SO_2Cl_2 has $k = 2.2 \times 10^{-5}$ s^{-1} (Example 13.4). What is $t_{1/2}$ for this reaction in seconds and in minutes?

    _____

25. At a certain temperature the half-life for the decomposition of N_2O_5 was 350 seconds when the N_2O_5 concentration was 0.200 mol/L. When the concentration was 0.400 mol/L, the half-life was 5.83 minutes. Is this decomposition reaction first- or second-order?

    _____

13.5 Molecular Basis of Collision Theory

Learning Objective

Explain the rate of chemical reactions based on a molecular view of collisions that includes frequency, energy, and orientation which make up collision theory.

Review

Collision theory postulates that the rate of a reaction is proportional to the number of effective collisions per second between the reactant molecules or ions. The number of *effective* collisions per second is less than the total number of collisions per second for two principal reasons:

1. For some reactions, it is important that the reactant molecules be in the correct *orientation* when they collide.

2. A minimum kinetic energy, called the *activation energy* (E_a), must be possessed by the reactant molecules in a collision to overcome the repulsions between their electron clouds and thereby permit the electronic rearrangements necessary for the formation of new product molecules.

13.6   Molecular Basis of Transition State Theory

Learning Objective

Describe the basics of transition state theory including activated complexes and potential energy diagrams.

Review

In *transition state theory*, we postulate the formation of a high-energy complex formed from the reactant particles by means of a collision, a complex called the *activated complex,* which can collapse into particles of the products as bonds break and form. We follow the energy of the reactants as they are transformed to the products by means of a graph on which the horizontal axis, the *reaction coordinate*, indicates the progress of the reaction and the vertical axis measures the changes in potential energy as the collision occurs. The potential energy at the *transition state* corresponds to a maximum in the curve, and the potential energy change needed to reach the maximum is the *activation energy, E_a*. Study Figures 13.12 to 13.16. On any progress-of-reaction diagram, you should be able to identify the activation energy for both the forward and reverse reactions, the potential energies of the reactants and products, and the *heat of reaction*. You should also be able to locate the transition state on the diagram.

Self-Test

26. Without peeking at the text, sketch and label the potential energy diagram for

   (a)   an exothermic reaction.

   (b)   an endothermic reaction.

27. Where is the transition state located on the energy diagram for a reaction?

   _____

28. One step in the reaction $NO_2(g) + CO(g) \longrightarrow NO(g) + CO_2(g)$ is believed to involve the collision of two NO_2 molecules to give NO and NO_3 molecules

   $$NO_2 + NO_2 \longrightarrow NO + NO_3$$

   What might be a reasonable structure for the activated complex in this collision?

29. On the basis of the transition state theory, explain why the temperature of a collection of molecules rises as an exothermic reaction occurs within it.

   _____

   _____

30. Why do reactions having a low activation energy usually occur faster than ones having a high activation energy (assuming no special requirements for molecular orientations)?

   _____

13.7   Activation Energies

Learning Objective

Use the Arrhenius equation to determine the activation energy of a reaction

Review

The activation energy is related to the rate constant by the *Arrhenius equation*,

$$k = Ae^{-E_a/RT}$$

where A, a proportionality constant, is called the *frequency factor* or *pre-exponential factor*. As described in Example 13.12, plotting the natural logarithm of the rate constant versus the reciprocal of the absolute temperature yields a straight line whose slope is equal to $-E_a/R$. After studying Example 13.12, work Question 34 of the Self-Test at the end of this section. To do that, set up a table with headings of "ln k" and "$1/T$" and compute these values from the data given. Then choose a piece of graph paper, plot the data, measure the slope, and calculate E_a, or set up a spreadsheet with your data and use the plotting function to obtain the graph and the slope.

To calculate E_a from rate constants at two different temperatures (which is actually less accurate than the graphical procedure) you will need to know Equation 13.14. Example 13.16 shows how this equation is used, and you should study them carefully before beginning the Practice Exercises.

In working with Equation 13.16, it is helpful to note that if a negative value is obtained for E_a when you calculate the activation energy, you probably interchanged the 1 and 2 subscripts on either the rate constants or the temperatures. The only effect that such an error will have on the computed E_a is to change its sign. Since the activation energy must be positive, just change its sign to positive.

When you use the activation energy to calculate a rate constant at some temperature, given the rate constant at some other temperature, keep in mind that the value of k is always larger at the higher temperature. After finishing the calculation, make sure your values of k fit this rule. If they don't, then you switched the subscripts 1 and 2 on the k's in the ratio of rate constants.

A final point to be especially careful about in these calculations is to use Kelvin temperatures in Equation 13.16, not Celsius temperatures.

Example 13.5   Estimating the activation energy

It is often said that the rate of a reaction doubles for each ten degree increase in temperature. What is the activation energy for a reaction that follows this maxim? A certain reaction is found to have a rate constant of 2.68×10^{-5} s^{-1} at 20 °C and a rate constant of 10.72×10^{-5} at 40 °C. What is the activation energy for this process? What is the activation energy of a process that doubles its reaction rate with each 10 °C increase in temperature?

Analysis:

You could make a small table in a spreadsheet program after converting the data to ln k and $1/T$. These data can be graphed and the slope, the activation energy divided by R, can be calculated by the program. In this case, we will use Equation 13.16.

Assembling the Tools:

The tool to use is the Arrhenius equation in either of the two forms presented in the text.

Solution:

Equation 13.12 is $\ln \dfrac{k_1}{k_2} = \dfrac{E_a}{R} \left(\dfrac{1}{T_2} - \dfrac{1}{T_1} \right)$. All we need to do is convert the Celsius temperatures to their Kelvin scale and insert the data to get (the appropriate value of R is 8.314 J mol^{-1} K^{-1} since we want the answer in energy units of Joules)

$$\ln \frac{2.68 \times 10^{-5}}{10.72 \times 10^{-5}} = \frac{E_a}{8.314 \text{ J mol}^{-1} \text{ K}^{-1}} \left(\frac{1}{313 \text{ K}} - \frac{1}{293 \text{ K}} \right)$$

$$-1.386 = (-2.62 \times 10^{-5} \text{ kJ}^{-1} \text{ mol}) E_a$$

$$E_a = 52.8 \text{ kJ mol}^{-1}$$

The reaction rate increased by a factor of four when the temperature was increased by 20 °C. This is the same as doubling every 10 °C. Therefore, reactions that double their rates with a ten–degree rise in temperature have activation energies of approximately 50 kJ.

Are the Answers Reasonable?

First, the activation energy must have a positive sign and our result does have a positive sign. Next, the units cancel as expected and the math checks out to be correct.

Self-Test

31. What is the natural logarithm of these quantities? (Express the answers to the correct number of significant figures.)

    (a)   ln 19.34 = _____   (b) ln 7.83 × 10^{-7} = _____

32. What is the natural antilogarithm of these quantities? (Express the answers to the correct number of significant figures.)

    (a)   antiln 6.445 = _____   (b) antiln −2.3450 = _____

33. A certain first-order reaction has a rate constant $k = 1.0 \times 10^{-2}$ s^{-1} at 30 °C. At 40 °C its rate constant is $k = 2.5 \times 10^{-2}$ s^{-1}. Calculate the activation energy for this reaction in kJ/mol.

    _____

34. A certain second-order reaction has an activation energy of 105 kJ/mol. At 25 °C the rate constant for the reaction has a value of 2.3×10^{-3} L mol^{-1} s^{-1}. What would its rate constant be at a temperature of 45 °C?

35. In the table below are tabulated values of the rate constant for a reaction at 5 °C intervals from 25 °C to 100 °C. Use these data to determine the activation energy by the graphical method.

Temperature(°C)	k (s^{-1})	Temp (°C)	k (s^{-1})
25	0.0240	65	0.206
30	0.0324	70	0.259
35	0.0433	75	0.325
40	0.0573	80	0.406
45	0.0751	85	0.502
50	0.0978	90	0.619
55	0.126	95	0.757
60	0.162	100	0.922

13.8 Mechanisms of Reactions

Learning Objective

Use the concepts of reaction mechanisms to recognize reasonable mechanisms and suggest plausible mechanisms given experimental data.

Review

Usually, a net overall reaction occurs by a *mechanism* involving a sequence of simple *elementary processes*, the slowest of which determines how fast the products are able to form. This is the *rate-determining step*, sometimes called the *rate-limiting step*, and the rate law of the overall reaction is the same as the rate law for the rate-determining step.

If we know the stoichiometry for an elementary process, we can predict its rate law; the coefficients of the reactants are equal to their exponents in the rate law. Remember, however, that this works *only* if the elementary process is known. When we first begin to study a reaction, we don't know what its mechanism is, so we can't predict with any hope of confidence what the exponents in the rate law will be.

Determining a mechanism involves guessing what the elementary processes are and then comparing the predicted rate law, based on the mechanism, with the rate law determined from experimental data. If the two rate laws match, the mechanism may be correct. If they don't match, the search for a mechanism must continue. Generally, successful collisions involving more than two "bodies" are so improbable that chemists almost never postulate them in devising possible mechanisms, which means that bimolecular collisions are highly probable but trimolecular collisions are very rare.

Self-Test

36. Suppose an elementary process is $2A + M \longrightarrow P + Q$. What is the rate law for this step?

_____

37. The following mechanism has been proposed for the reaction

$$(CH_3)_3CBr + OH^- \longrightarrow (CH_3)_3COH + Br^-$$

Step 1. $(CH_3)_3CBr \longrightarrow (CH_3)_3C^+ + Br^-$ (slow)

Step 2. $(CH_3)_3C^+ + OH^- \longrightarrow (CH_3)_3COH$ (fast)

If this mechanism is correct, what is the expected rate law for the overall reaction?

_____

38. What would be the rate law for the overall reaction in the preceding question if it occurred in a single step (i.e., if the overall reaction were actually an elementary process)?

_____

13.9   Catalysts

Learning Objective

Relate the properties of homogeneous and heterogeneous catalysts and how they act to increase reaction rates.

Review

Catalysts open alternative pathways (mechanisms) for reactions. These paths have lower activation energies than the uncatalyzed mechanisms, so catalyzed reactions occur faster. A *homogeneous catalyst* is in the same phase as the reactants; a *heterogeneous catalyst* is in a different phase than the reactants. A homogeneous catalyst, like the NO_2 used in the lead chamber process, is consumed in one step of the mechanism and then regenerated in a later step. A heterogeneous catalyst functions by *adsorption* of the reactants on its surface where the reactants are able to react with a relatively low activation energy. Be sure you know the difference between adsorption and absorption: adsorption occurs when a substance becomes bound to a surface; absorption occurs when a substance is taken into the absorbing medium (like a sponge absorbs water).

Self-Test

39. How do catalyst "poisons" work? _____

_____

Answers to Self-Test Questions

1. It would require the simultaneous collision of 27 molecules, which is an unlikely event.
2. By adjusting conditions, they can make their reactions go faster and more efficiently.
3. Nature of the reactants
4. Their biochemical reactions are slower in the cooler weather.
5. The possibility of a coal dust explosion.
6. Ability of reactants to meet; the effect of temperature.
7. The O_2 is less concentrated in air than in pure O_2.
8. Approximately 1.2×10^{-4} mol L^{-1} s^{-1} of HI is disappearing
9. 0.6×10^{-4} mol L^{-1} s^{-1} of H_2 is being produced
10. 2.0 mol L^{-1} s^{-1}
11. 0.100 mol C_8H_{18} L^{-1} s^{-1}, 1.25 mol O_2 L^{-1} s^{-1}
12. 225 mol C_8H_{18}
13. second-order
14. second-order with respect to both A and B, fourth-order overall.
15. L^3 mol^{-3} s^{-1}
16. third-order
17. (a) rate $= k[A]^1 [B]^2$ (b) $k = 4.9$ L^2 mol^{-2} s^{-1}
18. 0.014 mol SO_2Cl_2
19. 561 seconds or 9.4 minutes
20. 1.8×10^{-8} mol/L
21. (a) 384 seconds, (b) 0.098 mol/L
22. (a) 0.0625 M, (b) 6 half-lives = 3.0 hours
23. 2.6×10^{-12} s^{-1}
24. 3.2×10^4 s = 525 min =8.75 hr
25. second–order, only second–order half–lives depend on the initial concentration while first–order half–lives are constants.
26. (a) see Figure 13.12 (b) see Figure 13.14
27. At the highest point on the energy curve.
28.
```
    O                O
     \\              //
      N --- O --- N
     /
    O
```
29. As the potential energy decreases, the average kinetic energy rises, which means the temperature rises.
30. When E_a is small, a large fraction of molecules have at least this minimum energy that they need to react.
31. (a) 2.9621, (b) −14.060
32. (a) 605, (b) 0.09585
33. $E_a = 72.2$ kJ mol^{-1}
34. 3.3×10^{-2} L mol^{-1} s^{-1}

35. Data to be graphed:

ln k	1/T	ln k	1/T
−3.73	0.00335	−1.58	0.00296
−3.43	0.00329	−1.35	0.00291
−3.14	0.00324	−1.12	0.00287
−2.86	0.00319	−0.902	0.00283
−2.59	0.00314	−0.689	0.00279
−2.33	0.00309	−0.480	0.00275
−2.07	0.00305	−0.278	0.00272
−1.82	0.00300	−0.0809	0.00268

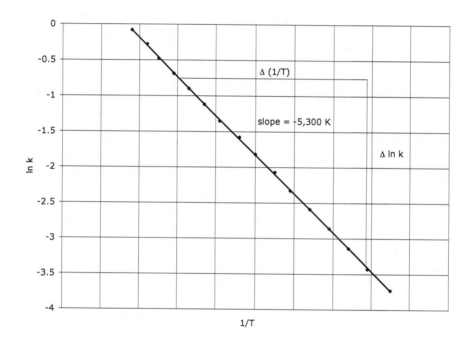

ARRHENIUS PLOT

$\Delta(1/T) = 0.00330 - 0.00280 = 0.00050$

$\Delta(\ln k) = -3.40 - (-0.75) = -2.65$

$$\text{slope} = \frac{\Delta(\ln k)}{\Delta(1/T)} = \frac{2.65}{5.0 \times 10^{-4} K^{-1}} = -5.3 \times 10^3 \text{ K} = \frac{E_a}{R} \text{ (Note units for the slope.)}$$

Using $R = 8.314$ J mol^{-1} K^{-1}, $E_a = -8.314(-5.3 \times 10^3) = 44 \times 10^3$ J/mol $= 44$ kJ/mol.

The activation energy is approximately 44 kJ/mol.

36. Rate = $k[A]^2 [M]$ 37. Rate = $k[(CH_3)_3CBr]$

38. rate = $k[(CH_3)_3CBr] [OH^-]$

39. The poison becomes attached to the catalyst's surface and prevents the reactants from reaching the catalytic site.

Tools for problem solving

In this chapter you learned to apply the following concepts as tools in solving problems. Study each one carefully so that you know what each is used for. When faced with solving a problem, recall what each tool does and consider whether it will be helpful in finding a solution.

You might want to tear these pages out to use along with solving problems in this chapter.

Factors that affect reaction rates (Section 13.1)

The five factors that affect the rate of reactions are (1) the chemical nature of the reactants; (2) the ability of the reactants to come in contact with each other; (3) the concentrations of the reactants; (4) the temperature of the reaction; and (5) the availability of catalysts.

Rate with respect to one reactant or product (Section 13.2)

$$\text{rate} = \frac{\Delta(\text{conc. of } X)}{\Delta t}$$

Rates are always expressed as a positive value, so if the concentration is decreasing for a reactant, the sign is changed to the positive.

Rate of a reaction (Section 13.2)

For the reaction $\qquad aA + bB \longrightarrow cC + dD$

The rate expression is

$$\text{rate} = -\frac{1}{a}\frac{\Delta[A]}{\Delta t} = -\frac{1}{b}\frac{\Delta[B]}{\Delta t} = \frac{1}{c}\frac{\Delta[C]}{\Delta t} = \frac{1}{d}\frac{\Delta[D]}{\Delta t}$$

Rate law of a reaction (Section 13.3)

$$\text{rate} = k\,[A]^n\,[B]^m$$

Determining rate laws (Section 13.3)

$$\frac{\text{rate}_2}{\text{rate}_1} = \frac{k[A]_2^n[B]_2^m}{k[A]_1^n[B]_1^m} = \frac{k}{k}\left(\frac{[A]_2}{[A]_1}\right)^n\left(\frac{[B]_2}{[B]_1}\right)^m$$

Integrated first-order rate law (Section 13.4)

$$\ln\frac{[A]_0}{[A]_t} = kt$$

Half-lives of a first–order reaction (Section 13.4)

$$\ln 2 = kt_{1/2}$$

Carbon-14 dating (Section 13.4)

The half-life for carbon-14 is 5730 years and the rate constant is 1.21×10^{-4} yr^{-1}. The age of an object can be found by comparing the ratio of $^{14}C/^{12}C$, r_t, from an object in question to the ratio of $^{14}C/^{12}C$ in a contemporary sample, r_0,

$$\ln\frac{r_0}{r_t} = (1.21 \times 10^{-4}\ yr^{-1})t$$

Integrated second-order rate law (Section 13.4)

For a second-order reaction of the form, Rate $= k[B]^2$, with known k, this equation is used to calculate the concentration of a reactant at some specified time after the start of the reaction, or the time required for the concentration to drop to some specified value.

$$\frac{1}{[B]_t} - \frac{1}{[B]_0} = kt$$

Integrated zero-order rate law (Section 13.4)

For a zero-order reaction of the form, Rate $= k$, the rate is independent of the concentration of the reactants. The integrated rate law is

$$[A_t] = -kt + [A]_0$$

Arrhenius equation (Section 13.7)

$$k = Ae^{E_a/RT}$$

Arrhenius equation alternate form (Section 13.7)

$$\ln\frac{k_2}{k_1} = \frac{-E_a}{R}\left(\frac{1}{T_2} - \frac{1}{T_1}\right)$$

Summary of Important Equations

Concentration versus time, first–order reaction:

$$\ln \frac{[A]_0}{[A]_t} = kt$$

Concentration versus time, second–order reaction:

$$\frac{1}{[B]_t} - \frac{1}{[B]_0} = kt$$

Half-life, first–order reaction:

$$t_{1/2} = \frac{\ln 2}{k}$$

Half-life, second–order reaction:

$$t_{1/2} = \frac{1}{k[B]_0}$$

Arrhenius equation:

$$k = Ae^{-E_a/RT}$$

$$\ln \left(\frac{k_2}{k_1} \right) = \frac{E_a}{R} \left(\frac{1}{T_2} \frac{1}{T_1} \right)$$

Remember, if you mix up the 1's and 2's, the only effect will be to change the sign of E_a. However, E_a must be positive. Also, k is always larger at the higher temperature.

Notes:

Chapter 14

Chemical Equilibrium

In Chapter 13 you learned the factors that determine how fast reactions are able to proceed. In this chapter we explore the fate of most chemical reactions, namely, dynamic equilibrium. For most reactions, the concentrations eventually level off at constant values, which are the equilibrium concentrations in the reaction mixture.

The principles we develop here will be used in the next several chapters as we study equilibria in solutions of acids and bases and solubility equilibria. For this reason, it is important that you learn well the topics discussed here, especially how we use the balanced chemical equation to construct the *equilibrium law* for the reaction. Also, study carefully the approach to solving equilibrium problems.

14.1 Dynamic Equilibrium in Chemical Systems

Learning Objective

Distinguish, describe, and explain the nature of a dynamic equilibrium.

Review

Equilibrium is established in a chemical system when the rate at which the reactants combine to form the "products" is equal to the rate at which the products react to form the "reactants." We call it a dynamic equilibrium because the reaction hasn't ceased; instead there are two opposing reactions occurring at equal rates.

When equilibrium is reached in a chemical system, the concentrations of the reactants and products attain steady, constant values that do not change with time. Because the reaction is proceeding in both directions simultaneously, the terms *reactants* and *products* no longer have their usual meanings. Instead, we use the term *reactants* to mean the substances on the left of the equilibrium arrows and the term *products* to mean the substances on the right of the arrows.

In the example shown in Figure 14.2 in the text 0.0350 mol N_2O_4 and 0.0700 mol NO_2 each contain the same total amount of nitrogen and oxygen. The two "initial" systems differ in how the nitrogen and oxygen atoms are distributed among the molecules. When these gases are allowed to come to equilibrium from either direction, the same equilibrium composition of NO_2 and N_2O_4 is reached. This demonstrates that for a given *overall* composition, the same equilibrium composition will always be reached regardless of whether we begin with reactants, products, or a mixture of them. The reaction can go in either direction, and the direction in which it proceeds is determined by how the concentrations must change in order to become equal to the appropriate equilibrium concentrations.

Self–Test

1. If we were able to follow a particular carbon atom in a solution of $HC_2H_3O_2$, part of the time it would exist in an acetate ion, $C_2H_3O_2^-$, and part of the time it would exist in a molecule of acetic acid, $HC_2H_3O_2$. Why is this so?

    _____

    _____

    _____

2. Explain how the concept of a dynamic equilibrium differs from the resonance concept in Section 8.8.

    _____

    _____

    _____

14.2   Equilibrium Laws

Learning Objective

Explain the basics of equilibrium laws.

Review

As you learned in Chapter 13, the molar concentration of a substance is represented symbolically by placing the formula for the substance between brackets. Thus, $[CO_2]$ stands for the molar concentration (in units of mol L^{-1}) of CO_2.

The *mass action expression* for a reaction is a fraction. Its numerator is constructed by multiplying together the molar concentrations of the *products*, each raised to a power that is equal to its coefficient in the balanced chemical equation for the equilibrium. The denominator is obtained by multiplying together the molar concentrations of the *reactants*, each raised to a power equal to its coefficient. For example, for the reaction

$$2CO(g) + O_2(g) \rightleftharpoons 2CO_2(g)$$

the mass action expression is

$$\frac{[CO_2]^2}{[CO]^2[O_2]}$$

The numerical value of the mass action expression is called the *reaction quotient*, Q, and at equilibrium, the reaction quotient has a value that we call the *equilibrium constant* K_c. Remember that the "c" in K_c means that the mass action expression is written with molar concentrations. Equating the mass action expression to the equilibrium constant gives the *equilibrium law* for the reaction. Thus, for the reaction above, the equilibrium law is

$$\frac{[CO_2]^2}{[CO]^2 [O_2]} = K_c$$

For any given reaction, the numerical value of K_c varies with temperature. But at a given temperature, the reaction quotient will always be the same when the system is at equilibrium, regardless of the individual concentrations. An important point is that there are no restrictions on individual equilibrium concentrations. They can have *any* values as long as they satisfy the equilibrium law when the system is at equilibrium.

Manipulating equations for chemical equilibria

The following rules apply to calculating the equilibrium constant when we manipulate chemical equilibria. Notice that the rules are different than the ones we used when we applied Hess's law. (Work Practice Exercises 14.4 and 14.5.)

- Change the direction of a reaction
  Take the reciprocal of K

- Multiply by a factor
  Raise K to a power equal to the factor

- Add two equations
  Multiply their K's

Self–Test

3. Write the equilibrium laws for the following reactions:

    (a) $CH_3OH(l) + CH_3CO_2H(l) \rightleftharpoons CH_3CO_2CH_3(l) + H_2O(l)$

    _____

    (b) $2HCrO_4^-(aq) \rightleftharpoons Cr_2O_7^{2-}(aq) + H_2O(l)$

    _____

    (c) $2N_2O(g) + 3O_2(g) \rightleftharpoons 4NO_2(g)$

    _____

    (d) $C_2H_4(g) + H_2(g) \rightleftharpoons C_2H_6(g)$

    _____

4. Write the equilibrium law for the following reaction:

    $$4NH_3(g) + 7O_2(g) \rightleftharpoons 4NO_2(g) + 6H_2O(g)$$

    _____

5. We saw that at 440 °C, $K_c = 49.5$ for the reaction

    $$H_2(g) + I_2(g) \rightleftharpoons 2HI(g)$$

    Which of the following mixtures are at equilibrium at 440 °C?

    (a) $[H_2] = 0.0122\ M$, $[I_2] = 0.0432\ M$, $[HI] = 0.154\ M$

    (b) $[H_2] = 0.708\ M$, $[I_2] = 0.0115\ M$, $[HI] = 0.635\ M$

    (c) $[H_2] = 0.728\ M$, $[I_2] = 0.0145\ M$, $[HI] = 0.435\ M$

    (d) $[H_2] = 0.0243\ M$, $[I_2] = 0.0226\ M$, $[HI] = 0.165\ M$

    (e) $[H_2] = 0.0708\ M$, $[I_2] = 0.00115\ M$, $[HI] = 0.0535\ M$

    _____

6. We saw that at 440 °C, $K_c = 49.5$ for the reaction

    $$H_2(g) + I_2(g) \rightleftharpoons 2HI(g)$$

    (a) What is K_c for the reaction $2HI(g) \rightleftharpoons H_2(g) + I_2(g)$?

    _____

    (b) What is K_c for the reaction $\tfrac{1}{2}H_2(g) + \tfrac{1}{2}I_2(g) \rightleftharpoons HI(g)$?

    _____

    (c) Write the equation for the reaction we would obtain if we added the equations in parts a and b. What is the value of the equilibrium constant for this final equation?

    _____

14.3   Equilibrium Laws Based on Pressures or Concentrations

Learning Objective

Write and convert between equilibrium laws based on molar concentration and gas pressures.

Review

The partial pressure of a gas in a mixture is proportional to its concentration. Therefore, the mass action expression for reactions involving gases can be written using partial pressures in place of concentrations. When partial pressures are used in the mass action expression, the equilibrium constant is designated as K_P. In general K_P does not have the same value as K_c.

For reactions in which there is a change in the number of moles of gas going from the reactants to the products, the numerical values of K_P and K_c are not the same, and in this section you learn how to convert between them. The equation that you need to remember in order to do this is Equation 14.10.

$$K_P = K_c(RT)^{\Delta n_{gas}}$$

In applying this equation, be careful to compute Δn_{gas} correctly; it is the difference in the total number of moles of *gas* between the product side and the reactant side of the equation. It is also important to remember to use $R = 0.0821$ L atm mol^{-1} K^{-1}. Any other value of R will give incorrect answers.

Example 14.1   Converting partial pressures to molarity

What is the molar concentration of nitrogen in air that has a total pressure of 775 torr at 25 °C? Air is composed of 79% N_2 by volume.

Analysis:

We need to go back to the ideal gas law We can rearrange this equation by dividing by the volume. One of the resulting terms is n/V that has the units of mol L^{-1}, which are the same units as molarity.

Assembling the Tools:

We need the ideal gas law $(PV = nRT)$ which is one of the tools we learned in Chapter 10.

Solution:

Rearranging the ideal gas law we get $P = (n/V)RT = MRT$. Solving this for molarity, M, we get

$$M = \frac{P}{RT}$$

To use this equation we need to calculate the partial pressure of nitrogen in units of atmospheres and the temperature must be converted to the Kelvin scale.

$$P = 775 \text{ torr} \times 0.79 \times \frac{1 \text{ atm}}{760 \text{ torr}} = 0.806 \text{ atm}$$

and $T = 25 + 273 = 298$ K, while $R = 0.0821$ L atm mol^{-1} K^{-1}. Entering these values in the equation above we get $M = 0.033$ mol N_2 L^{-1}

Is the Answer Reasonable?

The best indication that we did the problem right is that the units cancel to give us the units for molarity. The numerator had units of atm and the denominator had units of L atm mol^{-1} K^{-1} K. These result in the units of mol L^{-1}. We can estimate that the partial pressure of nitrogen is correct since it should be close to 80% of one atmosphere or 0.8 atm.

Example 14.2   Calculating K_P from K_c

The reaction $2H_2(g) + O_2(g) \rightleftharpoons 2H_2O(g)$ has $K_c = 9.1 \times 10^{80}$ at 25 °C. What is the value of K_P for the following reaction?

$$H_2(g) + \tfrac{1}{2}O_2(g) \rightleftharpoons H_2O(g)$$

Analysis:

Before starting to solve the equation for converting K_P to K_c we notice that the two equations given are different. The equation that we want K_P for has all of its coefficients one half of those given with the value of K_c. That means that we must manipulate the value of K_c first and then use our tool for converting K_P to K_c.

Assembling the Tools:

We need our tool that relates K_P to K_c (Equation 14.10 in the text) and our tools for manipulating reactions and their corresponding equilibrium laws.

Solution:

The square root of 9.1×10^{80} is 3.2×10^{40}. We now calculate the value of Δn_{gas} as

$$1 \text{ mol} - (1 \text{ mol} + 0.5 \text{ mol}) = -0.5 \text{ mol} = \Delta n_{gas}$$

We can now use Equation 14.10 above to obtain K_P.

$$K_P = K_c(RT)^{\Delta n_{gas}} = 3.2 \times 10^{40}(0.0821 \text{ L atm mol}^{-1} \text{ K}^{-1} \times 298 \text{ K})^{-0.5} = 6.47 \times 10^{39}$$

(NOTE: Equilibrium constants are dimensionless. If units were assigned they would be $P^{-1/2}$.)

Is the Answer Reasonable?

The best indication that the answer is correct is that K_P and K_c often have values close to each other. That is what our result shows. Perform the mathematical operations in a different order to check your work.

Self–Test

7. Write the K_P expression for the reaction

$$2N_2O(g) + 3O_2(g) \rightleftharpoons 4NO_2(g)$$

_____

8. A chemical reaction has the following equilibrium law.

$$K_P = \frac{P_{BrF_3}^2}{P_{F_2}^3 P_{Br_2}}$$

(a) What is the chemical equation for the reaction?

_____

(b) What is the expression for K_c?

_____

9. The partial pressure of O_2 in air is approximately 160 torr. What is the concentration of O_2 in air at 25 °C expressed in mol L^{-1}?

_____

10. For which of the following reactions will K_P be numerically equal to K_c?

    (a) $PCl_5(g) \rightleftharpoons PCl_3(g) + Cl_2(g)$

    (b) $H_2(g) + Cl_2(g) \rightleftharpoons 2HCl(g)$

    (c) $2CO(g) + O_2(g) \rightleftharpoons 2CO_2(g)$ _____

11. For the reactions in the preceding question, which will have

    (a) $K_P > K_c$ at 25 °C?   _____

    (b) $K_P < K_c$ at 25 °C?   _____

12. The reaction $2H_2(g) + O_2(g) \rightleftharpoons 2H_2O(g)$ has $K_c = 9.1 \times 10^{80}$ at 25 °C. What is the value of K_P for this reaction?

    _____

13. At 300 °C, the reaction $2SO_2(g) + O_2(g) \rightleftharpoons 2SO_3(g)$ has $K_P = 1.0 \times 10^8$. What is the value of K_c for this reaction at this temperature?

    _____

14.4   Equilibrium Laws for Heterogeneous Reactions

Learning Objective

Understand and explain why solids and pure liquids are not included in equilibrium laws.

Review

A heterogeneous equilibrium is one in which not all of the reactants and products are in the same phase. The mass action expression for a heterogeneous reaction is normally written without terms for the concentrations of pure liquids or solids. This is because the number of moles per liter for such substances does not depend on the amount of the substance in the reaction mixture. Here are two examples. For the reaction

$$Cl_2(g) + 2NaBr(s) \rightleftharpoons Br_2(l) + 2NaCl(s)$$

$$K_c = \frac{1}{[Cl_2]} \quad \text{and} \quad K_P = \frac{1}{P_{Cl_2}}$$

For the reaction

$$NH_4Cl(s) \rightleftharpoons NH_3(g) + HCl(g)$$

$$K_c = [NH_3][HCl] \quad \text{and} \quad K_P = P_{NH_3} P_{HCl}$$

Self–Test

14. Write the equilibrium law for the following reactions in terms of K_c.

    (a) $PCl_3(l) + Cl_2(g) \rightleftharpoons PCl_5(s)$

    _____

    (b) $CaCO_3(s) + SO_2 \rightleftharpoons CaSO_3(s) + CO_2(g)$

    _____

15. Write the equilibrium law, K_c, for the reaction $PbCl_2(s) \rightleftharpoons Pb^{2+}(aq) + 2Cl^-(aq)$.

    _____

16. Write the equilibrium law, K_c, for the reaction $3Zn(s) + 2Fe^{3+}(aq) \rightleftharpoons 2Fe(s) + 3Zn^{2+}(aq)$.

    _____

14.5   Position of Equilibrium and the Equilibrium Constant

Learning Objective

Interpret the value of the equilibrium constant as an indicator of the position of equilibrium.

Review

When the equilibrium constant (either K_P or K_c) is large, the reaction goes far toward completion by the time equilibrium is reached. We express this by saying that the "position of equilibrium lies far to the right." When K is small, only small amounts of products are formed when equilibrium is reached, so we say the position of equilibrium lies far to the left. See the summary in Section 14.5 of the text.

By comparing K's for reactions of similar stoichiometry, we are able to compare the extent to which the reactions proceed toward completion. Even if the stoichiometries differ, comparisons are still valid if the K's are vastly different.

Self–Test

17. For the reaction

$$CH_4(g) + H_2O(g) \rightleftharpoons CO(g) + 3H_2(g)$$

$K_c = 1.78 \times 10^{-3}$ at 800 °C, $K_c = 4.68 \times 10^{-2}$ at 1000 °C, and $K_c = 5.67$ at 1500 °C. From these data, should more or less CO and H_2 form as the temperature of this equilibrium is increased? Explain.

_____

_____

18. Hypochlorite ion, OCl^-, is the active ingredient in liquid laundry bleach. It has a tendency to react with itself as follows (although the reaction is slow at room temperature).

$$3OCl^- \rightleftharpoons 2Cl^- + ClO_3^- \qquad K_c = 10^{27}$$

If a bottle of bleach is allowed to come to equilibrium, how should the relative concentrations of the reactants and products compare?

_____

14.6   Equilibrium and Le Châtelier's Principle

Learning Objective

Use Le Châtelier's principle and the reaction quotient, Q, to interpret where a system lies in relation to equilibrium.

Review

Le Châtelier's principle states that if an equilibrium system is *stressed*, meaning it is subjected to an outside disturbance that upsets the equilibrium, the position of equilibrium shifts in a direction that counteracts the disturbance and, if possible, returns the system to equilibrium.

1. A reaction shifts in a direction away from the side of the reaction to which a substance is added. It shifts toward the side from which a substance is removed.

2. A decrease in the volume of a gaseous reaction mixture increases the pressure and shifts the equilibrium toward the side with the fewer number of molecules of *gas*. If both sides have the same number of molecules of gas, a volume change will not affect the equilibrium. Solids and liquids are virtually incompressible, so pressure changes have no effect on them.

3. An increase in temperature shifts an equilibrium in a direction that absorbs heat. The value of K increases with increasing temperature for a reaction that is endothermic in the forward direction. If you remember this, it's easy to figure out what happens for an exothermic reaction as well. Remember that temperature is the *only* thing that affects the value of K for a reaction.

4. Catalysts have absolutely no effect on the position of equilibrium. They only affect how fast a system gets to equilibrium.

5. Adding an inert gas, without simultaneously changing the volume, will have no effect on the position of equilibrium.

Self–Test

19. Question 17 gives values of K_c at three different temperatures for the reaction

$$CH_4(g) + H_2O(g) \rightleftharpoons CO(g) + 3H_2(g)$$

Is this reaction, as read from left to right, exothermic or endothermic? Explain.

_____

_____

_____

20. For the reaction in the preceding question, state how the amount of H_2 at equilibrium will be affected by

   (a)   adding CH_4 _____

   (b)   adding CO _____

   (c)   removing H_2O _____

   (d)   decreasing the volume _____

   (e)   adding helium at constant volume

                                _____

21. For the reaction $F_2 + Cl_2 \rightleftharpoons 2ClF + heat$, describe how the amount of Cl_2 will be affected by

   (a)   adding F_2 _____

   (b)   adding ClF _____

   (c)   decreasing the volume _____

   (d)   raising the temperature _____

   (e)   adding a catalyst   _____

22. How will the value of K_P change for the reaction in the preceding question if the temperature is lowered?

                                _____

14.7 Calculating Equilibrium Constants

Learning Objective

Describe experiments to determine the equilibrium constant and how to use the data from these experiments.

Review

As described in the text, the calculations that you learn how to perform in this section can be divided into two categories—calculating K_c from equilibrium concentrations, and calculating equilibrium concentrations from K_c and initial concentrations. These are not really very difficult *if you approach them systematically*. Until you become experienced at solving these problems, it's important that you not attempt to take shortcuts; that's where many students make mistakes or become lost and can't finish the problem.

In working equilibrium problems there are some important rules to follow, and the concentration table that we construct under the chemical equation helps you follow them.

1. The concentrations that you substitute into the equilibrium law *must* correspond to equilibrium concentrations. These are the *only* quantities that satisfy the equilibrium law.

2. Since we are working with K_c, quantities in the concentration table should have the units of molar concentration (mol L^{-1}). This means that if you are not given the molar concentration, but instead a certain number of moles in a certain volume, you must immediately change the data to mol L^{-1}. For example, if you are told that 5.0 mol of a certain reactant is in a volume of 2.0 L, change the data to 5.0 mol/2.0 L = 2.5 mol L^{-1} before entering it into the appropriate place in the concentration table.

3. An important difference between the two kinds of calculations that you are learning to perform is that when you are asked to calculate K_c, the equilibrium constant is the unknown and you must therefore find *numerical values* for *all* the entries in the "Equilibrium concentration" row of the table. On the other hand, when K_c is known and you are being asked to find an equilibrium concentration, then an *x* (or other symbol) appears in some way in the "Equilibrium concentration" row.

4. The initial concentrations in a reaction mixture are determined by the person doing the experiment. When you are given the composition of a mixture in a problem, these are initial concentrations, unless stated otherwise. In filling in this line of the table, we imagine that the reaction mixture can be prepared before any chemical changes take place. Then we let the reaction proceed to equilibrium and see how the concentrations change.

5. The changes in concentration are controlled by the stoichiometry of the reaction. When we want to compute K_c and have to calculate what the equilibrium concentrations are, the entries in this column are *numbers* whose ratios are the same as the ratios of the coefficients. When you are given K_c and initial concentrations, this is where the *x*'s first appear. The following points are useful to remember:

   - The coefficients of *x* can be the same as the coefficients in the balanced equation; this ensures that they are in the right ratio.

- The "changes" for the reactants all must have the same algebraic sign, and these signs must be the opposite of the algebraic signs of the "changes" for the products. (In other words, if the changes for the reactants are positive, the changes for the products are negative.)

- If the initial concentration of some reactant or product is zero, its change *must* be positive.

6. Equilibrium concentrations are obtained by algebraically adding the "Change in concentration" to the "Initial concentration." In cases where K_c is *very* small, the extent of reaction from left to right will also be *very* small. This allows equilibrium concentration expressions such as $(0.200 + x)$ or $(0.100 - x)$ to be simplified. The initial concentrations of the reactants hardly change at all as the reaction proceeds, so x can be expected to have a very small value. In this case, a quantity such as $(0.200 + x)$ will be very nearly equal to 0.200 after the very small value of x is added to 0.200.

$$(0.200 + x) \cong 0.200$$

This kind of simplifying assumption often makes a difficult problem very easy. If you are setting up a problem and find the algebra very complicated, it is likely you will be able to make simplifying assumptions.

7. Look for ways of simplifying the algebra in solving for x. Sometimes this isn't possible, as we see in Example 14.9, where the quadratic formula is used.

14.8   Using Equilibrium Constants to Calculate Concentrations

Learning Objective

Use tabulated values of equilibrium constants and experimental data to calculate the equilibrium concentrations (or pressures) of all species in an equilibrium mixture.

Review

We will review several of the methods used to determine concentrations from equilibrium data. The most common calculations involve a knowledge of the initial concentration of one or more of the substances shown in the chemical equation and the numerical value of the equilibrium constant that applies to that equation. We make extensive use of a concentration table to keep track of all the numbers. The following examples illustrate these calculations.

Example 14.3   Determining equilibrium concentrations by direct calculation

At a certain temperature the reaction $2HI(g) \rightleftharpoons H_2(g) + I_2(g)$ has $K_c = 8.4 \times 10^{-4}$. A mixture is prepared containing 0.200 M HI, 0.00500 M H_2, and 0.00500 M I_2. When this mixture reaches equilibrium, what will be the molar concentration of HI?

Analysis:

We are given a chemical equation (that implies we can write the correct equilibrium law) and the concentrations of all three substances in that equation. We are also given the value of K_c. We can enter the concentrations into the equilibrium law and calculate $Q = 6.25 \times 10^{-4}$. Since Q is smaller than the value of K_c we can tell that the system is not in equilibrium and furthermore it must proceed in the forward direction to reach equilibrium. This helps us set up the equilibrium table correctly.

Assembling the Tools:

We need our tools for writing equilibrium laws from chemical equations and the process for calculating Q. This information allows us to use the concentration tool to set up and solve for the concentrations of the solutes in the solution.

Solution:

We set up an equilibrium table to keep track of all the concentrations and their changes. First we must accurately transfer data from the problem to the first row of the table. For the change line we insert x's and the stoichiometric coefficients. Since the reaction goes in the forward direction, the $2x$ has a minus sign and the other x values are positive. The third line is the sum of the first two and are the terms inserted into the equilibrium law.

	$2HI(g) \rightleftharpoons$	$H_2(g) \quad +$	$I_2(g)$
Initial concentration (*M*)	0.200	0.00500	0.00500
Change in concentration (*M*)	$-2x$	$+x$	$+x$
Equilibrium concentrations (*M*)	$0.200 - 2x$	$0.00500 + x$	$0.00500 + x$

Setting up the equilibrium law we get

$$\frac{[H_2][I_2]}{[HI]^2} = \frac{[0.00500 + x][0.00500 + x]}{[0.200 - 2x]^2} = 8.4 \times 10^{-4}$$

We can take the square root of both sides of the equation to get

$$\frac{[0.00500 + x]}{[0.200 - 2x]} = 2.9 \times 10^{-2}$$

Multiplying both sides by $0.200 - 2x$ results in

$$0.00500 + x = 5.8 \times 10^{-3} - 5.8 \times 10^{-2}x$$

$$x + 0.058\,x = 0.00580 - 0.00500$$

$$x = 7.6 \times 10^{-4} = 0.00076$$

Using this value of x we calculate that

$$[HI] = 0.200 - 0.002 = 0.198 \text{ mol L}^{-1}$$

and

$$[H_2] = [I_2] = 0.00500 + 0.00076 = 0.00576 \text{ mol L}^{-1}$$

Is the Answer Reasonable?

The best test is to see if the values obtained will give us the value of the equilibrium constant. When we calculate K_c the value is $(0.00576)^2/(0.198)^2 = 8.46 \times 10^{-4}$. This is very close to the given K_c and we are satisfied that the answer is reasonable.

Example 14.4 Determining equilibrium concentrations using simplifying assumptions

At a certain temperature the reaction $2HBr(g) \rightleftharpoons H_2(g) + Br_2(g)$ has $K_c = 3.3 \times 10^{-5}$. A mixture is prepared containing 0.200 M HBr, 0.0 M H_2, and 0.0500 M Br_2. When this mixture reaches equilibrium, what will be the molar concentration of HBr?

Analysis:

We can evaluate this problem in the same way that we looked at Example 14.3 above. The only difference is that the concentrations of H_2 and Br_2 are not equal and we will not be able to solve this problem with a simple square root. Since the initial concentration of H_2 is zero, the reaction must go in the forward direction. However, considering the small equilibrium constant, we expect only a small amount of product.

Assembling the Tools:

We need the same tools and approaches as in Example 14.3.

Solution:

We will set up the equilibrium table as before.

	$2HBr(g) \rightleftharpoons$	$H_2(g) +$	$Br_2(g)$
Initial concentration (M)	0.200	0.0	0.0500
Change in concentration (M)	$-2x$	$+x$	$+x$
Equilibrium concentrations (M)	$0.200 - 2x$	$0.0 + x$	$0.0500 + x$

Setting up the equilibrium law we get

$$\frac{[H_2][Br_2]}{[HBr]^2} = \frac{[0.0+x][0.0500+x]}{[0.200-2x]^2} = 3.3 \times 10^{-5}$$

We cannot take the square root of both sides of the equation. We will assume that x << 0.0500 (<< means "much smaller") and that $2x$ << 0.200. This second assumption *must be true* if the first assumption is true. We can then write

$$\frac{[x][0.0500]}{[0.200]^2} = 3.3 \times 10^{-5}$$

Solving this equation gives us a value of $x = 6.6 \times 10^{-8}$. With this small value of x, both of our assumptions are true. Our result is that [HBr] = 0.200 M, [Br_2] = 0.0500 M, and [H_2] = 6.6×10^{-8} M.

Is the Answer Reasonable?

Because of the small size of the equilibrium constant we expect little forward reaction to occur. Formation of only 6.6×10^{-8} additional moles of H_2 and Br_2 agrees with this observation and our results are reasonable.

Self–Test

23. At 25 °C a mixture of Br_2 and Cl_2 in carbon tetrachloride reacted according to the equation

$$Br_2 + Cl_2 \rightleftharpoons 2BrCl$$

The following equilibrium concentrations were found: $[Br_2] = 0.124\ M$, $[Cl_2] = 0.237\ M$, $[BrCl] = 0.450\ M$. What is the value of K_c for this reaction?

24. Referring to Question 23, what is the value of K_c for the reaction

$$2BrCl \rightleftharpoons Br_2 + Cl_2$$

at 25 °C?

25. A mixture of $N_2O(g)$, $O_2(g)$, and $NO_2(g)$ was prepared in a 4.00–liter container and allowed to come to equilibrium according to the equation

$$2N_2O(g) + 3O_2(g) \rightleftharpoons 4NO_2(g)$$

The mixture originally contained 0.512 mol N_2O.

   (a) As the reaction came to equilibrium, the concentration of N_2O decreased by 0.0450 mol/L. What was the equilibrium concentration of N_2O?

   (b) By how much did the O_2 concentration change?

   (c) By how much did the NO_2 concentration change?

26. At a certain temperature, a mixture of $N_2O(g)$ and $O_2(g)$ was prepared having the following initial concentrations: $[N_2O] = 0.0972\ M$, $[O_2] = 0.156\ M$. The mixture came to equilibrium following the equation

$$2N_2O(g) + 3O_2(g) \rightleftharpoons 4NO_2$$

At equilibrium, the NO_2 concentration was found to be $0.0283\ M$. What is K_c for this reaction?

27. The reaction $2BrCl \rightleftharpoons Br_2 + Cl_2$ has $K_c = 0.145$ in CCl_4 at 25 °C. If 0.220 mol of BrCl is dissolved in 250 mL of CCl_4, what will be the concentrations of BrCl, Cl_2, and Br_2 when the reaction reaches equilibrium?

---

28. At 25 °C, $K_c = 1.6 \times 10^{-17}$ for the reaction

$$N_2(g) + 2O_2(g) \rightleftharpoons 2NO_2(g)$$

In air, the concentrations of N_2 and O_2 are $[N_2] = 0.0320 \ M$, $[O_2] = 0.00860 \ M$. Taking these as initial concentrations, what should be the equilibrium concentration of NO_2 in air?

Answers to Self–Test Questions

1. The $C_2H_3O_2^-$ ion that the carbon atom is in can pick up an H^+ and become an $HC_2H_3O_2$ molecule. Later this molecule can lose H^+ to become a $C_2H_3O_2^-$ ion again. This can be repeated over and over.

2. In a dynamic equilibrium the reactants and products actually exist as distinct entities that are alternating between each other as the forward and reverse reactions proceed at equal rates. In resonance, none of the individual resonance structures actually exist. The actual structure is a "blend" of the resonance structures.

3. (a) $\dfrac{[CH_3CO_2CH_3]}{[CH_3OH][CH_3CO_2H]} = K_c$ (b) $\dfrac{[Cr_2O_7^{2-}]}{[HCrO_4^-]^2} = K_c$

    (NOTE: For (a) and (b) we do not include water as part of the mass action expression, unless it is in the gas state. See Section 14.4 in the text on heterogeneous reactions.)

    (c) $\dfrac{[NO_2]^4}{[N_2O]^2[O_2]^3} = K_c$ (d) $\dfrac{[C_2H_6]}{[C_2H_4][H_2]} = K_c$

4. $\dfrac{[NO_2]^4[H_2O]^6}{[NH_3]^4[O_2]^7} = K_c$

5. Only (b) and (d) are at equilibrium.

6. (a) $K_c = 0.0202$
    (b) $K_c = 7.04$
    (c) $HI(g) \rightleftharpoons ½H_2(g) + ½I_2(g)$, $K_c = 0.142$

7. $K_P = \dfrac{P_{NO_2}^4}{P_{N_2O}^2 P_{O_2}^3}$

8. (a) $3F_2(g) + Br_2(g) \rightleftharpoons 2BrF_3(g)$

    (b) $K_c = \dfrac{[BrF_3]^2}{[F_2]^3[Br_2]}$

9. 8.60×10^{-3} mol L^{-1}

10. Reaction b

11. (a) Reaction a (b) Reaction c

12. $K_p = 3.7 \times 10^{79}$

13. $K_c = 4.7 \times 10^9$

14. (a) $K_c = \dfrac{1}{[Cl_2]}$ (b) $K_c = \dfrac{[CO_2]}{[SO_2]}$

15. $K_c = [Pb^{2+}][Cl^-]^2$

16. $K_c = \dfrac{[Zn^{2+}]^3}{[Fe^{3+}]^2}$

17. More CO_2 and H_2 should form, because K_c increases with increasing temperature.

18. At equilibrium the concentrations of Cl^- and ClO_3^- should be large compared to the concentration of OCl^-.

19. Endothermic, because K_c increases with increasing temperature.

20. (a) increase (b) decrease (c) decrease (d) decrease (e) no change

21. (a) decrease (b) increase (c) no change (d) increase (e) no change

22. K_P increases as the temperature is lowered.

23. $K_c = 6.89$

24. $K_c = 0.145$

25. (a) 0.083 M (The initial N_2O concentration was 0.512 mol/4.00 L = 0.128 M.)   (b) decreased by 0.0675 mol/L (c) increased by 0.0900 mol/L

26. $K_c = 0.0378$

27. [BrCl] = 0.500, $[Cl_2]$ = 0.190 M, $[Br_2]$ = 0.190 M

28. $[NO_2] = 3.1 \times 10^{-12} M$

Tools for problem solving

In this chapter you learned to apply the following concepts as tools in solving problems. Study each one carefully so that you know what each is used for. When faced with solving a problem, recall what each tool does and consider whether it will be helpful in finding a solution.

You might want to tear these pages out to use along with solving problems in this chapter.

The approach to equilibrium (Section 14.1)

Remember that the same equilibrium composition is reached regardless of whether it is approached from the direction of the reactants or from the direction of the products. When K_P or K_c is very large equilibrium problems can be greatly simplified if we first imagine the reaction going to completion (converting all the reactants to products) and then approaching the equilibrium from the direction of the products.

Requirement for equilibrium (Section 14.2)

For a system to be at equilibrium, the numerical value of Q must equal the equilibrium constant. Use this tool to determine whether a system is at equilibrium when you have a way to calculate Q. By comparing Q with K, you can determine the direction the reaction must proceed to reach equilibrium.

The equilibrium law (Section 14.2)

For a homogeneous reaction, an equation of the form

$$dD + eE \rightleftharpoons fF + gG$$

has the equilibrium law

$$\frac{[F]^f [G]^g}{[D]^d [E]^e} = K_c$$

Manipulating equilibrium equations (Section 14.2)

There are occasions when it is necessary to modify an equation for an equilibrium, or combine two or more chemical equilibria. The tools discussed here are used to obtain the new equilibrium constants for the new equations.

• When two equations are added, we multiply their Ks to obtain the new K.

• When an equation is multiplied by a factor n to obtain a new equation, we raise its K to the power n to obtain the K for the new equation.

• When an equation is reversed, we take the reciprocal of its K to obtain the new K.

Equilibrium law using partial pressures for gas phase reactions (Section 14.3)

$$dD + eE \rightleftharpoons fF + gG$$

$$K_P = \frac{(P_F)^f (P_G)^g}{(P_D)^d (P_E)^e}$$

Converting between K_P and K_c (Section 14.3)

To relate K_P to K_c, we use the equation

$$K_P = K_c(RT)^{\Delta n_g}$$

Remember to use $R = 0.0821$ L atm mol^{-1} K^{-1}. Also, remember that n_g ($= n_{products} - n_{reactants}$) is the change in moles of gas on going from reactants to products.

Equilibrium laws for heterogeneous reactions (Section 14.4)

Remember that pure solids and liquids do not appear in the mass action expression.

Magnitude of *K* (Section 14.5)

Use this tool to gain a rough estimate of the position of equilibrium.

- When *K* is very large, the position of equilibrium lies far to the right (toward the products).
- When *K* is very small, the position of equilibrium lies far to the left (toward the reactants).

Le Châtelier's principle (Section 14.6)

This tool lets us predict how disturbing influences shift the position of equilibrium. Factors to consider are:

- Adding or removing a reactant or product
- Changing the volume for gaseous reactions
- Changing the temperature
- Changing temperature changes *K*
- Catalysts or inert gases have no effect on the position of equilibrium.

Concentration table (Section 14.7)

This is a tool you will use in almost all equilibrium calculations. Remember the following points when constructing the table:

- All entries in the table must have units of molarity (mol L^{-1}).
- Any reactant or product for which an initial concentration or amount is not given in the statement of the problem is assigned an initial concentration of zero.
- Any substance with an initial concentration of zero must have a positive change in concentration when the reaction proceeds to equilibrium.
- The changes in concentration are in the same ratio as the coefficients in the balanced equation. When the changes are unknown, the coefficients of *x* can be the same as the coefficients in the balanced equation.
- Only quantities in the last row (Equilibrium Concentrations) satisfy the equilibrium law.

Quadratic formula (Section 14.8)

This equation allows us to analytically solve equations of the form $ax^2 + bx + c = 0$:

$$x = \frac{-b \pm \sqrt{b^2 - 4ac}}{2a}$$

Simplifications in equilibrium calculations (Section 14.8)

When the initial reactant concentrations are larger than $1000 \times K$, they will change only slightly as the reaction approaches equilibrium. You can therefore expect to be able to neglect the change when it is being added to or subtracted from an initial concentration. *If you use simplifying approximations, be sure to check their validity after obtaining an answer.* This tool is especially useful when working problems in which K is very small and the initial conditions are not far from the final position of equilibrium.

Summary of Important Equations

The equilibrium law from the coefficients of a balanced equation:

For the general chemical equation

$$dD + eE \rightleftharpoons fF + gG$$

The equilibrium law is

$$\frac{[F]^f [G]^g}{[D]^d [E]^e} = K_c$$

Relationship between K_p and K_c:

$$K_p = K_c(RT)^{\Delta n_{gas}}$$

Notes:

Chapter 15

Acids and Bases, A Molecular Look

This chapter continues our study, begun in Chapter 5, of *acid–base reactions*, which are among the most common and important kinds of chemical reactions. Chemists most often use the Brønsted view of acids and bases, because in the overwhelming majority of all acid–base reactions, *proton transfer* occurs. However, many reactions with the look and feel of an acid–base reaction do not involve proton transfers but involve nothing more, in the Lewis view, than the formation of a coordinate covalent bond.

The Brønsted view is nearly always used when working with aqueous solutions, but Brønsted acids (proton donors) vary widely in strength. We learn here how the periodic table and the concept of electronegativity help us to understand the trends in acid strengths and to organize them for study and learning.

The chapter includes a shorthand way for describing small molar concentrations (less than 1 *M*) of hydrogen ions in water, the pH concept. It's used throughout all of chemistry, biochemistry, and any fields of medicine that focus on the acid–base balance of a fluid of a living system. That's why it must be learned well.

15.1   Brønsted–Lowry Acids and Bases

Learning Objective

Identify Brønsted–Lowry acids and bases and explain the terms conjugate acid–base pair and amphoteric substances.

Review

Water is not a required solvent for acid–base reactions if we define an acid as any proton (H^+) donor and a base as any proton acceptor. Acids and bases thus defined are sometimes referred to as *Brønsted–Lowry acids and bases*, or simply *Brønsted acids and bases*, after the chemists who had the fundamental insight. A Brønsted acid–base reaction simply involves the transfer of a hydrogen ion (which is the same as a proton) from the acid to the base.

Brønsted acid–base reactions can always be viewed as reversible, so that in *both* the forward reaction and the reverse there is an acid and a base. An acid or proton donor occurring to the left of the equilibrium arrows becomes a base on the right; the proton acceptor or base on the left thereby becomes a Brønsted acid on the right.

Two chemical species that differ by only one proton are called an acid–base *conjugate pair*. The conjugate acid member has one more hydrogen and one additional positive charge (or one less negative charge) than the conjugate base. For example, consider the following pair,

$$H_2PO_4^- \longleftrightarrow HPO_4^{2-}$$

This is the conjugate acid because it has one more hydrogen and one less negative charge.	This is the conjugate base because it has one less hydrogen and one more negative charge.

If you need to write the formula for the conjugate acid of something, add a hydrogen and a positive charge (i.e., add one H^+). For example, to find the conjugate acid of PH_3, we add one H^+.

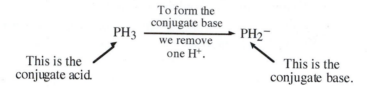

Similarly, to write the formula for the conjugate base of something, we take away one H^+. Suppose we want the formula for the conjugate base of PH_3:

Be sure to study Examples 15.1 and 15.2 in the text.

In any Brønsted acid–base reaction there are two conjugate pairs. Be sure to study text Example 15.2 so that you can quickly identify the conjugate pairs in an equation.

Two particularly important conjugate pairs are H_3O^+ and H_2O, and H_2O and OH^-. The hydronium ion is the conjugate acid of water; water is the conjugate base of the hydronium ion. Similarly, water is the conjugate acid of the hydroxide ion, and the hydroxide ion is the conjugate base of water. Notice that water is the base in one pair and the acid in the other. Any species that can be either an acid or a base, depending on the circumstances, is said to be *amphoteric* or *amphiprotic*. In the illustrations above, for instance, we would describe PH_3 as amphoteric if it is able to form both PH_2^- and PH_4^+.

Example 15.1   Identifying Brønsted acids and bases

(a) Nitrogen monoxide, NO, reacts with oxygen to give nitrogen dioxide, NO_2, but this reaction is not a Brønsted acid–base reaction. Why not?

(b) Consider the *possibility* that H_2O and S^{2-} might give a Brønsted acid–base reaction. If they do, what would be the likeliest products? Identify each of them as Brønsted acids or bases.

Analysis:

Both questions ask you to identify Brønsted acids and bases.

Assembling the Tools:

To answer these questions all we need to do is to recall the definition of Brønsted acids and bases. A Brønsted acid is a proton donor and a Brønsted base is a proton acceptor.

Solution:

(a) In this part we have reactants and products but there are absolutely no protons (hydrogen ions) involved. Since the Brønsted acid–base theory is based on proton acceptors and donors, this reaction is not a Brønsted acid–base reaction.

(b) In this part water does have protons to donate. In addition the sulfide ion, S^{2-} can accept a proton to form HS^-. We conclude that it is possible to have a Brønsted acid–base reaction. The reaction would involve water donating a proton (the Brønsted acid) and the sulfide ion accepting a proton (the Brønsted base). We write

$$H_2O + S^{2-}(aq) \longrightarrow OH^-(aq) + HS^-(aq)$$

The two Brønsted acids are H_2O and HS^- while the two Brønsted bases are S^{2-} and OH^-.

Are the Answers Reasonable?

Both answers are directly based on the definitions of Brønsted acids and bases and their reactions. They seem reasonable on that basis.

Self–Test

1. Write the formula of the conjugate acid of each of the following.

   (a) CH_3NH_2 _____

   (b) NH_2OH _____

   (c) ClO_3^- _____

   (d) $HC_2O_4^-$ _____

2. Write the formula of the conjugate base of each of the following.

   (a) $HC_2O_4^-$ _____

   (b) $H_2AsO_4^-$ _____

   (c) $HCHO_2$ _____

   (d) H_2S _____

3. Identify the conjugate acid–base pairs in the following reaction. In each pair, state which is the acid and which is the base.

   $$H_2SO_4 + Cl^- \longrightarrow HSO_4^- + HCl$$

   _____

4. Identify the conjugate acid–base pairs in the following reaction. In each pair, state which is the acid and which is the base.

   $$NH_3 + NH_3 \longrightarrow NH_4^+ + NH_2^-$$

   _____

5. Acetic acid, $HC_2H_3O_2$, is a Brønsted base in concentrated sulfuric acid and it is a Brønsted acid in water. Write chemical equations that illustrate these reactions.

   _____

   _____

   Explain why acetic acid is said to be *amphoteric*.

   _____

15.2 Strengths of Brønsted–Lowry Acids and Bases

Learning Objective

Compare the strengths of Brønsted–Lowry acids and bases, how they are classified as strong or weak, and the strengths of conjugate acid–base pairs.

Review

Measuring acid–base strength

To compare the strengths of a series of acids, we use a reference base and compare the extents to which the base is protonated by the acids. If we represent an acid by the formula HA and the base by B, we examine the equilibrium

$$HA + B \rightleftharpoons A^- + BH^+$$

The farther the reaction proceeds toward completion, the stronger is the acid.

In water, the strongest acid that can exist is H_3O^+; any stronger acid reacts with water to form H_3O^+ and the corresponding anion. This means we can't use water as a reference to compare the strengths of very strong acids such as HNO_3 and $HClO_4$. Both react completely with water, so differences in their strengths cannot be observed.

The strongest base that can exist in water is OH^-. Stronger bases, such as NH_2^-, react completely with water to give OH^- and the corresponding conjugate acid.

The position of equilibrium favors the weaker acid and base

This is the theme of the discussion that begins in Section 15.2. If we look at the position of equilibrium in a Brønsted acid–base reaction, we can tell the relative acid and base strengths of the substances involved. For example, consider the equilibrium

$$HCO_3^- + H_2O \rightleftharpoons H_3O^+ + CO_3^{2-}$$

Position of equilibrium lies to the left. ⟵

The position of equilibrium tells us that CO_3^{2-} is a stronger base than H_2O, and that H_3O^+ is a stronger acid than HCO_3^-.

Reciprocal relationships

Be sure to learn the inverse relationship between the strengths of acids and their conjugate bases. This is one of the most important concepts to come from this section.

The stronger the acid is, the weaker is its conjugate base. Thus, very strong acids, like HCl, $HClO_4$, and H_3O^+, have *very* weak conjugate bases, Cl^-, ClO_4^-, and H_2O, respectively.

The weaker the acid is, the stronger is its conjugate base. Thus, H_2O, a very weak acid, has a very strong conjugate base, namely, OH^-. Acetic acid, $HC_2H_3O_2$, is a stronger acid than H_2O, but still a weak acid among all the acids. Thus, its conjugate base, the acetate ion, $C_2H_3O_2^-$, is not as strong a base as OH^- but is much stronger as a base than Cl^-.

15.3   Periodic Trends in the Strengths of Acids

Learning Objective

Using the periodic table, describe the trends in the strengths of binary acids and oxoacids

Review

To develop a logical method for assessing the relative strengths of acids requires that we first divide acids into two groups, binary acids and oxoacids. Binary acids are composed of hydrogen and another element. Oxoacids are acids that contain oxygen, hydrogen, and one or more additional atoms. We also focus on the basic fact that acid strength is inversely related to the strength of the bond between a hydrogen atom and the rest of the molecule. In Chapters 8 and 9 the basics of bonding were discussed. From those discussions we draw the general notions that the greater the electron density between two atoms, the stronger the bond. Triple bonds are stronger than double bonds and double bonds are stronger than single bonds. The longer the bond, the lower the electron density and the weaker the bond. Additionally, polar bonds tend to be weaker than nonpolar bonds since electrons are withdrawn from the bond by the more electronegative atom. Lets see how these principles are applied.

Binary acids: Relative strengths and the periodic table

Binary acids have molecules made of just two elements, one being hydrogen joined to another atom. Examples are HCl and H_2S. Let's represent their formulas as H_nX. *The acids become stronger as the location of atom X in the periodic table moves from left to right in a period or from top to bottom in a group.*

The left–to–right variation is caused by increases in the electronegativity of the anion X, which also increases from left to right in a period. As the electronegativity increases, the electrons are drawn toward the anion thus weakening the bond with hydrogen. Recall that the weaker the bond with hydrogen, the stronger the acid.

The top to bottom trend, however, defies what we would expect on the basis of changes in electronegativity. Electronegativity *decreases* as we move *down* a vertical column (group), but the strengths of the binary acids *increase* as we move down. There is a weakening of the $H-X$ bond from top to bottom in a group. The atom X bonded to H increases in size as we move down a group, and the greater the radius of X, results in a longer bond length resulting in a weaker the $H-X$ bond. The decrease in bond strength due to increasing bond length outweighs the expected increase in bond strength due to a decrease in electronegativity, so the acids H_nX become stronger acids (better proton donors) as we descend a group.

Oxoacids: Relative strengths and the periodic table

Oxoacids are those in which a central atom, other than O or H, are bonded to one or more $-$OH groups. (Sometimes, extra O atoms are present; they are called *lone oxygens*.) Because O is very electronegative, the OH group is polar. It's made even more polar when it's attached to another electronegative atom, like a halogen, or S, or N. This combination makes the δ+ on H positive enough so that H can be transferred as H^+ from the oxoacid's OH group to an acceptor, like a molecule of H_2O. *All of the correlations in this section between the acidities of oxoacids and their structures or their locations in the periodic table ultimately come down to the magnitude of the partial positive charge, δ+, on H.* Whatever makes this δ+ greater makes the oxoacid stronger.

Oxoacids with central atoms in the same *group* bonded to identical numbers of O atoms. Examples are H_2SO_4 and H_2SeO_4. Such acids increase in strength as the central atom changes from bottom to top in the group. S is more electronegative than Se (which is below it in the table), and so the δ+ on H in an OH group of H_2SO_4 is greater than it is when in H_2SeO_4. As a result, H_2SO_4 is stronger than H_2SeO_4.

Oxoacids with central atoms in the same *period* bonded to identical numbers of O atoms. Examples are H_2SO_4 and H_3PO_4. Such acids increase in strength as the central atom moves from left to right in a period. This trend parallels the left–to–right trending increase in electronegativity. Thus, S is to the right of P in Period 3 of the table and S is more electronegative than P. As a result, H_2SO_4 is a stronger acid than H_3PO_4.

Oxoacids with the same central atom but different numbers of O atoms. Examples are H_2SO_3 and H_2SO_4. O atoms over and above those in OH groups add electron–withdrawing ability and increase the sizes of δ+ on the H atoms of the OH groups. So the more O atoms that are joined to the same central atom, the greater is the acidity of the oxoacid. Thus, H_2SO_4 is stronger than H_2SO_3.

When extra O atoms are *lone oxygens*, those not holding H atoms, they not only increase the electron withdrawal from OH groups but also help to disperse and stabilize the negative charge in the conjugate base.

Example 15.2   Using the relative strengths of oxoacids

An oxoacid of the general formula H_xZO_y has been discovered and found to be a stronger acid than H_2SeO_4. (a) If $y = 4$, which location in the periodic table is more likely for Z, to the *left* of Se or to the *right* of Se in the same row? (b) If $y = 4$ and $x = 2$, where does Z most likely lie in the periodic table, *above* or *below* Se in the same group?

Analysis:

We have learned the basic principles needed to assess the relative strengths of two oxoacids. Now we need to apply those principles using tools we've learned about in other chapters.

Assembling the Tools:

These principles (tools) are related to the electronegativity of the central atom, the number of oxygen atoms bound to the central atom, and the number of lone oxygen atoms. Finally, the ability of the anion to stabilize itself by delocalizing the electron left behind by the proton is important. For oxoacids we may need to draw Lewis structures, and knowing the periodic trends in atomic size and electronegativity can also be used.

Solution:

(a) The acid just to the right of H_2SeO_4 is $HBrO_4$, and the acid to the right of H_2SeO_4 is H_3AsO_4. Since the newly discovered oxoacid is stronger than H_2SeO_4 we would predict it is $HBrO_4$ because Br is more electronegative and the BrO_4^- anion has more lone oxygens to stabilize the anion.

(b) For this part each oxoacid will have the same number of oxygen atoms and lone oxygens, and the strength will be mostly dependent on the electronegativity of the central atom. In this case, since the atom higher up in a group is more electronegative, we predict that the acid we discovered is above H_2SeO_4.

Are the Answers Reasonable?

We check our reasoning to be sure we have not violated any of the principles used for assessing oxoacid strength. Since we have followed the principles, the answer is reasonable.

Self–Test

6. Consider the following equilibria and their equilibrium constants:

    $$HCHO_2 + H_2O \rightleftharpoons H_3O^+ + CHO_2^- \quad K = 1.8 \times 10^{-4}$$

    $$HOCl + H_2O \rightleftharpoons H_3O^+ + OCl^- \quad K = 3.0 \times 10^{-8}$$

    Which is the stronger acid? _____

7. The position of equilibrium in the following reaction lies to the left:

    $$HOCl + NO_2^- \rightleftharpoons HNO_2 + OCl^-$$

    (a) Which is the stronger acid? _____

    (b) Which is the stronger base? _____

8. NH_2^- is a stronger Brønsted base than OH^-. Which is the stronger acid, NH_3 or H_2O? _____

9. Which is the stronger acid, H_2Te or H_2Se? _____.

    Explain. _____

10. Which is the stronger acid, H_2Te or HI? _____.

    Explain. _____

11. Which is the stronger acid, H_3PO_3 or H_3PO_4? _____.

    Explain. _____

12. Which is the stronger acid, H_3AsO_4 or H_3PO_4? _____.

    Explain. _____

15.4 Lewis Acids and Bases

Learning Objective

Define Lewis acids and bases and compare them to the Arrhenius and Brønsted–Lowry definitions.

Review

Ordinarily, a covalent bond forms between two atoms when each atom supplies one electron for the shared pair. When both electrons for the pair come from one of the atoms, the bond is called a *coordinate covalent bond*. (Once formed, of course, it's just like any other covalent bond.)

A species furnishing an electron pair for a covalent bond is a *Lewis base*. The species accepting the pair is a *Lewis acid*. In the Lewis acid–base system, a neutralization reaction is the formation of a coordinate covalent bond. Hydrogen ions are not necessarily involved, although H^+ is a Lewis acid. (It's also what transfers in the Brønsted system.)

Lewis acids are species with less than an octet of electrons on a central atom (for example, BF_3), or they are species that can develop a vacancy open to more electrons by shifting electrons to other atoms around the central atom (like the C in CO_2).

Lewis bases are species that have unshared electron pairs on a central atom that can be donated into a coordinate covalent bond (like the O in H_2O or in OH^-).

A Brønsted acid–base reaction can be viewed as the transfer of a Lewis acid (H^+) from one Lewis base to another. The position of equilibrium favors the H^+ being attached to the stronger Lewis base.

Self–Test

13. Boron trifluoride, BF_3, can react with a fluoride ion to form the tetrafluoroborate ion, BF_4^-. Diagram this reaction using Lewis symbols to show the formation of a coordinate covalent bond.

14. BF_3 reacts with organic chemicals called ethers to form addition compounds. Use Lewis formulas to diagram the reaction of BF_3 with dimethyl ether. The structure of dimethyl ether is

    $$H_3C \overset{\cdot\cdot}{\underset{\cdot\cdot}{O}} CH_3$$

15. In the preceding question, which substance is the Lewis acid? _____

    Which is the Lewis base? _____

16. Which are the two Lewis bases in Question 7? Which is the stronger of them?

15.5   Acid–Base Properties of Elements and Their Oxides

Learning Objective

Using the periodic table, describe which elements are most likely to form acids or bases.

Review

As you learned in Section 4.3, some metal oxides, particularly those of metals in Groups IA and IIA of the periodic table, are called *base anhydrides*, because they react with water to give hydroxide ions. The oxides of most nonmetals react with water to give acidic solutions and so are *acid anhydrides*.

In water, metal ions with sufficiently high *charge-density*, which is the ratio of ionic charge to ionic volume, form proton–donating hydrates, and their solutions are acidic. A charge density large enough to yield an acidic species usually requires both a charge of at least 2+ and a small radius. Generally, the cations of Groups IA and IIA, except for Be^{2+}, do not have high enough charge densities to make their cations acidic. But the water–soluble cations of other groups form slightly acidic solutions.

Example 15.3 Evaluating the Acid–Base Properties of Oxides

Part I. A white solid readily dissolves in water to form a solution that turns red litmus paper blue. Which of the following compounds could this solid be?

$$CO_2, \ Na_2O, \ HNO_3, \ KOH, \ SO_3, \ HC_2H_3O_2$$

Consider the oxide of an element in Group IIA of the periodic table.

Part II (a) Write the formula of this oxide, using Z as the element's symbol. (b) Will a solution of this oxide in water give a blue or a red color to litmus paper? (c) Is this oxide very soluble in water or is it likely to be sparingly soluble?

Analysis:

We need to analyze the list of possibilities to determine those that white solids, of the white solids, which are soluble in water, and finally will the solution be acidic or basic? (Alternately we could ask which substances are soluble in water, then which is a white solid.)

Assembling the Tools:

We are asked to combine many of the principles (tools) learned so far in many of the previous chapters. Most important is recognizing which oxides are acidic and which are basic. We will also need skill in writing formulas, recalling solubility rules, and recalling specific properties of litmus paper.

Solution:

Part I. This question involves recognizing that oxides of metals tend to be basic and oxides of nonmetals tend to be acidic. We also need to recall that when litmus turns blue, the solution is basic. Thus we see that Na_2O

is a soluble basic oxide and KOH is a base in the list of substances given. The remaining substances are either acids or acid anhydrides.

Part II. For the second part we recall that metals in Group IIA form ions that can be written as Z^{2+} and as a result their formulas with the oxide ion, O^{2-}, will be ZO. These are metal oxides that are sparingly soluble in water. They will make solutions slightly basic, thus changing litmus paper blue.

Are the Answers Reasonable?

All we can do is check the facts of our answer with information from previous chapters and the information in Section 15.5 in the text. These answers do seem reasonable.

Self–Test

17. Why is Na_2O called a *base anhydride*? Write the equation for its reaction with water.

18. Why is SO_3 called an *acid anhydride*? Write the equation for its reaction with water.

19. Which salt, if either, will form the more acidic solution in water, $MgCl_2$ or $AlCl_3$? Explain.

20. Is the oxide OsO_4 more likely to be an acid or a base anhydride? Explain.

15.6   Advanced Ceramics and Acid–Base Chemistry

Learning Objective

Describe how the production of advanced ceramics depends on acid–base chemistry, including the sol–gel process.

Review

We are all familiar with ceramics. Bathroom sinks, toilets, ornamental figurines, mosaic tiles, and floor/wall tiles may be ceramic materials. Ceramics are very hard and durable and can be shaped in many ways that appeal to artists. Ceramics are formed by heating to high temperatures in a kiln, and ceramics themselves are resistant to heat. Perhaps the most publicized ceramics are those attached to the space shuttle to protect it from the high heat of re–entry after a mission.

One of the apparent disadvantages of ceramics is that high-temperature firing often does not melt all of the particles, leaving cracks and fissures that degrade its usefulness. Modern ceramics are precipitated from solution in an acid–base reaction resulting in extremely small particles that result in a more uniform, flaw-free product.

This section also discusses other ceramics that most people would not recognize. Aerogels are very lightweight ceramics that can be used as insulation. Figure 15.5 illustrates many ways that modern ceramics are formed or used. Chemistry Outside the Classroom 15.1 gives examples of modern uses of ceramics.

Self–Test

21. Describe the sol–gel process. _____

_____

_____

22. What are the advantages of ceramics materials? _____

_____

23. What is the main difference between sol–gel ceramics compared to traditional ceramics?

_____

_____

Answers to Self–Test Questions

1. (a) $CH_3NH_3^+$, (b) NH_3OH^+, (c) $HClO_3$, (d) $H_2C_2O_4$
2. (a) $C_2H_4^{2-}$, (b) $HAsO_4^{2-}$, (c) CHO_2^-, (d) HS^-
3. H_2SO_4 (acid), HSO_4^- (base); Cl^- (base), HCl (acid)
4. NH_3 (base), NH_4^+ (acid); NH_3 (acid), NH_2^- (base)
5. $HC_2H_3O_2 + H_2SO_4 \longrightarrow H_2C_2H_3O_2^+ + HSO_4^-$
   $HC_2H_3O_2 + H_2O \longrightarrow C_2H_3O_2^- + H_3O^+$
   In these reactions, acetic acid can be either an acid or a base.
6. $HCHO_2$ because it has the larger K_a
7. (a) HNO_2, (b) OCl^-
8. H_2O
9. H_2Te. The H—Te bond is weaker, because Te has a larger radius and the bond with H is longer and weaker than the bond between Se and H.
10. HI. The element I is more electronegative than Te. The larger electronegativity increases the bond polarity, which weakens the bond with H, thus making HI a stronger acid..
11. H_3PO_4 is stronger. It has one more O.
12. H_3PO_4 is stronger. P is more electronegative than As, standing above As in the table.

13.

$$\left[\ :\!\!\overset{\displaystyle ..}{F}\!:\ \right] \quad :\!F\!-\!B \quad :\!\overset{..}{F}\!:^- \longrightarrow \left[\ :\!F\!-\!B\!-\!\overset{..}{F}\!: \ \right]^-$$

14.

$$:\!F\!-\!B \quad :\!\overset{..}{O}\!\!\overset{CH_3}{\underset{CH_3}{}} \longrightarrow :\!F\!-\!B\!-\!\overset{..}{O}\!\!\overset{CH_3}{\underset{CH_3}{}}$$

15. BF_3 is the Lewis acid; $(CH_3)_2O$ is the Lewis base.

16. NO_2^- and OCl^-; the stronger base is OCl^-.

17. Na_2O reacts with water to give sodium hydroxide, NaOH.

$$Na_2O(s) + H_2O \longrightarrow 2NaOH(aq)$$

18. SO_3 reacts with water to give sulfuric acid, H_2SO_4

$$SO_3(g) + H_2O \longrightarrow H_2SO_4(aq)$$

19. $AlCl_3$. Al^{3+} has a greater charge density than Mg^{2+}, so the hydrated aluminum ion will be more acidic.

20. Acidic. The large charge on an Os^{8+} ion would give the ion a very large charge density.

21. The sol–gel process is an acid–base reaction. The first step is the replacement of an ethoxide ion with a hydroxide ion, which then reacts with another acid.

22. Ceramic materials have extremely high melting points and are also very hard substances with excellent heat resistance.

23. Sol-gel ceramics, prepared by a slow chemical reaction, have a very uniform surface that reduces chances of cracking and failure at high temperature and stress.

Tools for problem solving

In this chapter you learned to apply the following concepts as tools in solving problems. Study each one carefully so that you know what each is used for. When faced with solving a problem, recall what each tool does and consider whether it will be helpful in finding a solution.

You might want to tear these pages out to use along with solving problems in this chapter.

Brønsted–Lowry definitions (Section 15.1)

Acids are proton donors and bases are proton acceptors.

Conjugate acid–base pairs (Section 15.1)

Every conjugate acid–base pair consists of an acid plus a base with one less proton (H^+).

Conjugate acid–base equilibria (Section 15.1)

The conjugate acid is always on the opposite side of an equation from its conjugate base, and the Brønsted–Lowry acid–base equilibrium is typically written with two sets of conjugate acid–base pairs.

Acid–base strength and the position of equilibrium (Section 15.2)

In an acid–base equilibrium, the equilibrium favors the weaker acid and base. You can also use the position of equilibrium to establish the relative strengths of the acids or bases in an equilibrium.

Reciprocal relationship in acid–base strengths (Section 15.2)

The stronger an acid, the weaker is its conjugate base. This tool can help establish the position of equilibrium when dealing with more than one acid–base pair.

Periodic trends in strengths of binary acids (Section 15.3)

The acidities of H—X bonds increases from left to right across a period and from top to bottom in a group.

Trends in the strengths of oxoacids (Section 15.3)

As the electronegativity of the central atom increases, the acidity of the oxoacid increases, as long as the number of oxygen atoms on the atom remains the same. For the same central atom, as the number of oxygen atoms on the central atom increases, the strength of the oxoacid increases.

Lewis acid–base definitions (Section 15.4)

A Lewis acid accepts a pair of electrons and a Lewis base donates a pair of electrons. The bond formed in the acid–base neutralization reaction is a coordinate covalent bond.

Acidity of metal cations (Section 15.5)

Metal ions become better at polarizing H_2O molecules as their charge increases and their size decreases. Use this to compare the relative abilities of different metal ions to produce acidic solutions in water

Notes:

Chapter 16

Acid–Base Equilibria in Aqueous Solutions

The list of the strong acids and bases available in chemistry, as you now know, is very short. Most acids and bases are weak, and there are thousands of them. Those that are somewhat soluble in water do not come even close to being 100% ionized in solution. But they vary widely in how weak they are, so special equilibrium constants for these substances—acid or base ionization constants—have been devised whose values tell us at a glance how *relatively* strong or weak such acids and bases are.

Weak acids or bases and their conjugates have life-and-death roles in nature because they help to maintain the pH values of the fluids of living systems within very narrow limits. Our study of buffers explains how this all works. If you are planning to enter any one of the health sciences, buffers might be the most important single topic in this chapter, given their importance in human health and disease.

Learning how to do the various calculations in this chapter is of paramount importance. If you can do the calculations, and doing them is not just mechanical, you almost certainly understand the concepts. The large number of worked examples have been very carefully prepared. If you master one kind before going on to the next, the way to success will be smooth. The trickiest parts involve making simplifying assumptions about what we can safely ignore in certain calculations. It's particularly important that you understand these assumptions and can judge when to make them, because they really do simplify the calculations.

Good luck in your study of this vital chapter. There are very few places in either the chemical or the biological sciences where the equilibria of weak acids and bases in water are not important.

16.1 Water, pH, and "p" Notation

Learning Objective

Define pH and explain the use of "p" notation.

Review

Water undergoes *autoionization* as follows, and you should learn this equilibrium equation.

$$H_2O \rightleftharpoons H^+ + OH^-$$

Actually, because H^+ never exists in water unattached to H_2O, a better way to express water's autoionization is as follows, but chemists nearly always "shorten" the equation to the above.

$$2H_2O \rightleftharpoons H_3O^+ + OH^-$$

It is very common to use H^+ and $[H^+]$ as "stand-ins" for H_3O^+ and $[H_3O^+]$.

The *ion product* of the autoionization equilibrium is $[H^+]\,[OH^-]$, and at 25 °C, the value of this ion product, called the *ion product constant*, K_w, is 1.0×10^{-14}. It is very important that you learn the following.

$$[H^+][OH^-] = K_w = 1.0 \times 10^{-14} \text{ (at 25 °C)} \tag{1}$$

(Virtually all calculations involving the value of K_w in this chapter, as well as in quizzes and examinations, assume a temperature of 25 °C and a value of K_w of 1.0×10^{-14}.)

It's important to remember that Equation 1 relates $[H^+]$ and $[OH^-]$ *in any aqueous solution, regardless of the solutes*. You should be able to use the equation to calculate $[H^+]$ from $[OH^-]$ and vice versa. Work Practice Exercise 16.1 through 16.5.

Regardless of the solute and regardless of the temperature of the solution, $[H^+]$ and $[OH^-]$ *equal each other in a neutral solution*. In fact, *only* in a neutral solution do $[H^+]$ and $[OH^-]$ equal each other. At 25 °C, their values happen to equal $1.0 \times 10^{-7}\ M$, because at this temperature $K_w = 1.0 \times 10^{-14}$.

When $[H^+]$ is greater than $[OH^-]$, the solution is acidic. When $[H^+]$ is less than $[OH^-]$, the solution is basic. Figure 16.1 in the text illustrates the pH of common household items that you are familiar with. Notice that foods tend to be acidic and household cleaners tend to be basic solutions. Certain compounds, called indicators, such as phenolphthalein, bromothymol blue, and thymol blue, have different colors in acidic as compared to basic solutions (see Table 16.7).

When the value of $[H^+]$ is small (less than 1 M), it is convenient to represent acidity on a logarithmic scale called pH. Equations 16.4 and 16.5 in the text are alternative definitions of pH. Equation 16.4 is used to calculate pH from $[H^+]$ and Equation 16.5 is used to find $[H^+]$ from pH. Similar equations apply to pOH and values of $[OH^-]$.

$$pH = -\log [H+]$$

$$pOH = -\log [OH-]$$

$$pH + pOH = 14.00 \text{ (at 25 °C)}$$

Be careful to note that we use common logarithms in calculations involving pH and pOH, **not natural logarithms**.

Remember, the lower the pH, the more acidic the solution is. Similarly, the higher the pH, the more basic the solution is. Be sure to study and work Practice Exercises 16.1 through 16.5 before tackling the following problems. Once you've done a few and have gotten used to your pocket calculator for doing them, these problems are not hard. Get as good at doing them as you can.

Example 16.1   pH and [H⁺] relationships

An aqueous solution with a pH of 3.0 has how many times the molar concentration of H^+ as an aqueous solution with a pH of 6.0?

Analysis:

We need to determine the ratio of the hydrogen ion concentration in two solutions to have a measure of "how many times" one is compared to another. The ratio of the two concentrations will give us this answer. In this problem we are given two pH values and need to convert them into hydrogen ion (hydronium ion) concentrations to answer the question.

Assembling the Tools:

We need to use the equation that relates pH to $[H^+]$.

Solution:

The definition of pH is

$$pH = -\log[H^+] \text{ or } -\log[H_3O^+]$$

A pH = 3.0 is converted to $[H^+]$ by changing the sign and then taking the antilog (often the 10^x or 2nd function of the log key on your calculator). The result is that pH 3.0 has $[H^+] = 1 \times 10^{-3}\,M$ and pH 6.0 has $[H^+] = 1 \times 10^{-6}\,M$. The ratio of these concentrations is

$$\frac{1 \times 10^{-3}\,M}{1 \times 10^{-6}\,M} = 1000$$

So the solution with a pH of 3 has 1000 times the H^+ concentration as the solution with a pH of 6.

Is the Answer Reasonable?

We check the calculations and find they are done correctly. It is important to note that the ratio of the pH values is relatively meaningless.

Self–Test

1. When $[H_3O^+]$ equals $3.80 \times 10^{-5}\,M$ at 25 °C,

    (a) what is the value of $[OH^-]$? _____

    (b) Is the solution acidic, basic, or neutral? _____

2. At 37 °C, $K_w = 2.5 \times 10^{-14}$. If an aqueous solution has $[H_3O^+]$ equal to $1.6 \times 10^{-7}\,M$, is the solution acidic, basic, or neutral?

    _____

3. Calculate the concentration of H^+ in a solution in which the hydroxide ion concentration is

    (a) $2.0 \times 10^{-5}\,M$ _____ (b) $4.0 \times 10^{-8}\,M$ _____

4. An aqueous solution at 25 °C with a pOH of 10.00 has a hydrogen ion concentration equal to

    (a) $1.00 \times 10^{-10}\,M$ (b) $1.00 \times 10^{-4}\,M$ (c) $10.00\,M$ (d) $4.00\,M$

    _____

5. When $[H^+]$ equals 4.8×10^{-6}, the pH is

    (a) 5.20   (b) 6.48   (c) 5.32   (d) 4.80

    _____

6. When the pH equals 9.65 at 25 °C, the value of $[H^+]$ is

    (a) $2.24 \times 10^{-10}\,M$ (c) $4.35 \times 10^{-10}\,M$

    (b) $9.65 \times 10^{-14}\,M$ (d) $3.50 \times 10^{-9}\,M$

    _____

7. When pOH equals 7.35 at 25 °C, what is $[H^+]$? _____

8. What color does each of the following indicators have in a strongly acidic and in a strongly basic solution? (See Table 16.7 to check your answers.)

Indicator	Color in	
	Acid	**Base**
phenolphthalein		
bromothymol blue		
thymol blue		

16.2   pH of Strong Acid and Base Solutions

Learning Objective

Explain how to determine the pH of strong acids or bases in aqueous solution.

Review

When we have dilute solutions of strong acids and bases, the calculation of [H$^+$] or pH is particularly simple because these solutes are 100% ionized in solution. Remember, *we do not write the ionization reaction of a strong acid or a strong base as an equilibrium.* For strong, monoprotic acids, for example, the value of [H$^+$] is identical with the molar concentration of the acid. Thus, 0.10 *M* HCl has [H$^+$] = 0.10 *M*.

The autoionization of water contributes essentially nothing to the value of [H$^+$] except in extremely dilute solutions (10^{-6} *M* or less).

If you have not yet done so, be sure to memorize the names and formulas of the following strong, monoprotic acids:

HClO$_4$, perchloric acid	HCl, hydrochloric acid
HClO$_3$, chloric acid	HBr, hydrobromic acid
HNO$_3$, nitric acid	HI, hydriodic acid

Sulfuric acid, H$_2$SO$_4$, is also a strong acid, at least in its first ionization, but calculating the pH of its solutions is not as simple as for strong monoprotic acids. Also be sure to remember that all of the Group IA metal hydroxides are water-soluble strong bases and fully dissociated in water; the Group IIA metal hydroxides are strong bases too, but they are not as soluble as the hydroxides of the Group IA metals.

For a metal hydroxide from Group IIA, remember that you need to take into account the stoichiometry of the dissociation reaction when calculating the OH$^-$ concentration. For example,

$$Ba(OH)_2 \rightarrow Ba^{2+} + 2OH^-$$

Two moles of hydroxide ion are
released for each mole of Ba(OH)$_2$.

Self–Test

9. Calculate the pH of each of the following solutions.

    (a) 0.25 M HCl   _____

    (b) 0.25 M NaOH   _____

    (c) 0.0010 M Ca(OH)$_2$ _____

10. A solution of KOH has a pH of 12.45. What is the molarity of the KOH solution?

    _____

11. Suppose 25.0 mL of 0.45 M HCl is added to 30.0 mL of 0.35 M NaOH.

    (a) Write the equation for the chemical reaction that takes place in the solution.

    _____

    (b) After these substances have reacted, what is the pH of the solution? (*Hint*: Be sure to find the H$^+$ concentration in the final volume of the mixture.)

    _____

16.3   Ionization Constants, K_a and K_b

Learning Objective

Write expressions for the acid ionization constant, K_a, and base ionization constant, K_b, and explain how they are related to each other.

Review

Weak acids

The first thing you should learn here is the general chemical equation for the ionization of a weak acid.

$$HA \rightleftharpoons H^+ + A^- \tag{2}$$

The same equation applies regardless of the formula of the acid. Thus, HA could be a neutral species, like $HC_2H_3O_2$, an anion, like HSO_4^-, or a cation, like NH_4^+. From this equilibrium equation comes the corresponding *acid ionization constant*, K_a.

$$K_a = \frac{[H^+][A^-]}{[HA]} \tag{3}$$

Sometimes, it's convenient to express K_a in logarithmic form as pK_a, which is defined as follows.

$$pK_a = -\log K_a \tag{4}$$

You must be sure to master the skills for preparing Equations (2) through (4) for any specific acid before you go on to Section 16.4. You can test your progress by working the Self-Test below.

Weak bases

The general equation for the ionization of a weak base is

$$B + H_2O \rightleftharpoons BH^+ + OH^- \tag{5}$$

The corresponding *base ionization constant*, K_b is

$$K_b = \frac{[BH^+][OH^-]}{[B]} \tag{6}$$

The logarithmic expression for pK_b, is

$$pK_b = -\log K_b \tag{7}$$

Be sure that you can construct Equations (5) through (7) for any specific weak base before you go to Section 16.4. You can test your progress by working the Self-Test below.

Conjugate acid–base pairs

For any acid–base conjugate pair, the following relationships apply.

$$K_a \times K_b = K_w \tag{8}$$

$$pK_a + pK_b = 14.00 \text{ (at 25 °C)} \tag{9}$$

These useful equations mean that when we know K_a (or pK_a) for a weak acid, we can calculate the values of K_b (or pK_b) for its conjugate base. Most references supply tables only for molecular weak acids or bases, so when we want the K_a or K_b for an ionic species that's a weak acid or base, we usually have to calculate it from the corresponding values for its conjugate base or acid. For example, if we need the K_a for $N_2H_5^+$, we calculate it from the K_b for its conjugate base, N_2H_4. The K_b is tabulated for N_2H_4, but the K_a for $N_2H_5^+$ is not.

Be sure to work Practice Exercises 16.8 – 16.15; then try the Self–Test below. The following example reviews some of these principles.

Example 16.2 **Determining the pK_b, K_a, and pK_a from the K_b of a weak base**

The weak base, aniline, $C_6H_5NH_2$, is a starting material for the synthesis of many useful compounds including a large number of colored compounds called dyes. It has a $K_b = 1.5 \times 10^{-10}$. Write the ionization reaction for aniline and determine its pK_b. Determine the K_a and pK_a of its conjugate acid, the anilinium ion.

Analysis:

Aniline has been identified as a weak base. Since it is the only solute, we will use the K_b and start our problem by writing the chemical equation. Most weak bases we see in this course are related to ammonia and we see that if the C_6H_5 in aniline was replaced by a hydrogen atom it would have the formula of ammonia, NH_3. The advantage is that we know the equation for the reaction of ammonia with water so we can use that as the template for our ionization reaction, or any other nitrogen base.

For the second part of the problem, to convert from the K_b to the pK_b and then to the K_a and pK_a of the anilinium ion.

Assembling the Tools:

We need the tool that gives the general form of a weak nitrogen base ionizing in aqueous solution.

$$NH_3(aq) \;+\; H_2O \;\rightleftharpoons\; NH_4^+(aq) \;+\; OH^-(aq)$$

In addition, our tools for converting between K_a and K_b as well as pK_b and pK_a are needed. We recall that $K_aK_b = 1.0 \times 10^{-14}$ and that $pK_a + pK_b = 14.00$.

Solution:

Taking the ionization reaction as a sample we remove one H and add back the C_6H_5 group to get

$$C_6H_5NH_2(aq) \;+\; H_2O \;\rightleftharpoons\; C_6H_5NH_3^+(aq) \;+\; OH^-(aq)$$

where the $C_6H_5NH_3^+(aq)$ is the anilinium ion. We then calculate the pK_b as

$$pK_b \;=\; -\log(1.5 \times 10^{-10}) \;=\; 9.82$$

The K_a for $C_6H_5NH_3^+(aq)$ is calculated by dividing the K_b into 1.0×10^{-14} as

$$K_a = \frac{1.0 \times 10^{-14}}{1.5 \times 10^{-10}} = 6.7 \times 10^{-5}$$

The pK_a will be $-\log(6.7 \times 10^{-5}) = 4.17$. The pK_a can also be determined by subtracting the pK_b from 14.00. Using that method we get

$$K_a \;=\; 14.00 - 9.82 \;=\; 4.18$$

The difference between the two results is due to rounding errors and should not concern you.

Is the Answer Reasonable?

In writing the equation we can check that we have two conjugate acid-base pairs and we do. Counting charges is a quick way to check that our equation is balanced. To check the pK values we take the exponent of the equilibrium constant and make it a positive number we call x. Then the value of the pK must be between $x-1$ and x. In our problem the K_b has an exponent of -10, therefore $x = 10$. Then we know that the K_b must be in the range of $9-10$. In our case the pK_b was found to be 9.82, which is between 9 and 10 and we are confident the answer is correct.

Self–Test

12. For each of the following acids, write both the equilibrium equation for the ionization and the expression for K_a.

   (a)   HCN   _____

   (b) $HClO_2$ _____

   (c) $N_2H_5^+$ _____

13. For each of the following bases, write both the equilibrium equation for the ionization and the expression for K_b.

   (a) $(CH_3)_2NH$ _____

   (b) N_2H_4 _____

   (c) CN^- _____

14. What is the value of pK_a for the bicarbonate ion if its K_a is 4.7×10^{-11}?

   _____

15. Which is the stronger acid, nitrous acid ($K_a = 7.1 \times 10^{-4}$) or formic acid ($K_a = 1.8 \times 10^{-4}$)?

   _____

16. Which is the stronger acid, hydrocyanic acid, HCN ($pK_a = 9.20$) or the ammonium ion, NH_4^+ ($pK_a = 9.24$)?

   _____

17. The K_b for the fluoride ion is 1.5×10^{-11}. What is its pK_b?

   _____

18. Nitrous acid, HNO_2, has $K_a = 7.1 \times 10^{-4}$. What is the value of K_b for NO_2^-?

   _____

19. The azide ion, N_3^-, has $pK_b = 9.26$. What is K_a for hydrazoic acid, HN_3?

   _____

16.4   Determining K_a and K_b Values

Learning Objective

Describe how to determine acid and base ionization constants from experimental data.

Review

We deal with two kinds of calculations in this section.

- Calculating K_a or K_b from an initial concentration of a weak acid or base and either the pH of the solution or some other information about the extent of ionization, like the percent ionization.

$$\text{percentage ionization} = \frac{\text{moles ionized per liter}}{\text{moles available per liter}} \times 100\%$$

- Calculating the extent of ionization of a weak acid or base from K_a or K_b and the initial concentration of the solute. The results of such a calculation are often expressed as the calculated pH of the solution.

Example 16.1 in the text illustrates how to determine a pK_a of a weak acid when the pH of a solution of known molar concentration of the acid is measured. Example 16.2 shows how to determine the pK_b of a weak base when the molar concentration and percentage ionization are known. It is important to understand the reasoning involved in obtaining the entries in the concentration tables. We use such tables often.

16.5   pH of Weak Acid and Weak Base Solutions

Objective

Determine equilibrium concentrations and pH for weak acid or base solutions.

Review

Calculating equilibrium concentrations from K_a or K_b

The one thing all problems of this type have in common is the necessity *at the outset* of writing the correct equation for the equilibrium. Then you can write the appropriate equilibrium law. And then you'll have something into which to substitute quantities and to calculate equilibrium concentrations of products of the equilibrium. Writing the correct equilibrium equation is easy because there are only three options.

1. ***The only solute is a weak acid.*** Write the equilibrium equation for the ionization of the acid and from it write the equation for the acid ionization constant, K_a. Next, use this equation and the value of K_a (either given or from a table) to solve the problem. If the ionization constant you're given is not K_a but is K_b for the acid's conjugate base, then you must use Equation 7, above, to find K_a.

2. ***The only solute is a weak base.*** In this case, you must write the equation for the ionization of the base, modeling the equation after (4), above. Then you write the expression for K_b and use this equation and the value of K_b to solve the problem. If you've been given, instead, the value of K_a for the base's conjugate acid, you have to use Equation 7, above, to find K_b.

3. ***The solution contains both a weak acid and a salt of its conjugate base.*** Such solutions are called *buffers* and are discussed in Section 16.7. With a solution of this type, you can use either K_a or K_b to solve the problem. It depends on the information supplied. If you have a value of K_a, then work with the acid and write the equation for its ionization and for the equilibrium law, K_a. On the other hand, if you have a value of K_b for the acid's conjugate base, develop the ionization equation for the base, prepare the equilibrium law, K_b, and proceed to solve the problem.

Simplifications that apply to most weak acid–base calculations

This section of Chapter 16 takes you through a calculation where we're given the concentration of a weak acid and its K_a, and it illustrates a very important simplification: when K_a is very small, the change in the concentration of the weak acid or base that occurs when the ionization takes place doesn't appreciably change the initial concentration. As a result, the equilibrium concentration of the acid is effectively the same as the initial concentration. Let's review the reasoning.

Suppose we have a 0.10 *M* H*A* solution and K_a for the acid is very small. When the acid ionizes, a tiny amount of H*A* will be lost per liter, which we can call x. In other words, the concentration of H*A* decreases by x mol per liter, and at equilibrium the concentration of H*A* equals $(0.10 - x)$ *M*. But we can anticipate that x will be very small compared to the initial concentration, so we make the approximation

$$(0.10 - x) \, M \approx 0.10 \, M$$

In all the calculations you encounter in this section, this approximation is valid, so you can use the *initial* concentration of the acid or base as its *equilibrium* concentration. This makes these equilibrium computations much simpler.

The problem-solving strategy, then, boils down to the following.

- First, determine which of the three categories, above, fits the problem. (In other words, is the solute a weak acid, a weak base, or a mixture of an acid and its conjugate base?)

- Next, write the appropriate chemical equilibrium equation.

- Construct the equilibrium law (an equation either for K_a or K_b) from the equilibrium equation.

- If the only solute is a weak acid or a weak base, the algebra works out as follows:

$$\frac{x^2}{[\text{initial conc.}]} = K_a \quad \text{or} \quad \frac{x^2}{[\text{initial conc.}]} = K_b$$

depending on whether the solute is an acid or a base.

- If the solute is an acid, the value of x obtained is both the H^+ concentration and the concentration of the conjugate base formed by the ionization of the acid.

- If the solute is a base, the value of x obtained is both the OH^- concentration and the concentration of the conjugate acid formed by the ionization of the base.

Study Example 16.3 in the text carefully to see how this problem strategy works and how it can also be used to determine the percentage ionization. Then work Practice Exercises 16.22 through 16.24. When you've done this, you're ready for the Self-Test below.

Example 16.3   Calculations using initial concentrations and K_a

Consider a solution of propanoic acid with a concentration of 0.620 M. Calculate the $[H^+]$, $[C_3H_5O_2^-]$, $[HC_3H_5O_2]$, and the pH of the solution. (Use $K_a = 1.34 \times 10^{-5}$.)

Analysis:

We will use the sequence of steps described above to approach this problem. First we identify that the category that this problem fits into is one where a weak acid is the only solute. Second, we will write the ionization reaction for propanoic acid and third, we write the equilibrium law. We will solve the equilibrium law in its simplified format given above.

Assembling the Tools:

We need the tools that describe the ionization of a weak acid, the writing of the corresponding equilibrium law. Next the procedure for setting up the concentration table so that the appropriate mathematical relationships are evident is useful. Finally, the concept and tools for making reasonable approximations will be used.

Solution:

The first step is to write the ionization reaction as

$$HC_3H_5O_2 \rightleftharpoons H^+ + C_3H_5O_2^-$$

and the equilibrium law is

$$K_a = \frac{[C_3H_5O_2^-][H^+]}{[HC_3H_5O_2]} = \frac{x^2}{[\text{initial concentration}]}$$

We substitute the value of K_a (1.34×10^{-5}) and the initial concentration of propanoic acid ($0.620\ M$) into the equation

$$1.34 \times 10^{-5} = \frac{x^2}{(0.620\ M)}$$

The equation is then rearranged to solve for x

$$x^2 = (0.620\ M)(1.34 \times 10^{-5})$$

Next x is determined by taking the square root. Because we are using the K_a, x represents the hydrogen ion concentration and the concentration of propanoic acid's conjugate base.

$$x = 2.88 \times 10^{-3} = [H^+] = [C_3H_5O_2^-]$$

We can calculate the remaining, unionized propanoic acid by subtracting the amount that does ionize.

$$[HC_3H_5O_2] = 0.620 - 0.00288 = 0.617$$

Finally, the pH is

$$pH = -\log(2.88 \times 10^{-3}) = 2.54$$

Is the Answer Reasonable?

The first check is to be sure that this acid gave an acid pH (below pH 7), and it does. We should also check that the equilibrium law was set up correctly and that the correct numbers were used. We also have an exponent of −3 for the hydrogen ion concentration and therefore we expect the pH to be between 2 and 3.

Self−Test

20. A new organic acid was discovered and at 25 °C a 0.115 M solution of the acid in water had a pH of 3.42. Calculate the value of K_a for the acid.

   _____

21. A new drug obtained from the seeds of a Uruguayan bush was found to be a weak organic base. A 0.100 M solution in water of the drug had a pH of 10.80. What is the K_b of the drug?

   _____

22. Calculate the pH of a 0.425 M solution of NH_3 in water. (Use $K_b = 1.8 \times 10^{-5}$.)

   _____

23. Calculate the percentage ionization of HCN in a solution that is 0.10 *M* HCN. Obtain the K_a value from Table 16.2 in the text.

_____

Simplifications that apply to most weak acid–base calculations

Example 16.3 takes you through a calculation where we're given the concentration of a weak acid and its K_a, and it illustrates a very important simplification: when K_a is very small (this means that the K_a or K_b is more than 100 times *less* than the initial acid or base concentration), the change in the concentration of the weak acid or base that occurs when the ionization takes place doesn't appreciably change the initial concentration. As a result, the equilibrium concentration of the acid is effectively the same as the initial concentration. This means that the acid or base that ionizes is either (a) smaller than the last significant digit of the initial concentration or (b) less than 5% of the initial concentration. Let's review the reasoning.

Suppose we have a 0.10 *M* H*A* solution and K_a for the acid is very small. When the acid ionizes, a tiny amount of H*A* will be lost per liter, which we can call *x*. In other words, the concentration of H*A* decreases by *x* mol per liter, and at equilibrium the concentration of H*A* equals (0.10 – *x*) *M*. But we can anticipate that *x* will be very small compared to the initial concentration, so we make the approximation

$$(0.10 - x)\,M \approx 0.10\,M$$

so, if the first significant figure for *x* starts *after* the second decimal place, it will usually be dropped. In all the calculations you encounter in this section, this approximation is valid, so you can use the *initial* concentration of the acid or base as its *equilibrium* concentration. This makes these equilibrium computations much simpler.

16.6   Acid–Base Properties of Salt Solutions

Learning Objective

Explain how a salt solution can be acidic or basic.

Review

The questions in this section are "How can we tell if a salt, dissolved in water, affects the pH of the solution?" And, "How can we calculate the pH of such a solution?" If neither the anion nor the cation of a salt influences the pH, there is no calculation to carry out.

Does a salt affect the pH?

To answer this question, you need to know whether either (or both) of the ions of the salt are able to change the pH of the solution. In other words, are the ions weak acids or bases?

You can tell if a salt's anion affects a solution's pH very simply. First, determine the formula for the conjugate acid of the anion of the salt. Then, ask yourself whether this acid is on the list of strong acids. If it is not on the list, the conjugate acid is a weak acid, meaning that the anion in question is at least a weak base and it *will* affect the pH of the solution. On the other hand, if the conjugate acid of the salt's anion is strong, the anion will be such a weak base that it will not affect the pH. For example, suppose the anion is $C_3H_5O_2^-$.

$$C_3H_5O_2^- \xrightarrow{\text{Add } H^+} HC_3H_5O_2 \xrightarrow{\substack{\text{Is the acid on} \\ \text{the list of} \\ \text{strong acids?}}} \text{No}$$

Anion of salt   Conjugate acid

The acid isn't a strong acid, so the anion, $C_3H_5O_2^-$, must be a weak base that is capable of affecting the pH. This anion should tend to make the solution basic.

A somewhat similar procedure is applied to the cation of the salt. First, recall from Chapter 15 that the cations of the metals of Groups IA and IIA (except for Be^{2+}) do not affect the pH. The cations of other metals, however, form hydrates that can release protons to water molecules to form extra hydronium ions and thus make a solution acidic, *but we'll not deal with calculations involving them.* The only kinds of cations that could potentially affect pH and that we will work with are proton-donating cations, such as the ammonium ion and a few others like it. Therefore, if the cation of the salt is from Group IA or IIA, it won't affect the pH. However, *if the cation of the salt is not a metal, you can anticipate that it will be acidic and tend to lower the pH.*

With some salts, like ammonium cyanide, both the cation and the anion of a salt react to some extent with water, with the cation tending to lower the pH and the anion tending to raise it. But *we are not dealing with calculations of pH for such systems.*

Summary

1. If neither the cation nor the anion of a salt can affect the pH, the solution should be neutral. (Example: $NaNO_3$)

2. If only the anion is basic, the solution will be basic. (Example: $NaC_2H_3O_2$)

3. If only the cation is acidic, the solution will be acidic. (Example: NH_4Cl)

4. If the cation is acidic and the anion is basic, the pH outcome depends on their strengths relative to each other. We'll not encounter calculations dealing with such systems. (Example: NH_4CN)

Work Practice Exercises 16.25 and 16.26.

How can we calculate the pH of a salt solution when the salt does affect the pH?

If you determine that a salt's anion (but not its cation) affects pH, the pH calculation is simply the same kind that you do for any Brønsted base, as you just learned.

If the salt involves a proton-donating cation, like the ammonium salt of a strong acid, the pH calculation is nothing more than what you do for any weak acid, just as you learned in Section 16.5.

The calculations, in other words, involve essentially nothing new, once it's determined that a calculation is warranted. The problem-solving strategy mirrors the one described in the previous section, above:

1. Determine the nature of the salt. Which ion, if either, affects the pH?

2. Write the appropriate chemical equilibrium equation for the reaction of that ion with water.

3. Construct the equilibrium law (an equation either for K_a or K_b) from the equilibrium equation.

4. Get the value of the ionization constant, K_a or K_b. Seldom for any *ion* will this value be found directly in a table. If you need the K_b of an *anion*, you'll have to use the K_a of its conjugate acid and than calculate K_b from $K_a \times K_b = K_w$. If you need the K_a of a cation, like NH_4^+, you may have

to find the K_b of its conjugate base, NH_3, and use the equation $K_a \times K_b = K_w$ to calculate the K_a that you need.

From here on, the calculation is just like that for weak molecular acids or molecular bases. Once again, the algebra works out to be

$$\frac{x^2}{[\text{initial conc.}]} = K_a \quad \text{or} \quad \frac{x^2}{[\text{initial conc.}]} = K_b$$

depending on whether the reactant is the cation (an acid) or the anion (a base).

With the above before you, study Example 16.4 and work Practice Exercises 16.27 through 16.31. Then tackle the Self–Test that follows.

Example 16.4   Calculating the pH of a salt solution

Is a 0.100 M solution of KNO_2 acidic, basic, or neutral? If it is not neutral, what is its calculated pH?

Analysis:

The first step in solving this problem is to identify the ions that this compound dissociates into when dissolved in water. Next we decide if these ions came from strong or weak acids and bases. The easiest way to do this is to add H^+ to the negative ion and OH^- to the positive ion to make the conjugate acid and base. From there we can decide if they are strong or weak. If an ion comes from a strong acid or strong base, it will be too weak to change the pH of water. If on the other hand the ion came from a weak acid or base, that ion will in turn be a weak conjugate base or acid that will affect the pH of water. If either ion has an effect on the pH, we can calculate the necessary K_a or K_b to determine the pH using a calculation similar to what was used in solving Practice Exercises 16.14 and 16.15.

Assembling the Tools:

We need our tools for writing the appropriate ions from a salt that dissolves in water. We also need our methods of identifying weak acids and bases as well as strong acids and bases. We then use the concept that the ion that is the stronger conjugate acid (or base as the case may be) will dictate if the solution is acidic or basic.

Solution:

When dissolved in water KNO_2 undergoes the reaction

$$KNO_2 \longrightarrow K^+(aq) + NO_2^-(aq)$$

If we add H^+ to the NO_2^-, we get HNO_2 and recognize it as a weak acid. If we add OH^- to K^+, we get KOH that is recognized as a strong base. Therefore K^+ will not affect the pH of the solution because it is a very weak conjugate acid, but the NO_2^- will. It will act as a base (it must be a conjugate base since it came from an acid, HNO_2). We then write the reaction of NO_2^- with water as

$$NO_2^- + H_2O \rightleftharpoons HNO_2 + OH^-$$

The equilibrium law for this reaction is

$$K_b = \frac{[HNO_2][OH^-]}{[NO_2^-]}$$

The initial concentration of NO_2^- is 0.100 M, and a small amount, x, reacts with water. We then conclude that :

$$[NO_2^-] = 0.100 - x \quad \text{and} \quad [HNO_2] = [OH^-] = x$$

To simplify the calculations we assume that x is very small so that $[NO_2^-] = 0.100$.

Next, we can look up the K_a for HNO_2, which is 7.1×10^{-4}. From this, the $K_b = 1.4 \times 10^{-11}$ (obtained by dividing K_w by K_a). We then enter these variables into the equilibrium law and solve for x to get

$$K_b = \frac{x^2}{0.100} = 1.4 \times 10^{-11}$$

$$x^2 = (1.4 \times 10^{-11}) \times (0.100) = 1.4 \times 10^{-12}$$

$$x = 1.2 \times 10^{-6} = [OH^-] \quad \text{and} \quad pOH = 5.93$$

The pH $= 14.00 - pOH = 14.00 - 5.93 = 8.07$

Is the Answer Reasonable?

We decided that the NO_2^- was a conjugate base of the weak acid HNO_2. Therefore the solution should be basic (pH greater than 7), and it is.

We always check to be sure that any simplifying assumptions are true before declaring a problem solved. It is disheartening to find that after all the calculations, an assumption turned out to be false. We do have a "rule-of-thumb" that tells us that if our initial concentration is more than one hundred times the value of the equilibrium constant, the assumption will surely work. So we say that if $C_{HA} > 100 \times K_a$, then the assumption will work. For weak bases we say that $C_B > 100 \times K_b$ assures that the assumption is true for weak bases.

Self–Test

24. Decide if each ion is a weak base that's capable of affecting the pH of a solution.

    (a) $C_6H_5O^-$ _____

    (b) OCN^- _____

    (c) NO_3^- _____

25. Predict the behavior of each salt in water by stating if it will cause an aqueous solution to be acidic, basic, or neutral.

    (a) KNO_3 _____   (d) $NaClO_4$ _____

    (b) $NaClO_2$ _____   (e) C_5H_5NHCl _____

    (c) KF _____   (f) $NaNO_2$ _____

26. Will either ion from potassium butyrate, $KC_4H_7O_2$, react with water? _____

    If so, calculate the pH of a 0.120 M solution. (For $HC_4H_7O_2$, $K_a = 1.5 \times 10^{-5}$.) _____

27. Will either ion from ammonium bromide, NH_4Br, react with water? _____

    If so, calculate the pH of a 0.240 *M* solution. _____

28. Will either ion from potassium iodide, KI, react with water? _____

    If so, calculate the pH of a 0.100 *M* solution. _____

29. What is the minimum concentration of chloroacetic acid, $HC_2H_2O_2Cl$, in an aqueous solution that would allow us to reasonably use the simplifying assumption about the equilibrium concentration of the acid? The K_a of this acid is 1.4×10^{-3}.

    _____

30. Calculate the pH of a 0.010 *M* solution of chloroacetic acid. (Refer to the previous Question for its formula and K_a value.) _____

    Did the assumption work? _____

16.7 Buffer Solutions

Learning Objectives

Describe what a buffer solution is, how it works, and how to calculate pH changes in a buffer.

Review

A *buffer* is a solution that maintains a nearly constant pH even when small amounts of strong acid or strong base are added to it. Be sure to note, however, that a buffer doesn't necessarily hold a solution to a *neutral* pH of 7.00 (at 25 °C). Rather, the pH is held reasonably steady at whatever pH value the buffer is prepared. Thus, the buffered solution could be acidic, basic, or neutral.

Note further that a buffer doesn't necessarily maintain an absolutely *constant* pH. It only prevents large changes of pH when strong acids or bases challenge the system.

A buffer is a solution that contains both a weak acid and a weak base. Often, the *buffer pair* consists of a weak acid, H*A*, and one of its salts (e.g., Na*A*). The anion A^- is the Brønsted base in the buffer; its conjugate acid, H*A*, is the acid component of the buffer. A typical pair is sodium acetate and acetic acid. The base, acetate ion, has the ability to react with extra protons that enter. The acid component, acetic acid, can neutralize extra hydroxide ions that get into the solution. This prevents large changes in the concentrations of H^+ or OH^- ions, and therefore prevents large changes in pH.

Another kind of buffer pair consists of a weak base and its salt – for example, ammonia plus the ammonium salt of a strong acid, such as ammonium chloride. The cation, NH_4^+, can neutralize any OH^- ion that enters the solution, and NH_3 can react with any protons that get in.

To perform buffer calculations, you must understand how the components of a buffer function. In other words, you should be able to write equations that show how they neutralize H^+ or OH^-. Here's a summary:

When a strong acid is added to a buffered solution, it supplies H^+, which is neutralized by the reaction

$$\text{(conjugate base)} + H^+ \longrightarrow \text{(conjugate acid)}$$

$$C_2H_3O_2^- + H^+ \longrightarrow HC_2H_3O_2$$

$$NH_3 + H^+ \longrightarrow NH_4^+$$

When a strong base is added to a buffered solution, it supplies OH^-, which is neutralized by the reaction

$$\text{(conjugate acid)} + OH^- \longrightarrow \text{(conjugate base)} + H_2O$$

$$HC_2H_3O_2 + OH^- \longrightarrow C_2H_3O_2^- + H_2O$$

$$NH_4^+ + OH^- \longrightarrow NH_3 + H_2O$$

Study these equations as they apply to the acetic acid/acetate buffer and the ammonium ion/ammonia buffer.

Factors that determine the pH of a buffer pair

There are two factors.

1. The pK_a of the Brønsted acid of the pair. (If the buffer is the ammonium ion/ammonia buffer or one like it, then this factor is the pK_b of the Brønsted base.)

2. The logarithm of the *ratio* of the *initial* values of the molarities of the buffer's components.

It's safe to use *initial* molarities for *equilibrium* molarities in all of our buffer calculations.

Buffer calculations start with the chemical equations and equilibrium expressions that define K_a or K_b. When the buffer consists of a weak acid, HA, and a salt, like NaA or KA, we use the equation for K_a.

$$K_a = \frac{[H^+]\,[A^-]}{[HA]}$$

The text explains how, in buffer calculations, we can use *initial* molar concentrations as equilibrium concentrations for both $[A^-]$ and $[HA]$. Solving this equation for $[H^+]$ and switching to logarithms gives us the pH of the buffer in terms of the value of pK_a and the log of the ratio of acid to anion. Assuming the acid is monoprotic and can be represented by HA, we have

$$pH = pK_a - \log \frac{[HA]_{init}}{[A^-]_{init}} \tag{10}$$

This equation is known as the Henderson-Hasselbalch equation. It is frequently used in biology and other life sciences.

When the buffer is made up of a weak base, B, and its conjugate acid, BH^+, (e.g., the ammonia/ammonium ion buffer), then we use the defining equation for K_b for the weak base.

$$K_b = \frac{[BH^+]\,[OH^-]}{[B]}$$

Again, *initial* molar concentrations of cation and base can be used as equilibrium values. Solving for [OH⁻] and taking the negative logarithm gives pOH from which pH is easily obtained.

Factors that determine the choice of a specific buffer pair

There are two factors, and they are essentially identical with the two given above that are involved in the pH of a buffer pair. As Equation 10 above shows, for a buffer with an acidic pH, these factors are the pK_a of the weak acid and the ratio of the molar concentrations of the weak acid and its conjugate base. In experimental work, these factors govern the selection of a buffer pair.

When we want a buffer for a pH not identical to, but still not too far from the acid's pK_a, we calculate the ratio of the molarities of anion to acid that will adjust the system to the desired pH. But the choice of the acid for its pK_a value is of first importance. For maximum effectiveness, the ratio of the initial molar concentrations of the anion to the acid should be somewhere between 10:1 and 1:10.

The text does not deal with choosing buffer pairs for basic pH values in general; it only develops the ammonium ion/ammonia buffer. However, the same principles would apply.

Calculating the effect of strong acid or base on the pH of a buffer

In calculations that deal with the results of adding strong acid or base to a given buffer, we need to apply the equations given above that describe how a buffer works. Remember that when H⁺ is added to the buffer, it reacts with the conjugate base. This reduces the amount of conjugate base by the number of moles of H⁺ added, and increases the amount of conjugate acid by the identical amount. Similarly, when OH⁻ is added to a buffer, by reaction it reduces the amount of conjugate acid by the number of moles of OH⁻ added, and increases the amount of conjugate base by the same amount. Study Example 16.7 in the text to see how these calculations proceed.

Factors that determine actual quantities of the members of the buffer pair

This consideration has to do with buffer *capacity*, that is, with just how much acid or base we expect the prepared buffer will have to absorb during some experiment while holding the pH within a predetermined range. For a large capacity buffer, we would want relatively high concentrations of each component, remembering that it is the *ratio* of the two molarities that, together with the pK_a, determines the initial pH. In biological applications of buffers, however, the potential toxicities of the buffer components at high concentrations must be considered. These principles are also covered in Examples 16.5 to 16.7 in the text.

Example 16.5 Preparation of a buffer solution

You are a member of a team of scientists that decides that the lab should have on hand 1.0 L of an aqueous buffer for a pH of 4.50. How will you prepare this buffer if the conjugate acid must be 0.15 *M*?

Analysis:

This question has no single answer. Any acid that has a pK_a within one pH unit of 4.5 will work. However, the best buffering capacity toward both acids and bases will be found if the acid we choose has a pK_a as close as possible to the desired pH 4.5. We look in the lists of acids to make our selection. From the pH we can calculate [H⁺] and we can rearrange the equilibrium law to

$$\frac{K_a}{[H^+]} = \frac{[A^-]}{[HA]}$$

Since we know three of the four terms in this equation (K_a for the acid we will select, [H⁺] from the pH, and 0.15 *M* = [HA] from the statement of the problem), we can determine [A⁻]. Finally, since we know we want one liter of buffer, we can determine the moles of the H*A* and *A*⁻ and then the mass of each we need to weigh.

Assembling the Tools:

We need a list of weak acids and their $K_a$values (or pK_a values). The tools for writing the equilibrium equation for the weak acid and its equilibrium law are also needed.

Solution:

First we select an acid. Benzoic acid, acetic acid and hydrazoic acid all have pK_a values close to 4.5. Since acetic acid is the most common of all these, we will choose to use it and its salt, sodium acetate, will provide the conjugate base. The chemical equation is

$$HC_2H_3O_2 \rightleftharpoons H^+ + C_2H_3O_2^-$$

The K_a for acetic acid is 1.8×10^{-5}. The pH of 4.5 can be converted into an $[H^+] = 3.2 \times 10^{-5}$, and the concentration of acetic acid, $[HC_2H_3O_2]$, is 0.15 *M*.

The equilibrium law is

$$K_a = \frac{[H^+][C_2H_3O_2^-]}{[HC_2H_3O_2]}$$

and substituting the known values gives us

$$1.8 \times 10^{-5} = \frac{(3.2 \times 10^{-5})[C_2H_3O_2^-]}{(0.15)}$$

Rearranging and solving for $[C_2H_3O_2^-]$ results in

$$[C_2H_3O_2^-] = \frac{0.15}{3.2 \times 10^{-5}} \times 1.8 \times 10^{-5} = 0.084\ M\ C_2H_3O_2^-$$

Now that we know the molarities of the acetic acid and the sodium acetate (0.084 *M*) that are needed, we convert them to the mass needed.

$$1\ L\ HC_2H_3O_2 \times \frac{0.15\ mol\ HC_2H_3O_2}{1\ L\ HC_2H_3O_2} \times \frac{60.1\ g\ HC_2H_3O_2}{1\ mol\ HC_2H_3O_2} = 9.0\ g\ HC_2H_3O_2$$

and

$$1 \text{ L NaC}_2\text{H}_3\text{O}_2 \times \frac{0.084 \text{ mol NaC}_2\text{H}_3\text{O}_2}{1 \text{ L NaC}_2\text{H}_3\text{O}_2} \times \frac{82.0 \text{ g NaC}_2\text{H}_3\text{O}_2}{1 \text{ mol NaC}_2\text{H}_3\text{O}_2} = 6.9 \text{ g NaC}_2\text{H}_3\text{O}_2$$

Finally, to prepare the buffer we weigh 9.0 g of acetic acid and 6.9 g of sodium acetate into a 1 liter flask. We dissolve them in a small amount of distilled water and then dilute to one liter and mix well.

Is the Answer Reasonable?

Since the pK_a of the acid and the pH of the buffer are close to each other, we expect the ratio of the moles of the conjugate acid to conjugate base to be close to one, and it is. Since the desired pH is lower than the pK_a of the acid, we also expect that the mixture should have more moles of acid than conjugate base, and it does. Recheck your math, but we are already confident that the answer is correct.

Self–Test

31. Calculate the pH of a buffered solution made up of 0.125 M $KC_2H_3O_2$ and 0.100 M $HC_2H_3O_2$. The pK_a of acetic acid is 4.74.

         _____

32. The pK_a of formic acid, $HCHO_2$, is 3.74. What mole ratio of sodium format, $NaCHO_2$, to formic acid is needed to prepare a solution buffered at a pH of 3.00?

         _____

33. How many grams of NH_4Cl have to be added to 200 mL of 0.25 M NH_3 to make a solution that is buffered at pH 9.10? For NH_3, $K_b = 1.8 \times 10^{-5}$.

         _____

Calculating pH change of a buffer

Buffers minimize the change in pH when a small amount of a strong acid or strong base is added, This is because the conjugate acid of the buffer reacts with OH^- ions and the conjugate base of the buffer will react with added acid. This is actually a simple limiting reactant calculation that is based on how much of the conjugate acid will be neutralized by adding a base and how much conjugate base is formed. Example 16.7 in the text illustrates how this type of calculation is approached. When you feel confident with the process, try Practice Exercises 16.38 and 16.39 and the Self-Test below.

Self–Test

34. Suppose 0.050 mol of HCl were added to 1.00 L of the buffer in Question 31. By how much and in which direction will the pH change?

         _____

16.8   Polyprotic Acids

Learning Objective

Define polyprotic acids and describe how to calculate concentrations of all species in a solution of a polyprotic acid.

Review

Be sure you learn to write the equations for the stepwise ionizations of polyprotic acids. Follow Example 16.8 in the text for carbonic acid, H_2CO_3. Also learn to write the acid ionization constant for each step. These are designated as K_{a_1}, K_{a_2}, and so forth.

The heart of this section is in the simplifications that greatly reduce the work of calculations. They are possible because the numerical values of the K_a's generally differ very much from one ionization step to the next. The simplifications are as follows.

1.   We can use just the first ionization step and its K_{a_1} to calculate $[H^+]$ of a solution of the acid.

2.   The value of the anion formed in the first step of the ionization is equal to $[H^+]$.

3.   The equilibrium concentration of the polyprotic acid equals its initial concentration.

4.   The molarity of the ion formed in the second ionization step numerically equals K_{a_2} for the acid, provided that there is no other solute.

In other words, by simplifications 1 and 2 we can treat the acid as if it were monoprotic to find $[H^+]$. The third simplification is made possible by the very small sizes of the first ionization constants of weak polyprotic acids; very little ionization occurs to lower the concentration of the un-ionized species.

The anions of polyprotic acids are weak Brønsted bases. Those of diprotic acids are able to react with water in two steps, each step producing some OH^-. Section 16.6 in the text illustrates this using the carbonate ion.

Just as with diprotic acids, there are two equilibrium constants, one for each of the two steps, designated as K_{b_1} and K_{b_2}. In general, K_{b_1} is much larger than K_{b_2}. Therefore, almost all of the OH^- produced by the two successive reactions with water comes from the first step. We are able to ignore the contribution of OH^- by the second step. Therefore, we can use K_{b_1} and its associated equilibrium expression to calculate $[OH^-]$ and ultimately the pH of the solution.

You may have already noticed the parallel between the simplifications that apply to solutions of bases, such as CO_3^{2-}, and those that apply to solutions of diprotic acids, such as H_2CO_3. In fact, the simplifications are identical, except that in one case we are calculating $[OH^-]$ and in the other $[H^+]$.

The one complicating factor in dealing with problems involving salts of polyprotic acids is that we must calculate the K_b values from the corresponding K_a values for the conjugate acids. But we now know the relationship among K_a, K_b, and K_w for any acid–base conjugate pair. With polyprotic acids, this relationship works itself out with slightly different labels.

$$K_{b_1} = \frac{K_w}{K_{a_2}} \quad \text{and} \quad K_{b_2} = \frac{K_w}{K_{a_1}}$$

Note carefully the subscripts: K_{b_1} comes from K_{a_2} and K_{b_2} comes from K_{a_1}.

After studying Examples 16.8, and 16.9 in the text and working Practice Exercises 40 through 44, take the Self-Test below.

Example 16.6   Calculations involving polyprotic acids

Selenous acid is a weak diprotic acid. Write the two ionization steps for selenous acid along with the equilibrium law for each step and the value of the equilibrium constant for each step. Calculate the concentration of all species present in a 0.50 molar solution of selenous acid.

Analysis:

The key to this problem is that we need to find the formula for selenous acid so that we can deduce the equation for the two ionization steps. Each step is simply the loss of one hydrogen ion. Once we have written the equations the equilibrium laws can be constructed. In addition we need to calculate the concentrations of all species in solution. Let's identify what they are. We will have selenous acid H_2SeO_3 and its two salts, $HSeO_3^-$ ion and the SeO_3^{2-} ion. Finally we will have hydrogen ions, hydroxide ions, and, of course, water molecules in this solution.

Assembling the Tools:

We need to use the tools defining the ionization of polyprotic acids. The pH is calculated using the techniques similar to those described above.

Solution:

We look in the textbook or a handbook for the formula. Removing one hydrogen ion we get the first ionization reaction.

$$H_2SeO_3 \rightleftharpoons HSeO_3^- + H^+$$

The second ionization is

$$HSeO_3^- \rightleftharpoons SeO_3^{2-} + H^+$$

Next we write the equilibrium laws and K_a values.

$$K_{a_1} = \frac{[HSeO_3^-][H^+]}{[H_2SeO_3]} = 4.5 \times 10^{-3}$$

and

$$K_{a_2} = \frac{[SeO_3^{2-}][H^+]}{[HSeO_3^-]} = 1.1 \times 10^{-8}$$

Now that we have all of the equations we can use the concepts that we learned about polyprotic acids. First, if the two values differ by a ratio of 10^4 or more the two steps are independent. This means that we can quickly calculate the $[H^+]$ and $[HSeO_3^-]$ from

$$[H^+] = [HSeO_3{}^{2-}] = \sqrt{K_{a_1} \times (\text{initial conc.})} = \sqrt{4.5 \times 10^{-3} \times 0.50} = 4.7 \times 10^{-2}\,M$$

We also recall that the concentration of $SeO_3{}^{2-}$ is equal to K_{a_2} so we can write

$$[SeO_2{}^{2-}] = K_{a_2} = 1.1 \times 10^{-8}\,M$$

and the concentration of the selenous acid is

$$[H_2SeO_3] = 0.50\,M - 0.047\,M = 0.45\,M$$

We can calculate the concentration of hydroxide ions as

$$[OH^-] = (1.0 \times 10^{-14})/(4.7 \times 10^{-2}) = 2.1 \times 10^{-13}\,M$$

Is the Answer Reasonable?

We have calculated many things in this problem. First, we check the equilibrium equations to be sure the conjugate acids and bases are written correctly. Second, we check the equilibrium laws for proper format. Finally, we see that the K_{a_1} is fairly large as far as weak acids go and the hydrogen ion concentration is correspondingly high.

Self–Test

35. On a separate sheet of paper, write the equilibrium equations and the expressions for K_{a_1} and K_{a_2} for the step-wise ionization of H_2SeO_3.

36. What are the values of $[H^+]$, pH, and $[Asc^{2-}]$ in a solution of ascorbic acid, which we represent as H_2Asc, with a concentration of 0.100 M? $K_{a_1} = 6.8 \times 10^{-5}$ and $K_{a_2} = 2.7 \times 10^{-12}$.

     $[H^+] = $ _____   pH $ = $ _____ $[Asc^{2-}] = $ _____

37. What is the concentration of $[HAsc^-]$ in the solution of the previous question?

     _____

38. Calculate the pH of a 0.20 M solution of sodium oxalate, $Na_2C_2O_4$. Use Table 16.3 in the text for the successive ionization constants of oxalic acid.

     _____

39. Calculate the pH of a 0.10 M solution of Na_2X at 25 °C. The first ionization constant of H_2X is 9.5×10^{-8} and the second is 1.3×10^{-14}. (In this case, $[X^{2-}]_{initial} < 400 \times K_{b_1}$, so you will have to use either the quadratic equation or the method of successive approximations.)

_____

40. In the solution described in Question 38, what is the concentration of $H_2C_2O_4$ at equilibrium?

_____

41. In the solution described in Question 38, what is the concentration of $C_2O_4^{2-}$?

_____

16.9   Acid–Base Titrations

Learning Objective

Draw and explain titration curves for reactions of strong or weak acids and bases.

Review

The nature and appearance of an acid–base titration curve is a function of the relative strengths of the acid and the base, because these determine how the ions being produced by the neutralization reaction react (if at all) with water as Brønsted acids or bases.

- A strong acid is titrated with a strong base. Example: HCl and NaOH

    The titration curve is symmetrical about pH 7, the equivalence point.

- A weak acid is titrated with a strong base. Example: $HC_2H_3O_2$ and NaOH

    The equivalence point is greater than 7 (because the newly forming anion, a Brønsted base, reacts with water to give some OH^- and a slightly alkaline pH).

- A weak base is titrated with a strong acid. Example: NH_3 and HCl

    The equivalence point is less than 7 (because the newly forming cation, a Brønsted acid, reacts with water to give some H^+ and a slightly acidic pH).

The calculations required to obtain points for a titration curve are identical in kind to those studied earlier in the chapter.

- Determine the pH before any titrant is added.

    Use molarity and an acid (or base) ionization constant to find $[H^+]$ (or $[OH^-]$, depending on what exactly is the substance about to be titrated).

- As soon as some titrant is added, you have to take into account three factors.

    (1)   The *total* volume changes, so concentrations change for this reason alone.

(2) The concentration of the substance being titrated, which is needed in the calculation of pH, decreases because it is being consumed in the neutralization.

(3) *If the salt being formed is one with an ion that is basic or acidic*, there is a buffer effect requiring a buffer-type calculation.

Choice of Indicator Ideally, the midpoint of the pH range of the indicator's color change should be the same as the pH of the equivalence point for the titration. What this means is that if the acid ionization constant of the indicator is K_{In}, then pK_{In} should equal the pH of the titration's equivalence point.

Example 16.7 Titration curve calculations for a strong acid and strong base

We want to titrate 25.00 mL of 0.0800 *M* propanoic acid with 0.1200 M KOH. Calculate the volume of KOH needed to reach the equivalence point. Calculate the pH at the start, after 10.00 mL of KOH has been added, and at the equivalence point of this titration. Suggest a good indicator for this titration.

Analysis:

We have five things to accomplish in this rather complex question:

1. Determine the equivalence point volume

2. Determine the starting pH

3. Determine the pH after some base is added

4. Determine the pH at the equivalence point determined in item #1

5. Use the above information to select an indicator

Now that we have made this list, the problem seems more manageable. Since we have solved problems like these before, let's solve each in turn. However, first let's write the chemical equation and the equilibrium law that applies.

Assembling the Tools:

We need to review the tool that breaks titration curves into four parts for easy solution. Then we can use our tools for calculating the end point volume, pH of a weak acid solution, a buffer solution, and a salt solution.

Solution:

The formula for propanoic acid is $HC_3H_5O_2$ and we can write the ionization equation as

$$HC_3H_5O_2 \rightleftharpoons C_3H_5O_2^- + H^+$$

The K_a expression is

$$K_a = \frac{[C_3H_5O_2^-][H^+]}{[HC_3H_5O_2]} = 1.34 \times 10^{-5}$$

The reaction of propanoic acid with KOH is

$$KOH + HC_3H_5O_2 \rightleftharpoons C_3H_5O_2^- + H_2O$$

With our equations in place we can now do the five sets of calculations

1. To determine the equivalence point volume, we need to convert mL of propanoic acid to mL of KOH using our basic stoichiometric steps.

$$25.00 \text{ mL } HC_3H_5O_2 = ? \text{ mL KOH}$$

The conversions are

$$25.00 \text{ mL } HC_3H_5O_2 \frac{0.0800 \text{ mol } HC_3H_5O_2}{1 \text{ L } HC_3H_5O_2} \frac{1 \text{ mol KOH}}{1 \text{ mol } HC_3H_5O_2} \frac{1 \text{ L KOH}}{0.1200 \text{ mol KOH}} = 16.7 \text{ mL KOH}$$

and the equivalence point is expected at 16.7 mL.

2. The pH at the start of the titration is simply a solution containing a weak acid and the hydrogen ion concentration is calculated, using the assumptions as we did before.

$$[H^+] = \sqrt{K_a \times (\text{initial conc.})} = \sqrt{1.34 \times 10^{-5} \times 0.080} = 1.0 \times 10^{-3} M$$

The pH = $-\log(1.0 \times 10^{-3})$ = 3.00.

3. This part of the problem is actually a simple limiting reactant problem. Let's start by calculating the moles of each substance that we have. This can be done by multiplying the molarity of each solution by its volume in liters so that

$$\text{mol KOH} = (0.01000 \text{ L KOH})(0.1200 \text{ mol KOH L}^{-1}) = 1.200 \times 10^{-3} \text{ mol KOH}$$

$$\text{mol } HC_3H_5O_2 = (0.02500 \text{ L } HC_3H_5O_2)(0.0800 \text{ mol } HC_3H_5O_2 \text{ L}^{-1}) = 2.00 \times 10^{-3} \text{ mol } HC_3H_5O_2$$

We have more moles of $HC_3H_5O_2$ than KOH so that all of the KOH is used up while forming 1.200×10^{-3} of moles of $C_3H_5O_2^-$. Subtracting the consumed KOH from the original moles of $HC_3H_5O_2$ leaves us with 0.80 $\times 10^{-3}$ mol $HC_3H_5O_2$. We now have enough information to solve the equilibrium law for the hydrogen ion concentration.

$$K_a = \frac{[C_3H_5O_2^-][H^+]}{[HC_3H_5O_2]} = 1.34 \times 10^{-5} = \frac{[1.200 \times 10^{-3}][H^+]}{[0.80 \times 10^{-3}]}$$

Solving this for $[H^+] = 8.9 \times 10^{-6} M$ and the pH = $-\log(8.9 \times 10^{-6})$ = 5.05.

4. To calculate the pH at the equivalence point we need to know what the solution contains. Since we have reacted exactly equal moles of KOH and $HC_3H_5O_2$, neither is present in any major amount. The solution is simply one that contains the conjugate base of propanoic acid. We need to calculate K_b for this salt as

$$K_b = K_w/K_a = (1.0 \times 10^{-14})/(1.34 \times 10^{-5}) = 7.5 \times 10^{-10}$$

The concentration of the conjugate base is calculated by dividing the moles of the original acid by the total volume at the equivalence point.

$$[C_3H_5O_2^-] \quad = \quad (2.00 \times 10^{-3} \text{ mol HC}_3\text{H}_5\text{O}_2)/(0.02500 \text{ L} + 0.0167 \text{ L}) = 0.048 \ M$$

The hydroxide ion concentration can now be determined as

$$[OH^-] = \sqrt{K_b \times (\text{initial conc.})} = \sqrt{7.5 \times 10^{-10} \times 0.048} = 6.0 \times 10^{-6} M$$

Finally, the pOH = 5.22 and the pH is 8.78 at the equivalence point.

5. To select an indicator we need it to change color right at the equivalence point. Therefore we look at the table of indicators (Table 16.7 in the text) and try to find one that has a good color change right around pH 8.78. In other words, the pK_{in} should be close to 8.78. There are several in this range including thymol blue and phenolphthalein.

Are the Answers Reasonable?

One of the main features of a titration is that as base is added to the acid, the pH must increase. Our results increased in pH from the start to a point somewhere before the equivalence point and then to the equivalence point. This occurred and we can check our math for minor errors.

Self-Test

42. Calculate the pH of the solution that exists in a titration after 22.00 mL of 0.10 M NaOH has been added to 25.00 mL of 0.10 M HC$_2$H$_3$O$_2$.

   _____

43. Calculate the pH of the solution that forms in a titration in which 24.00 mL of 0.20 M KOH is added to 25.00 mL of 0.20 M HCl.

   _____

44. Calculate the pH of the solution that forms during a titration when 26.00 mL of 0.10 M NaOH has been added to 25.00 mL of 0.10 M HC$_2$H$_3$O$_2$.

   _____

45. Calculate the pH of the solution that has formed during a titration when 28.00 mL of 0.10 M HCl has been added to 25.00 mL of 0.10 M NH$_3$.

   _____

46. If you know that there are two indicators, X and Y, that undergo their color changes in the same pH range, and that X changes from colorless to blue and Y from orange to red, which indicator would be the better choice and why?

   _____

47. If you were asked to carry out a titration of dilute hydrobromic acid with potassium hydroxide, which indicator would be the better choice (and why)—thymol blue or bromothymol blue?

_____

Answers to Self-Test Questions

1. (a) $2.63 \times 10^{-10}\ M$ (b) acidic because $[H_3O^+]$ is greater than $[OH^-]$.
2. The solution is neutral because $[H_3O^+] = [OH^-] = 1.6 \times 10^{-7}\ M$.
3. (a) $5 \times 10^{-10}\ M$ (b) $2.5 \times 10^{-7}\ M$
4. b. When pOH = 10, pH = 4, and $[H^+] = 1.0 \times 10^{-4}\ M$.
5. c. When $[H^+] = 4.68 \times 10^{-6}\ M$, pH $= -\log 4.68 \times 10^{-6}$.
6. a. When pH = 9.65, $[H^+] = 1 \times 10^{-9.65}\ M = 2.24 \times 10^{-10}\ M$.
7. $2.2 \times 10^{-7}\ M$
8.

	Color in	
Indicator	**Acid**	**Base**
phenolphthalein	none	red
bromothymol blue	yellow	blue
thymol blue	yellow	blue

9. (a) 0.60, (b) 13.40, (c) 11.30
10. 0.028 M
11. (a) $HCl(aq) + NaOH(aq) \rightarrow NaCl(aq) + H_2O$
    (b) pH = 1.90

12. (a) $HCN \rightleftharpoons H^+ + CN^-$, $K_a = \dfrac{[H^+]\,[CN^-]}{[HCN]}$

    (b) $HClO_2 \rightleftharpoons H^+ + ClO_2^-$, $K_a = \dfrac{[H^+]\,[ClO_2^-]}{[HClO_2]}$

    (c) $N_2H_5^+ \rightleftharpoons H^+ + N_2H_4$, $K_a = \dfrac{[H^+]\,[N_2H_4]}{[N_2H_5^+]}$

13. (a) $(CH_3)_2NH + H_2O \rightleftharpoons (CH_3)_2NH_2^+ + OH^-$, $K_b = \dfrac{[(CH_3)_2NH_2^+]\,[OH^-]}{[(CH_3)_2NH]}$

    (b) $N_2H_4 + H_2O \rightleftharpoons N_2H_5^+ + OH^-$, $K_b = \dfrac{[N_2H_5^+]\,[OH^-]}{[N_2H_4]}$

    (c) $CN^- + H_2O \rightleftharpoons HCN + OH^-$, $K_b = \dfrac{[HCN]\,[OH^-]}{[CN^-]}$

14. 10.33

15. Nitrous acid

16. HCN

17. 10.82

18. 1.4×10^{-11}

19. 1.8×10^{-5}

20. $K_a = 1.26 \times 10^{-6}$

21. $K_b = 4.01 \times 10^{-6}$

22. pH = 11.44

23. 0.0079%

24. (a) $C_6H_5O^-$ is a Brønsted base, (b) OCN^- is a Brønsted base, (c) NO_3^- is not a Brønsted base.

25. (a) neutral, (b) basic, (c) basic, (d) neutral, (e) acidic (contains $C_5H_5NH^+$ and Cl^- ions), (f) basic

26. Yes. $C_4H_7O_2^-$. pH = 8.95

27. Yes. NH_4^+. pH = 4.93

28. No

29. 0.14 M

30. $[H^+] = 3.1 \times 10^{-3}$ M; pH = 2.51. No, the method of successive approximations was used.

31. pH = 4.84

32. $[CHO_2^-]/[HCHO_2] = 0.18$, so the desired mole ratio is $(0.18 \text{ mol } NaCHO_2)/(1 \text{ mol } HCHO_2)$.

33. 3.7 g NH_4Cl

34. The pH will decrease by 0.40 units.

35. $H_2SeO_3 \rightleftharpoons H^+ + HSeO_3^-$ $K_{a_1} = \dfrac{[H^+][HSeO_3^-]}{[H_2SeO_3]}$

    $HSeO_3^- \rightleftharpoons H^+ + SeO_3^{2-}$ $K_{a_2} = \dfrac{[H^+][SeO_3^{2-}]}{[HSeO_3^-]}$

36. $[H^+] = 2.6 \times 10^{-3}$ M; pH = 2.58; $[Asc^{2-}] = 2.7 \times 10^{-12}$ M

37. $[HAsc^-] = 2.6 \times 10^{-3}$ M

38. pH = 8.76

39. pH = 12.95

40. $[H_2C_2O_4] = K_{b_2} = 1.5 \times 10^{-13}$ M

41. 0.20 M

42. pH = 5.61

43. pH = 2.39

44. pH = 11.30 (caused solely by the OH^- ion provided by the excess NaOH)

45. pH = 2.25 (caused solely by the H^+ ion provided by the excess HCl)

46. Indicator X, because a very dramatic color change is easier to notice.

47. Bromothymol blue, because the equivalence point for the titration of a strong acid and a strong base is at pH 7, so we need an indicator whose color change occurs at or near pH 7.

Tools for problem solving

In this chapter you learned to apply the following concepts as tools in solving problems. Study each one carefully so that you know what each is used for. When faced with solving a problem, recall what each tool does and consider whether it will be helpful in finding a solution.

You might want to tear these pages out to use along with solving problems in this chapter.

Ion product constant of water, K_w (Section 16.1)

$$K_w = [H^+][OH^-] \quad \text{or} \quad K_w = [H_3O^+][OH^-]$$

At 25 °C the value of K_w is 1.0×10^{-14}.

Definition of pH (Section 16.1)

$$pH = -\log[H^+]$$

Defining the p-function, pX (Section 16.1)

$$pX = -\log X$$

Relating pH and pOH to pK_w (Section 16.1)

$$pH + pOH = pK_w \qquad pK_w = 14.00 \text{ at } 25\ °C$$

General equation for the ionization of a weak acid (Section 16.3)
$$HA + H_2O \rightleftharpoons H_3O^+ + A^- \quad \text{or} \quad HA \rightleftharpoons H^+ + A^-$$

General equation for the ionization of a weak base (page 646)
$$B + H_2O \rightleftharpoons BH^+ + OH^-$$

Inverse relationship between K_a and K_b (Section 16.3)

$$K_a \times K_b = K_w$$

Relationship between pK_a and pK_b (Section 16.3)

$$pK_a + pK_b = pK_w \qquad pK_w = 14.00 \text{ at } 25\ °C$$

Percentage ionization (Section 16.4)

$$\text{percentage ionization} = \frac{\text{(moles per liter ionized)}}{\text{(moles per liter available)}} \times 100\%$$

Criterion for predicting if simplifying assumptions will work (Section 16.5)

$$C_{HA} > 100 \times K_{a} \quad \text{or} \quad C_{B} > 100 \times K_{b}$$

Identification of acidic cations and basic anions (Section 16.6)

We use the concepts developed here to determine whether a salt solution is acidic, basic, or neutral. This also is the first step in calculating the pH of a salt solution.

How buffers work (Section 16.7)

These reactions determine how the concentrations of conjugate acid and base change when a strong acid or strong base is added to a buffer. Adding H^+ decreases $[A^-]$ and increases $[HA]$; adding OH^- decreases $[HA]$ and increases $[A^-]$. These reactions keep the pH of the solution from changing drastically.

Henderson–Hasselbalch equation (Section 16.7)

$$pH = pK_{a} + \log\frac{[A^-]}{[HA]}$$

Polyprotic acids ionize stepwise (Section 16.8)

Polyprotic acids have more than one ionizable proton in their formulas. Each proton will ionize sequentially and the chemical equation will have a form similar to the general equation for the ionization of a weak acid tool above.

Titration curves have four parts (Section 16.9)

To plot the graph of pH as a function of volume of titrant added, that is, a titration curve, there are four different stages of calculation. They are (a) the starting point, (b) the stage from the start to the equivalence point, (c) the equivalence point, and (d) the stage after the equivalence point.

How acid–base indicators work (Section 16.9)

Titration indicators are weak acids that have one color for the conjugate acid and a different color for the conjugate base. And the change in color depends on the pH of the solution.

Summary of Important Equations

Ion product constant of water:

$$[H^+]\,[OH^-] = K_w$$

pH and pOH and the relationship between pH and pOH:

$$pH = -\log[H^+] \qquad pOH = -\log[OH^-] \qquad pH + pOH = 14.00 \ \text{(at 25 °C)}$$

Acid ionization constant: K_a and pK_a for a weak acid: $HA \rightleftharpoons H^+ + A^-$

$$K_a = \frac{[H^+]\,[A^-]}{[HA]} \qquad pK_a = -\log K_a$$

Base ionization constant: K_b and pK_b for a weak base: $B + H_2O \rightleftharpoons BH^+ + OH^-$

$$K_b = \frac{[BH^+]\,[OH^-]}{[B]} \qquad pK_b = -\log K_b$$

Relationship among K_a, K_b, and K_w for conjugate acid–base pairs:

$$K_a \times K_b = K_w$$

$$pK_a + pK_b = 14.00 \ \text{(at 25 °C)}$$

Buffers:

$$pH = pK_a - \log \frac{[HA]_{init}}{[A^-]_{init}}$$

For the best buffer action:

$$pH = pK_a \pm 1$$

Successive acid ionization constants, polyprotic acids, illustrated by H_2A:

$$H_2A \rightleftharpoons H^+ + HA^- \qquad\qquad K_{a_1} = \frac{[H^+][HA^-]}{[H_2A]}$$

$$HA^- \rightleftharpoons H^+ + A^{2-} \qquad\qquad K_{a_2} = \frac{[H^+][A^{2-}]}{[HA]}$$

Concentration of A^{2-} in a solution of H_2A (when H_2A is the *only* solute)

$$[A^{2-}] = K_{a_2}$$

Chapter 17

Solubility and Simultaneous Equilibria

In this chapter we extend our discussion of ionic equilibria to deal with salts that until now we have considered insoluble. As you learn here, such compounds do dissolve slightly and their saturated solutions represent situations in which the solid salt is in equilibrium with its ions in solution. Among the topics we examine are the effects of one salt on the solubility of another and how relative degrees of insolubility permit separation of metal ions by selective precipitation. We also study equilibria in the formation of substances called complex ions and how such substances can influence the solubilities of salts. This is a topic that has many practical applications.

17.1   Equilibria in Solutions of Slightly Soluble Salts

Learning Objective

Perform solubility calculations using neutral salts.

Review

The equilibrium law for the solubility equilibrium of a salt is written exactly the same way as all other chemical equilibrium laws. However, since the sole reactant is always in the solid phase, the denominator of the equilibrium law is always one (if this is unclear, review Section 14.4 on heterogeneous equilibria). The result is that this type of equilibrium law involves only the salt's ion product, which is the product of the ion molarities raised to powers obtained from the subscripts in the salt's formula. The sparingly soluble salt lead(II) chloride, $PbCl_2$, for example, gives the following equilibrium in its saturated aqueous solution.

$$PbCl_2(s) \rightleftharpoons Pb^{2+}(aq) + 2Cl^-(aq)$$

The ion product for $PbCl_2$ is $[Pb^{2+}][Cl^-]^2$. Notice that the exponents come from the coefficients of the ions in the equilibrium. When the solution is saturated, and only then, the value of the ion product is a constant (at a given temperature) and is called the solubility product constant, K_{sp}. Thus, for lead(II) chloride, the equation for its K_{sp} is

$$K_{sp} = [Pb^{2+}][Cl^-]^2 \text{ (saturated solution)}$$

This first section deals with the simplest situation, salts that are composed of ions that do not affect the acidity of a solution by reacting with water (hydrolysis). The calculations in this section fall into three categories.

- Calculating K_{sp} from the known or measured solubility of a salt.

- Calculating solubility from K_{sp}.

- Using the calculated ion product, Q, for a solution of a salt to determine whether a precipitate will form.

The first and third calculations are readily done by determining the molarity of each ion in a saturated solution from the experimental solubility data. These data are either the molarity of the dissolved salt or the g/L of the solute that dissolves. The latter is readily converted to the molarity of the ions in the solution.

For the second calculation, the concentration table is helpful in organizing our thinking. We imagine the saturated solution being formed in a stepwise fashion. First, we examine what's in the solvent to which the sparingly soluble salt will be added and determine concentration values for any of the salt's ions that might already be present (say, from some other compound already dissolved). These go in the "initial concentration" row. If the solvent is pure water, then all the values in this top row are zero. The next step is to consider how the concentrations change as the salt dissolves. When we're calculating solubility from K_{sp}, the quantities in the "change" row will be our unknown quantities expressed as multiples of s (we got tired of using x for our variables and switched to s in this chapter) as we did in the previous chapters on equilibrium. As usual, the equilibrium concentrations in the last row are obtained by adding the changes to the initial values.

In the worked examples in this section, you will notice that we place no entries (NE) under the formula for the sparingly soluble salt because the concentration of the solid doesn't appear in the mass action expression (the ion product). But we must always realize that solubility product equilibria *always requires some solid in contact with the solution.*

Calculating a salt's K_{sp} from its solubility

Examples 17.1 and 17.2 in the text illustrate how the measured solubility in terms of grams per liter or moles per liter can be used to determine the value of the solubility product. Since we just have to determine the concentrations of the ions in solution (in moles per liter) to determine the K_{sp}, there is no need for a concentration table.

Calculating a salt's molar solubility from its K_{sp}

To set up a concentration table for a specific salt, we must first obtain three pieces of information.

1. The equation for the equilibrium in the saturated solution of the solute.

2. The ion product equation for the salt's K_{sp} (which we figure out from the salt's equilibrium equation).

3. The value of K_{sp} for the salt.

Now it is useful to set up a concentration table as shown in Examples 17.3 and 17.4. When the solvent does not contain any of the ions in the slightly soluble salt, the row for the initial concentrations of the ions has only zeros. The change in the concentrations of the salt's ions depends on the coefficients of the ions produced as the salt dissociates. (For example, 1 mol of AgCl can give one mole each of Ag^+ and Cl^-, but 1 mol of Ag_2CrO_4 yields two moles of Ag^+ ion and one of CrO_4^{2-} ion.) In this usage, the row for the changes in concentration uses s to stand for the molar solubility, but for each ion, s must be multiplied by the number of ions each formula unit releases, as illustrated in Examples 17.3 and 17.4. This assures us that the changes in concentration will be in the same ratio as the coefficients in the balanced equilibrium equation. In the third row the individual equilibrium concentrations are the sums, column by column, of the initial concentration and the change in concentration already developed (that is, rows 1 and 2).

The equilibrium values expressed in units of s are finally substituted into the K_{sp} equation, and then the equation is solved for s, the molar solubility. The molar concentration of each ion is obtained by multiplying s by the stoichiometric factor used in the table.

Example 17.1 Determining solubility from the K_{sp} value

Determine the molar solubility and solubility in grams/100 mL of silver iodide in H_2O ($K_{sp} = 8.5 \times 10^{-17}$).

Analysis:

First, we will need to determine the formula for silver iodide and then write the equation for its dissolution in water. Next we write the K_{sp} law (equation) for silver iodide and construct the equilibrium table using s to symbolize the molar solubility of silver iodide. Last, we write the K_{sp} in terms of s and solve for s, which is our desired molar solubility. The second part of the problem asks us to convert molar solubility to solubility with units of grams per 100 mL. If our molar solubility is low, we can assume that the density of pure water and the solution are the same (generally 1.00 g/mL for aqueous solutions).

Assembling the Tools:

We need nomenclature tools to get the formula and solution tools, and periodic relationships to predict the ions produced. We need stoichiometry tools to calculate the concentrations and convert one concentration unit to another.

Solution:

Silver iodide has the formula AgI (review Chapter 2 if needed) and its solubility equation is

$$AgI(s) \rightleftharpoons Ag^+(aq) + I^-(aq)$$

	AgI(s) $\rightleftharpoons$	Ag$^+$(aq)	+	I$^-$(aq)
Initial molar concentrations	Solids are not part of the equilibrium law	0		0
Change in concentration (M)	s mol/L will dissolve	$+s$		$+s$
Equilibrium concentration (M)	No entry here	$+s$		$+s$

$$K_{sp} = 8.5 \times 10^{-17} = [Ag^+][I^-] = (s)(s) = s^2$$

We have written the equilibrium law and inserted the variables deduced in the equilibrium table. Taking the square root we find that $s = 9.2 \times 10^{-9}$ M for the molar solubility. For the second part, since the mass of 9.2×10^{-9} moles of AgI is small, we can assume that the density of the solution, to two significant figures, will be 1.0 g mL^{-1}. We can then set up the following conversion:

$$\frac{9.2 \times 10^{-9} \text{ mol AgI}}{1000 \text{ mL solution}} \times \frac{1 \text{ mL solution}}{1 \text{ g solution}} \times \frac{235 \text{ g AgI}}{1 \text{ mol AgI}} = \frac{2.2 \times 10^{-6} \text{ g AgI}}{1000 \text{ g solution}}$$

Since we want grams per 100 mL we divide the numerator and denominator by 10 to get 2.2×10^{-7} g AgI per 100 mL.

Are the Answers Reasonable?

The low solubilities we found is a good indication that our answers are correct. We can go back and check our setup of the solubility reaction, the equilibrium law, and equilibrium table. Finally, if we square the molar solubility we should obtain the solubility product, and we do.

Common ion effect

A salt is less soluble in a solution that contains one of its ions than it is in pure water, a phenomenon called the common ion effect. This is really nothing more than an application of Le Châtelier's principle.

To calculate a salt's solubility from its K_{sp} under a common ion situation, the only thing that changes in the strategy described above is that the initial concentration row will have a non–zero entry, namely, the molar concentration of one of the ions already provided by some other solute. The unknown is still the molar solubility, s. Entries into the change row are made exactly as before, with the coefficients of s equal to the coefficients in the equation for the solubility equilibrium. The sums that you next enter into the equilibrium concentration row might include an expression such as $(0.10 + 2s)$, as in Example 17.5 in the text. Because we're dealing with *sparingly* soluble salts, we can safely assume that s or $2s$ is very small compared to 0.10, so we can simplify a $(0.10 + 2s)$ term to 0.10. This makes the arithmetic much simpler without sacrificing accuracy.

Example 17.2 Determining solubility from the K_{sp} value when a common ion is present

Determine the molar solubility and solubility in grams/100 mL of AgI in an aqueous solution of 0.14 M NaI ($K_{sp} = 2.4 \times 10^{-5}$).

Analysis:

This problem is identical to Example 17.1 above except that a common ion has been added.

Assembling the Tools:

We need the tools mentioned for Example 17.1 and the concept of the common ion.

Solution:

Silver iodide has the formula AgI (review Chapter 2 if needed) and its solubility equation is

$$\text{AgI}(s) \rightleftharpoons \text{Ag}^+(aq) + \text{I}^-(aq)$$

	AgI(s) $\rightleftharpoons$	Ag$^+$(aq)	+	I$^-$(aq)
Initial molar concentrations	Solids are not part of the equilibrium law	0		0.14
Change in concentration (*M*)	*s* mol/L will dissolve	+*s*		+*s*
Equilibrium concentration (*M*)	No entry here	+*s*		*s* + 0.14

$$K_{sp} = 8.5 \times 10^{-17} = [Ag^+][\, I^-]$$

$$= (s)(s + 0.14) = s^2 + 0.14s$$

We have written the equilibrium law and inserted the variables deduced in the equilibrium table. This time we have a quadratic equation to solve. However, if you recall that the solubility in the presence of a common ion is always less than in distilled water, we can safely assume that $(s + 0.14) = 0.14$. We will then solve

$$8.5 \times 10^{-17} = (s)(0.14)$$

$$s = 6.1 \times 10^{-16} \, M$$

Since the second part of this problem is also the same as before except that the mass of 6.1×10^{-16} moles of AgI is even smaller, we can safely assume that the density of the solution, to two significant figures, will be 1.0 g mL^{-1}. We can then set up the following conversion:

$$\frac{6.1 \times 10^{-16} \text{ mol AgI}}{1000 \text{ mL solution}} \times \frac{1 \text{ mL solution}}{1 \text{ g solution}} \times \frac{235 \text{ g AgI}}{1 \text{ mol AgI}} = \frac{1.4 \times 10^{-13} \text{ g AgI}}{1000 \text{ g solution}}$$

Since we want grams per 100 mL we divide the numerator and denominator by 10 to get 1.4×10^{-14} g AgI per 100 mL.

Are the Answers Reasonable?

The lower solubility we found in the presence of a common ion is a good indication that our answers are correct. We can go back and check our setup of the solubility reaction, the equilibrium law, and equilibrium table.

Predicting if a precipitate will form

A precipitate can form only if a solution is supersaturated, and this condition is fulfilled only when the ion product of the salt in question exceeds the value of K_{sp} for the salt. This simple fact will help you to remember the summary just preceding Example 17.6 in the text. Be sure you know this summary.

When trying to predict what might happen when two solutions of different salts are mixed, you first have to consider which combinations of their ions might produce a sparingly soluble salt. If you find one, you have to do a solubility–product type of calculation. When the problem describes the mixing of two solutions, be sure to take into account the dilution of the ion concentrations (see Equation 4.6), as illustrated in Example 17.6. Then, with the correct molarities, compute the value of the ion product, Q, of the salt. Finally, compare the ion product with the salt's K_{sp} to see if a precipitate should form.

Example 17.3 Using the K_{sp} to determine if a precipitate will form

Rhubarb leaves contain a high concentration of oxalate ions, so much so that if the leaves are used along with the stalks, this delicious pie can become lethal. One reason for this is that the calcium ions in the blood precipitate as calcium oxalate, CaC_2O_4. In the laboratory, if we mix the contents of a beaker containing 75.0 mL of a 0.045 millimolar solution of $Ca(NO_3)_2$ with another beaker containing 115 mL of 0.0013 millimolar $Na_2C_2O_4$, should we expect a precipitate to form? (Since these are both aqueous solutions we can safely assume that volumes will be additive.)

Analysis:

We will first have to find the correct formula for calcium oxalate and then write the correct dissolution equation and the equilibrium law for the K_{sp}. We need to compare Q to the K_{sp} to determine if a precipitate will form. If Q is larger than the K_{sp} a precipitate will form, otherwise it will not. In this case we need to calculate the concentrations of our ions in the final solution.

Assembling the Tools:

We can use the analysis to determine the tools needed. We need nomenclature tools to get the formula and solution tools, and periodic relationships to predict the ions produced. The concept of equilibrium is used in calculating and using Q. We need stoichiometry tools, including the dilution equation and perhaps limiting reactant calculations to calculate the concentrations.

Solution:

The formula for calcium oxalate is CaC_2O_4 and we find its $K_{sp} = 2.3 \times 10^{-9}$. The chemical equation and equilibrium law are written as

$$CaC_2O_4(s) \rightleftharpoons Ca^{2+} + C_2O_4^{2-}$$

and

$$K_{sp} = [Ca^{2+}][C_2O_4^{2-}] = 2.3 \times 10^{-9}$$

Our final solution will have a total volume of 75.0 mL + 115 mL = 190 mL. We can calculate the final concentration of calcium ions as

$$(4.5 \times 10^{-5} M)(75.0 \text{ mL}) = C_f (190 \text{ mL})$$
$$C_f = 1.78 \times 10^{-5} M_{\text{Ca ions}}$$

For the oxalate ions we calculate

$$(1.3 \times 10^{-6} M)(115.0 \text{ mL}) = C_f (190 \text{ mL})$$
$$C_f = 7.87 \times 10^{-7} M_{\text{oxalate ions}}$$

Now we can enter the two molarities into the K_{sp} expression to get

$$Q = (1.78 \times 10^{-5} M_{\text{Ca ions}})(7.87 \times 10^{-7} M_{\text{oxalate ions}}) = 1.4 \times 10^{-11} \text{ (properly rounded)}$$

We compare Q to the K_{sp} and find that Q is much less than the value of the K_{sp} and we conclude that a precipitate will not form.

Is the Answer Reasonable?

If we forgot to convert the given millimolar concentrations into molar concentrations we would come to the opposite conclusion. First we look at the calculated diluted concentrations. Since the volumes are almost equal, the concentrations should be about half of what we started with, and they are. Finally, a quick estimate at multiplying the two concentrations shows that we should end with a value close to 10^{-11}. We therefore conclude that our answer is correct.

Self–Test

In the following questions assume that the conjugate base (the anion) does not hydrolyze significantly.

1. The molar solubility of $Mg_3(PO_4)_2$ in pure water is 3.57×10^{-6} mol/L. What is the K_{sp} for $MnCO_3$?

    _____

2. The molar solubility of lead iodate, $Pb(IO_3)_2$, is 4.0×10^{-5} M. What is the K_{sp} of this salt?

    _____

3. Lead chromate has been a favorite red–orange pigment of painters for centuries. In 100.0 mL of a saturated solution of $PbCrO_4$ in water there are 4.33 µg of $PbCrO_4$. What is the K_{sp} for this salt?

    _____

4. The molar solubility of $Cr(OH)_3$ in 0.00010 M NaOH is 2.0×10^{-18} mol/L. What is the K_{sp} of $Cr(OH)_3$?

    _____

5. For $PbCO_3$, $K_{sp} = 7.4 \times 10^{-14}$. What is the molar solubility of $PbCO_3$ in pure water?

    _____

6. What is the molar solubility of $PbCO_3$ in 0.020 M Na_2CO_3? For $PbCO_3$, $K_{sp} = 7.4 \times 10^{-14}$.

    _____

7. What is the molar solubility of $PbCl_2$ in 0.30 M $CaCl_2$? For $PbCl_2$, $K_{sp} = 1.7 \times 10^{-5}$.

    _____

8. Referring to Question 7, what is the molar solubility of $PbCl_2$ in 0.30 M NaCl solution?

    _____

9.   How many grams of Ag_2CrO_4 will dissolve in 100 mL of water to give a saturated solution?

_____

10. Will a precipitate of $PbCl_2$ form in a solution that contains 0.20 M Pb^{2+} and 0.030 M Cl^-? For $PbCl_2$, $K_{sp} = 1.7 \times 10^{-5}$.

_____

11. Will a precipitate of $CaSO_4$ form in a solution that contains 0.0030 M Ca^{2+} and 0.0010 M SO_4^{2-}? Check a table in Chapter 17 or Appendix C in the text for the needed K_{sp}.

_____

17.2   Solubility of Basic Salts Is Influenced by Acids

Learning Objective

Understand how hydrogen ion concentration affects the solubility of basic salts.

Review

A basic salt is one that will produce a basic solution when dissolved in water. Often a basic salt contains an anion from a weak acid and a cation from a strong base. Calcium phosphate is one example of a basic salt. The solubility equation is

$$Ca_3(PO_4)_2 \rightleftharpoons 3Ca^{2+} + 2PO_4^{3-}$$

and the solubility product equation is

$$K_{sp} = [Ca^{2+}]^3[PO_4^{3-}]^2$$

If the PO_4^{3-} concentration can be reduced, Le Châtelier's principle informs us that more calcium phosphate will dissolve to reestablish equilibrium. Addition of acid will protonate the phosphate ion to make HPO_4^{2-}, thus reducing the phosphate concentration. The overall reaction is the sum of the reaction of acid with two phosphate ions

$$2PO_4^{3-} + 2H^+ \rightleftharpoons 2HPO_4^{2-}$$

and the solubility equation above, which gives

$$Ca_3(PO_4)_2 + 2H^+ \rightleftharpoons 3Ca^{2+} + 2HPO_4^{2-}$$

The equilibrium law for this reaction is

$$K = \frac{[Ca^{2+}]^3[HPO_4^{2-}]^2}{[H^+]^2} = \frac{K_{sp}}{K_{a_3}^2}$$

This illustrates that the solubility of a salt, that has a basic anion, can be increased by having acid present. The Analyzing and Solving Multi–Component Problems example in the text shows how equations of this type may be applied.

Self–Test

12. Write the equilibrium equations for dissolving the following salts in acidic solutions.

    (a) $CaCO_3$ (b) PbC_2O_4 (c) Ag_2CrO_4

    _____

    _____

13. Will the addition of acid to a saturated solution of $MgCO_3$ increase or decrease the concentration of magnesium ions? What about $MgBr_2$? Justify your answers with equations.

    _____

    _____

17.3   Equilibria in Solutions of Metal Oxides and Sulfides

Learning Objective

Describe why oxides and sulfides are special cases and need unique methods to determine solubilities.

Review

The sparingly soluble oxides and sulfides of metal ions cannot be treated like other sparingly soluble ionic compounds because the O^{2-} and S^{2-} ions cannot exist in water. Once either is released from a crystalline substance, it instantly reacts essentially 100% with water to give OH^- and HS^-, respectively. The rest of this section deals only with metal sulfides.

For a metal sulfide of the general formula MS, the solubility equilibrium is

$$MS(s) + H_2O \rightleftharpoons M^{2+}(aq) + HS^-(aq) + OH^-(aq)$$

and the solubility product constant is

$$K_{sp} = [M^{2+}][HS^-][OH^-]$$

If the solution is made acidic, we have to rewrite the equilibrium, because both HS^- and OH^- are neutralized by acids. In dilute acid, the equilibrium involving the general sulfide, MS, is

$$MS(s) + 2H^+(aq) \rightleftharpoons M^{2+}(aq) + H_2S(aq)$$

The solubility product expression based on this equilibrium gives a different constant, the *acid solubility product constant*, or K_{spa}.

$$K_{spa} = \frac{[M^{2+}][H_2S]}{[H^+]^2}$$

(When the metal sulfide is not of the form MS, these equations must be altered accordingly.)

Acid–insoluble sulfides

One group of metal sulfides, the *acid–insoluble sulfides*, have values of K_{spa} so small that no acid exists that can be made concentrated enough in water to bring them into solution simply by converting the sulfide ion to H_2S. Their K_{spa} values range from about 10^{-32} to 10^{-5}.

Acid–soluble sulfides

When K_{spa} of a metal sulfide is about 10^{-4} or greater, then it will dissolve in acid. These are called the *acid–soluble sulfides*.

Self–Test

14. Consider lead(II) sulfide.

    (a) Write its solubility equilibrium and the K_{sp} equation for a saturated aqueous solution.

    _____

    (b) Write its solubility equilibrium and the K_{spa} equation for a saturated solution in aqueous acid.

    _____

17.4 Selective Precipitation

Learning Objective

Use selective precipitation to separate ions for qualitative and quantitative analysis.

Review

Because of the huge differences in K_{spa} values among the metal sulfides, the adjustment of the pH of an aqueous solution containing two metal ions that can form insoluble sulfides can sometimes permit the precipitation of one metal ion but not the other when H_2S is bubbled into the solution.

Some metal carbonates also differ widely enough in solubility that an adjustment of the pH of a solution of metal ions that are able to form insoluble carbonates can sometimes permit the precipitation of one metal ion but not the other when CO_2 is bubbled into the solution.

This section uses equations developed in the previous section to illustrate how selective precipitation works.

Selective precipitation of a metal sulfide

To achieve a separation of metal ions that can form insoluble sulfides, hydrogen sulfide is bubbled into their solution until it is saturated in hydrogen sulfide (0.1 M in H_2S). But before this is done, the pH of the solution is adjusted according to the results of a calculation, as explained in Example 17.9 in the text.

Selective precipitation of a metal carbonate

The solubility products of the sparingly soluble metal carbonates do not differ nearly as widely as do those of the metal sulfides. Yet it's still possible to use pH to control the carbonate ion concentration so that one of two metal carbonates will precipitate and the other will not as CO_2 is bubbled into the solution of the ions. Using MCO_3 as a general formula, the relevant ionization equilibrium is

$$MCO_3(s) \rightleftharpoons M^{2+}(aq) + CO_3^{2-}(aq)$$

How the molarity of CO_3^{2-} responds to pH control is seen in the ionization equilibrium of carbonic acid, shown next in a form that combines the two stepwise ionizations. Adding the two ionization equilibria gives the combined equation.

$$H_2CO_3(aq) \rightleftharpoons H^+(aq) + HCO_3^-(aq)$$
$$\underline{HCO_3^-(aq) \rightleftharpoons H^+(aq) + CO_3^{2-}(aq)}$$
$$\text{Sum:} \quad H_2CO_3(aq) \rightleftharpoons 2H^+(aq) + CO_3^{2-}(aq)$$

The combined equilibrium constant, K_a, equals the product of K_{a_1} and K_{a_2}.

$$K_{a_1} K_{a_2} = K_a = \frac{[H^+]^2[CO_3^{2-}]}{[H_2CO_3]} = 2.4 \times 10^{-17}$$

Recalling Le Châtelier's principle, you can see that if we decrease the pH by adding acid to this system, the combined equilibrium will shift to the left, reducing the availability of CO_3^{2-}. And if we increase the pH by adding something to neutralize H^+, like OH^-, the equilibrium will shift to the right, increasing the availability of CO_3^{2-}.

The source of carbonic acid for the selective precipitation of carbonates is simply the bubbling into the solution of CO_2 until the solution is saturated, where its concentration becomes approximately 0.030 M. We use this value for $[H_2CO_3]$ in calculations. Example 17.10 in the text describes a calculation for separating ions by selective precipitation of their carbonates.

Example 17.4   Using the K_{spa} to determine conditions for a selective precipitation

A solution contains both $Fe(NO_3)_2$ and $Cu(NO_3)_2$ at concentrations of 0.010 M each. It is to be saturated with hydrogen sulfide. Is it possible to adjust the pH of the solution so as to prevent one of the cations from precipitating when hydrogen sulfide gas is bubbled into the system? If so, which cation remains dissolved and what pH should be used? Consult Table 17.2 in the text as needed.

Analysis:

We need to determine the conditions where Q for one of the ions is less than, or equal to, its K_{spa} (no precipitate forms) while Q is larger than its K_{spa} for the other ion (precipitate forms). We also see that since $[H^+]$ is in the denominator of the K_{spa}, lowering $[H^+]$ will increase the value of Q. We know that when a solution is saturated with $H_2S(g)$ the concentration is approximately $0.10\ M\ H_2S$, which means that two of the three variables in the K_{spa} expression are known. We can therefore calculate, for each ion, the $[H^+]$ needed so that $Q = K_{spa}$. Selecting the smaller of the two hydrogen ion concentrations will allow us to calculate the pH at which one ion will definitely precipitate while the other will still remain in solution.

Assembling the Tools:

We can use the tools used in previous examples as well as the concept that predicts whether or not a precipitate forms.

Solution:

For the sulfide of each of our ions the solubility reactions are

$$FeS(s) + 2H^+ \rightleftharpoons Fe^{2+}(aq) + H_2S(aq); \qquad K_{spa} = 6 \times 10^2$$

$$CuS(s) + 2H^+ \rightleftharpoons Cu^{2+}(aq) + H_2S(aq); \qquad K_{spa} = 6 \times 10^{-16}$$

The equilibrium laws for these reactions are written as

$$K_{spa} = \frac{[Cu^{2+}][H_2S]}{[H^+]^2}$$

$$K_{spa} = \frac{[Fe^{2+}][H_2S]}{[H^+]^2}$$

We now calculate $[H^+]$ for the FeS.

$$[H^+]^2 = \frac{(0.01)(0.10)}{6 \times 10^2} \quad \text{and} \quad [H^+] = 1.7 \times 10^{-6} \quad \text{or pH}\ \ 5.78$$

For CuS we calculate

$$[H^+]^2 = \frac{(0.01)(0.10)}{6 \times 10^{-16}} \quad \text{and} \quad [H^+] = 1.2 \times 10^6 \quad \text{or pH}\ \ -6.11$$

Notice that a molarity of more than one million is impossible as is a pH of -6.11. However, this means that CuS will precipitate from all solutions, even the most acidic. However, if the pH is kept below a pH of 5.78, the FeS will not precipitate.

Is the Answer Reasonable?

The ridiculously large concentration of hydrogen ions needed to make CuS soluble should give us some pause. However, rechecking our math shows it is correct and the interpretation is that CuS is extremely insoluble. Our equation shows that Q increases as the $[H^+]$ decreases and therefore we want high $[H^+]$ to keep FeS soluble. This corresponds to lowering the pH, therefore it should be below pH 5.78 so that Q is less than the K_{spa} for FeS.

Self–Test

15. What value of pH permits the selective precipitation of the sulfide of just one of the two metal ions in a solution that is 0.020 molar in Pb^{2+} and 0.020 molar in Fe^{2+}? Assume $[H_2S] = 0.100\ M$.

_____

16. A solution contains $Mg(NO_3)_2$ and $Sr(NO_3)_2$, each at a concentration of 0.050 M. Can one of the metal ions be precipitated as its insoluble carbonate without the other? If so, which one? The pH of the solution must be within what range of values for this to work as CO_2 is bubbled into the solution? Assume $[CO_2] = 0.030\ M$.

_____

17.5   Equilibria Involving Complex Ions

Learning Objective

Describe and use equilibria related to reactions of complexes

Review

This section delves into equilibria involving complex ions and how the ability to form complex ions can be used to affect the solubility of a salt. Only an overview is given in this chapter. Sections 21.5 through 21.9 discuss complex ions in more detail, including the rules of nomenclature and the description of bonding in these substances.

Metal ions other than those in Groups IA and IIA are strong enough Lewis acids to draw electron–rich species, whether neutral or negatively charged, into Lewis acid–base interactions, forming coordinate covalent bonds. The products are called *complex ions* or simply *complexes*. Compounds made up of complex ions and ions of opposing charge are *coordination compounds* (or *coordination complexes*). The metal ion is called the *acceptor*, and a Lewis base to which the metal becomes attached is called a *ligand*. The specific atom within the ligand that provides the electron pair for the new bond is called the *donor atom*. Common ligands include neutral species, like H_2O and NH_3, and anions, like OH^- or any halide ion.

Complex ion equilibria

There is no uniform agreement among chemists concerning the way to represent the equilibrium between a complex ion and the metal and ligands that form it. Two approaches to writing the equilibria are used, one being for the formation of the complex from its constituents and the other being for the decomposition of the complex. Fortunately, one is simply the inverse of the other.

Formation constants

When the equilibrium equation shows the formation of the complex ion (rather than its decomposition), the associated equilibrium constant is called the *formation constant* or the *stability constant* and is given the symbol K_{form}. The equation corresponding to K_{form} has the complex in the numerator and the ligand and donor in the denominator. For example, the *formation* of the complex of the cadmium ion with the cyanide ion is represented as follows:

$$Cd^{2+}(aq) + 4CN^-(aq) \rightleftharpoons Cd(CN)_4^{2-}(aq) \qquad K_{form} = \frac{[Cd(CN)_4^{2-}]}{[Cd^{2+}][CN^-]^4}$$

When K_{form} is relatively large, it must mean that the value of $[Cd(CN)_4^{2-}]$ is high, relative to the terms in the denominator. In other words, the larger that K_{form} is, the more stable is the complex, which is why K_{form} is sometimes called the *stability constant* of a complex ion. Some chemists prefer to view complexes from such a perspective.

Instability constants

Other chemists prefer the reciprocal view. They write the equilibrium equation to show the decomposition of the complex and the equation for its associated equilibrium constant as follows.

$$Cd(CN)_4^{2-}(aq) \rightleftharpoons Cd^{2+}(aq) + 4CN^-(aq) \qquad K_{inst} = \frac{[Cd^{2+}][CN^-]^4}{[Cd(CN)_4^{2-}]}$$

The equilibrium constant is now called the *instability constant* because this view puts the emphasis on the instability of a complex. Thus, the larger the value of K_{inst}, the more unstable is the complex and the greater its tendency to undergo decomposition into the metal ion and free ligands. Notice that the equilibrium equation for K_{form} is the reverse of that for K_{inst}. This means that the numerical value of K_{inst} is merely the reciprocal of K_{form}. To prove this to yourself, take any value for K_{form} in Table 17.3 in the text and take its reciprocal to show it is equal to the tabulated K_{inst}.

Self–Test

17. The zinc ion forms a complex with four OH^- ions. Write the formula of the complex ion.

   _____

18. Fe^{3+} and SCN^- form a complex ion with a net charge of 3−. Write the formula of this complex ion.

   _____

19. Write the equilibria that are associated with the equations for K_{form} for each of the following complex ions. Write also the equations for the K_{form} of each.

   (a) $Hg(NH_3)_4^{2+}$ _____

   (b) SnF_6^{2-} _____

   (c) $Fe(CN)_6^{3-}$ _____

20. Cobalt(II) ion, Co^{2+}, forms a complex with ammonia, $Co(NH_3)_6^{2+}$. Write the chemical equation involving this complex for which the equilibrium constant would be referred to as

   (a) K_{form}

   _____

   (b) K_{inst}

   _____

21. Silver ion forms a complex ion with cyanide, $Ag(CN)_2^-$. Its formation constant equals 5.3×10^{18}. What is the value of the instability constant?

   _____

17.6 Complexation and Solubility

Learning Objective

Enhance solubility by using complexation reactions.

Review

Complex ions and the solubilities of salts

Example 17.11 in the text shows how AgBr is 4000 times more soluble in 1.0 M NH_3 than it is in pure water. Let's look at another example to be sure you understand the thinking involved.

Suppose we wish to anticipate the effect of adding sodium cyanide on the solubility of cadmium hydroxide. We saw above that Cd^{2+} forms a complex with cyanide ion. When CN^- is added to a saturated solution of $Cd(OH)_2$, the following two equilibria must be considered.

$$Cd(OH)_2(s) \rightleftharpoons Cd^{2+}(aq) + 2OH^-(aq) \qquad K_{sp} = [Cd^{2+}] [OH^-]^2$$

$$Cd^{2+}(aq) + 4CN^-(aq) \rightleftharpoons Cd(CN)_4{}^{2-}(aq) \qquad K_{form} = \frac{[Cd(CN)_4{}^{2-}]}{[Cd^{2+}][CN^-]^4}$$

Thus any Cd^{2+} ion that is in the solution from the first equilibrium will begin to form the complex ion when CN^- is added, which upsets the first equilibrium. The lowering of $[Cd^{2+}]$ as its complex forms must cause the first equilibrium to shift to the right; it's the Le Châtelier's principle response. This is just another way of saying that forming a complex with the cadmium ion increases the solubility of $Cd(OH)_2$. It's as if we're dealing with the following overall equilibrium.

$$Cd(OH)_2(s) + 4CN^-(aq) \rightleftharpoons Cd(CN)_4{}^{2-}(aq) + 2OH^-(aq) \qquad K_c = K_{sp} \times K_{form}$$

Note: It is assumed when working problems involving a system like this that *all* of the Cd^{2+} ion is tied up either in $Cd(OH)_2$ or in $Cd(CN)_4{}^{2-}$; that is, there is a negligible amount of free $Cd^{2+}(aq)$.

The text shows how the equilibrium constant, K_c, is the product of two other equilibrium constants.

$$K_c = K_{sp} \times K_{form} = [Cd^{2+}] [OH^-]^2 \frac{[Cd(CN)_4{}^{2-}]}{[Cd^{2+}][CN^-]^4} = \frac{[Cd(CN)_4{}^{2-}][OH^-]^2}{[CN^-]^4}$$

Example 17.5 Using the K_{sp} and K_{form} to determine conditions for dissolving a precipitate

What molarity of sodium cyanide, NaCN, solution needs to be used to dissolve 0.10 moles of iron(II) oxalate in 500 mL of the solution? The $K_{sp} = 2.1 \times 10^{-7}$ and the formation constant $K_{form} = 1.0 \times 10^{24}$. Ignore the reaction of oxalate ions with water.

Analysis:

We need to write two chemical reactions and combine them. At the same time we will combine their equilibrium constants to obtain the overall equilibrium constant.

Assembling the Tools:

We can use the tools from Chapter 14 to manipulate chemical equations and their equilibrium constants. From there we use the tools developed, such as the concentration table and making approximations, to calculate the desired concentration.

Solution:

For the sulfide of each of our ions the solubility reactions are

$$FeC_2O_4(s) \rightleftharpoons Fe^{2+}(aq) + C_2O_4{}^{2-}(aq)$$
$$Fe^{2+}(aq) + 6CN^-(aq) \rightleftharpoons Fe(CN)_6{}^{4-}(aq)$$

The overall equilibrium constant $K_c = K_{sp} K_{form} = 2.1 \times 10^{17}$ and the equilibrium law is

$$K_c = \frac{[Fe(CN)_6{}^{4-}][C_2O_4{}^{2-}]}{[CN^-]^6}$$

If all of the iron(II) oxalate dissolves, we will have 0.20 M $Fe(CN)_6{}^{4-}(aq)$ and the same concentration of $C_2O_4{}^{2-}(aq)$. We can solve for the concentration of $CN^-(aq)$.

$$[CN^-]^6 = \frac{[Fe(CN)_6{}^{4-}][C_2O_4{}^{2-}]}{K_c} = \frac{(0.2)(0.2)}{2.1 \times 10^{17}} = 1.9 \times 10^{-19}$$

which turns out to be 7.6×10^{-4} M after taking the sixth root. The total $[CN^-]$ is the sum of the free cyanide ion we just calculated (7.6×10^{-4} M) plus six times the concentration of the iron cyanide complex ($0.2 \times 6 = 1.2$ M). The sum of these is 1.2 M when correctly rounded.

Is the Answer Reasonable?

The very large value of the overall equilibrium constant suggests that the reaction goes virtually to completion. The fact that our answer agrees with that assessment (there is very little free CN^- in the solution) indicates that our calculation is correct. All we need to do is to be sure that the molarity of the cyanide solution starts out as six times the molarity that the iron would produce, and it does.

Self–Test

22. Write the chemical equation and the equation for K_{sp} for the following slightly soluble salts. Write the chemical equation and the equation for K_{form} for the complex formation reaction with the specified number of ligands. Write the overall equation for the reaction that occurs upon dissolving the slightly soluble salt in a solution of the ligand. Write the equilibrium constant, K_c, for this process.

    (a) $Fe(OH)_3$ and $6CN^-$ ions

    (b) $CoCO_3$ and $6NH_3$ molecules

    (c) $Al(OH)_3$ and $6F^-$ ions

Answers to Self–Test Questions

1. $K_{sp} = 6.26 \times 10^{-26}$
2. $K_{sp} = 2.6 \times 10^{-13}$
3. $K_{sp} = 1.8 \times 10^{-14}$
4. $K_{sp} = 2 \times 10^{-30}$
5. $2.7 \times 10^{-7} \, M$
6. $3.7 \times 10^{-12} \, M$
7. $4.7 \times 10^{-5} \, M$
8. $1.9 \times 10^{-4} \, M$
9. $2.2 \times 10^{-3} \, g$
10. Yes. The ion product is equal to 1.8×10^{-4}, which is greater than K_{sp}.
11. No. The ion product is less than K_{sp}.
12. a. $CaCO_3(s) + H^+(aq) \rightleftharpoons Ca^{2+}(aq) + HCO_3^-(aq)$
    b. $PbC_2O_4 + H^+(aq) \rightleftharpoons Pb^{2+}(aq) + HC_2O_4^-(aq)$
    c. $Ag_2CrO_4 + H^+(aq) \rightleftharpoons 2Ag^+(aq) + HCrO_4^-(aq)$
13. Yes, the equations answering the previous question (Self–Test question 12) indicate that adding acid increases solubility. $MgBr_2$ is not affected by acids.
14. (a) $PbS(s) + H_2O \rightleftharpoons Pb^{2+}(aq) + OH^-(aq) + HS^-(aq)$
    (b) $PbS(s) + 2H^+(aq) \rightleftharpoons Pb^{2+}(aq) + H_2S(aq)$
15. $pH = 2.7$
16. Yes. The Sr^{2+} ion can be selectively precipitated. The pH of the solution must be in the range of 5.0 and 6.0.
17. $Zn(OH)_4^{2-}$
18. $Fe(SCN)_6^{3-}$

19. (a) $Hg^{2+}(aq) + 4NH_3(aq) \rightleftharpoons Hg(NH_3)_4^{2+}(aq)$ $\qquad K_{form} = \dfrac{[Hg(NH_3)_4^{2+}]}{[Hg^{2+}][NH_3]^4}$

    (b) $Sn^{4+}(aq) + 6F^-(aq) \rightleftharpoons SnF_6^{2-}(aq)$ $\qquad K_{form} = \dfrac{[SnF_6^{2-}]}{[Sn^{4+}][F^-]^6}$

    (c) $Fe^{3+}(aq) + 6CN^-(aq) \rightleftharpoons Fe(CN)_6^{3-}(aq)$ $\qquad K_{form} = \dfrac{[Fe(CN)_6^{3-}]}{[Fe^{3+}][CN^-]^6}$

20. (a) $Co^{2+}(aq) + 6NH_3(aq) \rightleftharpoons Co(NH_3)_6^{2+}(aq)$
    (b) $Co(NH_3)_6^{2+}(aq) \rightleftharpoons Co^{2+}(aq) + 6NH_3(aq)$

21. $K_{inst} = 1.9 \times 10^{-19}$

22. (a) $Fe(OH)_3(s) \rightleftharpoons Fe^{3+}(aq) + 3OH^-(aq)$ $\qquad K_{sp} = [Fe^{3+}][OH^-]^3$

    $Fe^{3+}(aq) + 6CN^-(aq) \rightleftharpoons Fe(CN)_6^{3-}(aq)$ $\qquad K_{form} = \dfrac{[Fe(CN)_6^{3-}]}{[Fe^{3+}][CN^-]^6}$

    $Fe(OH)_3(s) + 6CN^-(aq) \rightleftharpoons Fe(CN)_6^{3-}(aq) + 3OH^-(aq)$ $\qquad K_c = \dfrac{[Fe(CN)_6^{3-}][OH^-]^3}{[CN^-]^6}$

    (b) $CoCO_3(s) \rightleftharpoons Co^{2+}(aq) + CO_3^{2-}(aq)$ $\qquad K_{sp} = [Co^{2+}][CO_3^{2-}]$

    $Co^{2+}(aq) + 6NH_3(aq) \rightleftharpoons Co(NH_3)_6^{2+}(aq)$ $\qquad K_{form} = \dfrac{[Co(NH_3)_6^{2+}]}{[Co^{2+}][NH_3]^6}$

$$CoCO_3(s) + 6NH_3(aq) \rightleftharpoons Co(NH_3)_6{}^{2+}(aq) + CO_3{}^{2-}(aq)$$

$$K_c = \frac{[Co(NH_3)_6{}^{2+}][CO_3{}^{2-}]}{[NH_3]^6}$$

(c) $\quad Al(OH)_3(s) \rightleftharpoons Al^{3+}(aq) + 3OH^-(aq) \qquad\qquad K_{sp} = [Al^{3+}][OH^-]^3$

$\quad\quad\quad\quad Al^{3+}(aq) + 6F^-(aq) \rightleftharpoons Al(F)_6{}^{3-}(aq) \qquad\qquad K_{form} = \dfrac{[Al(F)_6{}^{3-}]}{[Al^{3+}][F^-]^6}$

$\quad\quad Al(OH)_3(s) + 6F^-(aq) \rightleftharpoons Al(F)_6{}^{3-}(aq) + 3OH^-(aq) \qquad K_c = \dfrac{[Al(F)_6{}^{3-}][OH^-]^3}{[F^-]^6}$

Tools for problem solving

In this chapter you learned to apply the following concepts as tools in solving problems. Study each one carefully so that you know what each is used for. When faced with solving a problem, recall what each tool does and consider whether it will be helpful in finding a solution.

You might want to tear these pages out to use along with solving problems in this chapter.

Solubility product constant, K_{sp} (Section 17.1)

The K_{sp} is the name of an equilibrium constant that applies when the equilibrium law describes a saturated solution.

Molar solubility (Section 17.1)

This is defined as the moles of solute dissolved in a liter of a saturated solution.

Predicting precipitation (Section 17.1)

$\qquad\qquad\qquad Q > K_{sp}$, a precipitate forms

$\qquad\qquad\qquad Q < K_{sp}$, a precipitate does not form

Acid solubility product constant, K_{spa} (Section 17.3)

The K_{spa} data are used to calculate the solubility of a metal sulfide at a given pH.

Combined K_a expression for a diprotic acid (Section 17.4)

Combining expressions for K_{a_1} and K_{a_2} for a diprotic acid, H_2A, yields the equation, for ex?

$$K_a = K_{a_1} K_{a_2} = \frac{[H^+]^2[CO_3{}^{2-}]}{[H_2CO_3]}$$

Formation constants of complex ions (Section 17.5)

We can use them to make judgments concerning the relative stabilities of complexes and the solubilities of salts in solutions containing a ligand.

Summary of Important Equations

Equilibrium equation for a sparingly soluble salt, M_mA_a:

$$M_mA_a(s) \rightleftharpoons mM^{a+}(aq) + aA^{m-}(aq)$$

Solubility product constant for a sparingly soluble salt, M_mA_a:

$$K_{sp} = [M^{a+}]^m[A^{m-}]^a$$

For a sparingly soluble metal sulfide, MS, in acid:

The solubility product equilibrium:

$$MS(s) + 2H^+(aq) \rightleftharpoons M^{2+}(aq) + H_2S(aq)$$

The solubility product constant, K_{spa}:

$$K_{spa} = \frac{[M^{2+}][H_2S]}{[H^+]^2}$$

Notes:

Chapter 18

Thermodynamics

In this chapter we turn our attention to what it is that causes events of any kind, whether they be physical or chemical, to occur spontaneously—that is, by themselves without outside assistance. Spontaneous events are crucial for our existence, because without them nothing would ever happen and we wouldn't exist. Understanding the factors that contribute to spontaneity is important because such events provide the driving force for all natural changes. You will also learn that there is a close connection between thermodynamics and chemical equilibrium.

18.1   First Law of Thermodynamics

Learning Objective

State the first law of thermodynamics and explain how the change in energy differs from the change in enthalpy.

Review

Thermodynamics is the study of energy changes, with a focus on the way energy flows between a system and the surroundings. Studying thermodynamics allows us to understand why spontaneous events occur and gives us insight into answers to many questions that plague modern society.

In this section, we review some topics introduced in Chapter 6. The first law of thermodynamics is given by the equation

$$\Delta E = q + w$$

where q is the heat absorbed by the system and w is the work done on the system. Study the signs of q and w in relation to the direction of energy flow.

A system can do work by expanding against an opposing pressure, and the amount of work can be calculated as $w = -P\Delta V$, where ΔV is the change in volume. (Here w is expressed as a negative quantity because the system loses energy when it does work.) If a system cannot change volume during a reaction, then the heat of reaction is equal to ΔE for the change.

$$\Delta E = q_v \text{ (where } q_v \text{ is the heat of reaction at constant volume)}$$

The enthalpy, H, is defined to deal with changes that occur at constant pressure. If the only kind of work a system can do is expansion work against the opposing atmospheric pressure, then ΔH is equal to the heat of reaction, and it is called the heat of reaction at constant pressure.

$$\Delta H = q_p$$

where the subscript p indicates "constant pressure," thus q_p is the heat of reaction at constant pressure,

For most reactions, the difference between ΔE and ΔH is very small and can be neglected. To convert between ΔE and ΔH, we can use the equation

$$\Delta H = \Delta E + \Delta n_{gas}RT$$

where Δn_{gas} is the change in the number of moles of gas on going from the reactants to the products. (If the energy is to be in joules or kilojoules, remember to use $R = 8.314$ J mol^{-1} K^{-1}.) Study Example 18.1 in the text to see how this equation is applied.

Example 18.1 Converting from ΔH to ΔE

When propane is burned at 356 °C, the heat of reaction is –2003 kJ. What is ΔE for this process under the same conditions?

Analysis:

To do this we need the formulas and a balanced chemical equation. Then we need to calculate Δn_{gas} and apply the relationships that were just reviewed.

Assembling the Tools:

Our tool is the equation for converting between ΔH and ΔE. We will need tools from previous chapters to write the formula for propane and the balanced equation for its combustion. The concept for determining Δn_{gas} is also used.

Solution:

The balanced chemical reaction for the combustion of propane is:

$$C_3H_8(g) \ + \ 5O_2(g) \longrightarrow 3CO_2(g) \ + \ 4H_2O(g)$$

From this we calculate Δn_{gas} as

7 mols of gaseous products – 6 moles of gaseous reactants = change of +1 mol of gas

Using the equation: $\Delta H = \Delta E + \Delta n_{gas}RT$, we obtain

$$-2003 \times 10^3 \text{ J} = \Delta E + (+1 \text{ mol}) (8.314 \text{ J mol}^{-1} \text{ K}^{-1}) (629 \text{ K}) = \Delta E + 5.23 \times 10^3 \text{ J}$$

$$\Delta E = -2008 \times 10^3 \text{ J} = -2008 \text{ kJ}$$

Is the Answer Reasonable?

We see that the difference between ΔE and ΔH is very small (less than 0.25% difference) as we would expect. We also expect ΔE to be more negative than ΔH, and it is. Review the math to be sure calculations were done correctly, and be especially careful that you don't add joules and kJ together without converting one into the other first.

Self-Test

1. Give the definition of enthalpy in terms of the internal energy, the pressure, and the volume.

    _____

2. The product of pressure in pascals (Pa, which has the units newtons per square meter, N/m^2) times volume in cubic meters (m^3) gives work in units of joules. In Chapter 8, the standard atmosphere was defined in terms of the pascal as 1 atm = 101,325 Pa. With this information, calculate the amount of work, in joules, done by a gas when it expands from a volume of 500 mL to 1500 mL against a constant opposing pressure of 5.00 atm.

    (Remember that 1 mL = 1 cm^3.)   _____

3. Suppose that during an exothermic reaction at constant volume, a gas is produced, so that the pressure increases. How would ΔE for this reaction compare to its ΔH? Why?

    _____

    _____

4. How would the values of ΔE and ΔH compare for the reaction

    $$CO_2(g) + H_2(g) \rightarrow CO(g) + H_2O(g)$$

    Explain your answer.   _____

    _____

5. At 25 °C and a pressure of 1 atm, the reaction

    $$2NO_2(g) \rightarrow N_2O_4(g)$$

    has $\Delta H = -57.9$ kJ. What is the value of ΔE for this reaction at the same temperature?

    _____

18.2 Spontaneous Change

Learning Objective

Explain what is meant by the direction of spontaneous change.

Review

A spontaneous change is one that takes place all by itself without continual assistance. Once conditions are right, it proceeds on its own. A nonspontaneous change needs continual help to make it happen, and some spontaneous change must occur first to drive the nonspontaneous one. Everything we see happen is ultimately caused by a spontaneous event of some kind.

When a change is accompanied by a decrease in the potential energy of the system (i.e., when the change is exothermic), it tends to occur spontaneously. Some examples are shown in the photographs in Figure 18.5. A potential energy decrease is a factor in favor of spontaneity. It is not the sole factor, however, because there are changes that are endothermic and nevertheless spontaneous.

In chemical reactions at constant pressure and temperature, the energy change is given by ΔH. For a pressure of 1 atm and a temperature of 25 °C, it is $\Delta H°$, the standard enthalpy change. When $\Delta H°$ is negative, the process occurs by a decrease in its potential energy and it tends to be spontaneous.

Self-Test

6. Which of the following do you recognize as spontaneous events?

    (a) Ice melts on a hot day.   _____

    (b) Water boils in a home in Alaska.   _____

    (c) A scratched fender on an automobile rusts.   _____

    (d) Dirty clothes become clean.   _____

7. Which of the following chemical changes *tend* to be spontaneous at 25 °C and 1 atm, based on their enthalpy changes? (Refer to Table 6.2 in the text.)

    (a) $2PbO(s) \longrightarrow 2Pb(s) + O_2(g)$

    (b) $NO(g) + SO_3(g) \longrightarrow NO_2(g) + SO_2(g)$

    (c) $Fe_2O_3(s) + 3CO(g) \longrightarrow 3CO_2(g) + 2Fe(s)$

    Answer: _____

8. Give an example based on observations you have made in your daily life of a change that is endothermic and spontaneous.

18.3   Entropy

Learning Objective

Describe the factors that affect the magnitude of entropy and the sign of an entropy change in a chemical reaction.

Review

In this section we see that statistical probability is an important factor to consider in analyzing spontaneous events. In general, when a change is accompanied by an increase in the statistical probability of an energy distribution, the change has a tendency to occur spontaneously. Such a change increases the number of equivalent ways of distributing the energy, so a spontaneous process tends to disperse energy.

The thermodynamic quantity related to probability is the entropy, S. The more probable the state, the higher is the entropy, and the greater is the number of ways of distributing the energy in the system. Thus, an increase in entropy corresponds to a dispersal (spreading out) of energy over more possible arrangements. It also corresponds to an increase in the freedom of molecular motion — any change that increases the freedom of motion also increases the entropy. Mathematically, a change tends to occur spontaneously if $\Delta S > 0$ (ΔS is positive and $S_{final} > S_{initial}$).

The entropy change can often be predicted for a change. In general, the entropy of a system increases when

   • the temperature of the system increases
   • there is an increase in the volume of the system

- there is a change from solid → liquid, liquid → gas, or solid → gas

For a chemical reaction, ΔS will be positive when

- there is an increase in the number of moles of gas in the system
- there is an increase in the number of independent particles in the system

---

Example 18.2 Estimating entropy change

The reaction $N_2O_4(g) \longrightarrow N_2(g) + 2O_2(g)$ is exothermic. Is this reaction expected to be spontaneous? (Base your answer on the enthalpy and entropy changes involved.)

Analysis:

In this chapter we learned that both entropy change and enthalpy change contribute to a spontaneous process. We are given the fact that the given reaction is exothermic. We need to evaluate the entropy change. Factors we will look for are, in order, (1) a change in the moles of gas, Δn_{gas}, (2) a change in the number of molecules, and (3) a change in the physical state of the molecules.

Assembling the Tools:

Our tool is the group of factors that can be used to assess the sign and magnitude of the entropy change.

Solution:

We find an increase in the number of moles of gas, from one to two. This is a major factor and we could stop here and declare that the entropy increases for this reaction. However, considering the second factor, this also suggests an increase in entropy since we have two smaller molecules. The third factor does not apply since the entire reaction takes place in the gas phase.

The result is that the enthalpy is exothermic and the entropy change is positive; therefore, we conclude the reaction is spontaneous.

Is the Answer Reasonable?

About all we can do is check our logic with the principles developed in the text.

Self-Test

9. If two containers holding different gases are connected together, the gases will gradually diffuse until the composition of the mixture is the same in both containers. Explain this in probability terms.

---

10.   What is the sign of ΔS for each of the following changes?

   (a)   Solid iodine vaporizes.   _____

   (b)   Waste oil is spilled into the ground.   _____

   (c)   A gas is compressed into a tire.   _____

   (d)   A liquid is cooled.   _____

   (e)   Sugar dissolves in water to form a solution.   _____

11.   What is the sign of ΔS for each of the following reactions?

   (a) $2HgO(s) \longrightarrow 2Hg(g) + O_2(g)$ _____

   (b) $CaO(s) + CO_2(g) \longrightarrow CaCO_3(s)$ _____

   (c) $H_2(g) + I_2(s) \longrightarrow 2HI(g)$ _____

   (d) $2C_2H_6(g) + 7O_2(g) \longrightarrow 4CO_2(g) + 6H_2O(g)$ _____

18.4   Second Law of Thermodynamics

Learning Objective

State the second law of thermodynamics in terms of entropy, and explain the significance of

Gibbs free energy.

Review

In regard to favoring spontaneity, the enthalpy change and entropy change sometimes work in the same direction, but at other times one favors spontaneity while the other does not. In the latter situations, the temperature becomes the controlling factor. All of these concepts are brought together by the second law of thermodynamics.

   The *second law of thermodynamics* tells us that any spontaneous event is accompanied by an overall increase in the entropy of the universe. Earlier we saw that everything that happens is the net result of a spontaneous change of some sort, so every time something happens in the world there is an increase in entropy and some energy is dispersed and made unavailable for future use.

   The Gibbs free energy is a composite function of the enthalpy and entropy. It allows us to gauge quantitatively how important these individual factors are in determining the spontaneity of an event.

   At constant temperature and pressure,

$$\Delta G = \Delta H - T\Delta S$$

A change will be spontaneous when the enthalpy term (ΔH) and the entropy term ($T\Delta S$) combine to give a negative value for ΔG, that is, when the free energy decreases.

You should be able to analyze how (or if) temperature will affect the spontaneity of an event when you know the signs of ΔH and ΔS. Study the summary in Figures 18.9 and 18.10.

Example 18.3 Judging the expected sign for ΔG

In Example 18.2 above, you were told that the reaction

$$N_2O_4(g) \longrightarrow N_2(g) + 2O_2(g)$$

is exothermic. What is the expected sign of ΔG for this reaction and how do we expect the algebraic sign of ΔG to be affected by the temperature?

Analysis:

Reviewing our results, this reaction is exothermic and the entropy change is positive. We need to deduce the signs for ΔH and ΔS so that we can predict the change in ΔG, if any, with changes in temperature.

Assembling the Tools:

Our tools include the concepts for estimating the signs and magnitudes of ΔH and ΔS. Then we can use the Gibbs free energy equation to determine the sign of in ΔG.

Solution:

The free energy equation is $\Delta G = \Delta H - T\Delta S$. Since the ΔH is negative, it will remain negative if temperature changes. Since the ΔS term is positive, the $-T\Delta S$ term in the equation must always be negative too (remember that T, the Kelvin temperature, is always positive). Therefore we conclude that the sum of two negative numbers will also be negative and ΔG will be negative for this reaction at all temperatures.

Is the Answer Reasonable?

Once again there is no mathematical operation to check, just our logic. Review the logic to be sure it makes sense.

Self-Test

12. For which of the following is ΔG *always* positive, regardless of the temperature?

    (a) ΔH positive, ΔS negative

    (b) ΔH negative, ΔS negative   _____

18.5   Third Law of Thermodynamics

Learning Objective

Explain how the third law of thermodynamics leads to a standard entropy of formation.

Review

At 0 K the entropy of any pure crystalline substance is zero. This is a statement of the *third law of thermodynamics*. Because the zero point on the entropy scale is known, the absolute amount of entropy that a substance possesses can be determined. (This can be compared with energy or enthalpy, where it is impossible to figure out exactly how much energy a substance has because it is impossible to know when a substance has zero energy.)

Standard entropies—entropies at 25 °C and 1 atm—can be used to calculate standard entropy changes by a Hess's law type of calculation using Equation 18.6. Be sure to study Example 18.3 in the text.

Self-Test

13. Calculate the standard entropy change, in J/K, for the following reactions.

(a) $2NaCl(s) \longrightarrow 2Na(s) + Cl_2(g)$ _____

(b) $CO(g) + 2H_2(g) \longrightarrow CH_3OH(l)$ _____

(c) $CH_4(g) + 2O_2(g) \longrightarrow CO_2(g) + 2H_2O(g)$ _____

(d) $2KCl(s) + H_2SO_4(l) \longrightarrow K_2SO_4(s) + 2HCl(g)$ _____

14. If entropy changes were the *only* factors involved, which of the reactions in the preceding question would occur spontaneously at 25 °C and a pressure of 1 atm?

_____

18.6   The Standard Free Energy Change, $\Delta G°$

Learning Objective

Calculate standard free energy changes.

Review

As with other thermodynamic quantities, the standard free energy change is the free energy change at 25 °C and 1 atm. We can compute it in two ways. One is from $\Delta H°$ and $\Delta S°$.

$$\Delta G° = \Delta H° - (298\text{ K})\Delta S°$$

The other is from tabulated values of standard free energies of formation.

$$\Delta G° = (\text{sum } \Delta G_f° \text{ products}) - (\text{sum } \Delta G_f° \text{ reactants})$$

These are not difficult calculations. Study Examples 18.4 and 18.5 in the text and work Practice Exercises 18.14 through 18.17; then try the Self-Test below.

Example 18.4 Calculation of ΔG using two methods

Calculate the ΔG for the combustion of butane at 298 K by using $\Delta H_f°$ and $S°$ and then by using $\Delta G_{298}°$.

Analysis:

We return to Table 6.2 to find data to calculate the $\Delta H_{298}°$ and we use Table 18.1 in this chapter to find the data to calculate $\Delta S_{298}°$. These two values are combined using the Gibbs free energy equation to calculate $\Delta G_{298}°$. We will then use the standard free energies of formation in Table 18.2 to find the data to calculate $\Delta G_{298}°$ directly.

Assembling the Tools:

Our tools include the concepts for writing formulas and balanced chemical equations. Then we apply the tools for calculating enthalpy, entropy, and free energy changes from tabulated data. The Gibbs free energy equation gives us a second way to determine $\Delta G_{298}°$. Recall that all of these are similar to Hess's law.

Solution:

To calculate the heat of reaction we need the balanced chemical equation

$$2C_4H_{10}(g) \ + \ 13O_2(g) \longrightarrow \ 8CO_2(g) \ + \ 10H_2O(g)$$

$$\Delta H_{298}° = [8(-393.5\text{ kJ}) \ + \ 10(-241.8\text{ kJ})] - [2(-126\text{ kJ})]$$
$$= -5314\text{ kJ}$$
$$\Delta S_{298}° = [8(213.6) \ + \ 10(188.7)] - [2(310.2) + 13(205)] \text{ (all J mol}^{-1}\text{ K}^{-1})$$
$$= +310.4\text{ J mol}^{-1}\text{ K}^{-1}$$
$$= +0.3104\text{ kJ mol}^{-1}\text{ K}^{-1}$$

Now we calculate $\Delta G°$.

$$\Delta G_{298}° \ = -5314\text{ kJ} - (298\text{ K})(+0.3104\text{ kJ mol}^{-1}\text{ K}^{-1})$$

$$= -5406 \text{ kJ}$$

Using the standard free energies in Table 18.2 we calculate ΔG°_{298} a different way.

$$\Delta G^{\circ}_{298} = [8(-394.4 \text{ kJ}) + 10(-228.6 \text{ kJ})] - [2(-17 \text{ kJ})]$$

$$\Delta G^{\circ}_{298} = -5427 \text{ kJ}$$

Are the Answers Reasonable?

Our answers, -5406 kJ and -5427 kJ, can be checked by being sure the correct values are transferred from the tables and that the calculations are correct. Our answers are not identical, but considering the number of steps, experimental error, and rounding errors, they are acceptably close. They are within $\pm 0.4\%$ of each other.

Self-Test

15. Using Table 6.2 and Table 18.1 in the text, calculate ΔG° (in kilojoules) for the reaction

$$CaO(s) + H_2O(l) \longrightarrow Ca(OH)_2(s)$$

_____

16. Use standard free energies of formation in Table 18.2 to calculate ΔG° (in kilojoules) for the following reactions:

    (a) $2C_2H_5OH(l) + 6O_2(g) \longrightarrow 4CO_2(g) + 6H_2O(l)$

_____

    (b) $2HNO_3(l) + 2HCl(g) \longrightarrow Cl_2(g) + 2NO_2(g) + 2H_2O(l)$

_____

18.7 Maximum Work and ΔG

Learning Objective

Explain the connection between the Gibbs free energy change and the maximum energy produced by a reaction.

Review

The free energy change for a reaction is equal to the maximum amount of energy that, theoretically, can be recovered as useful work. In all real situations, however, somewhat less than this maximum is actually obtained. This is because the maximum work can only be produced if the process occurs reversibly. A reversible process takes an infinite length of time and consists of an infinite number of small changes in which the driving "force" is very nearly balanced by an opposing "force." No real process from which work is extracted occurs reversibly, so we always obtain less than the maximum. Nevertheless, ΔG gives us a goal to aim at, and allows us to gauge the efficiency of the way we are using a reaction to obtain work.

Example 18.5   Maximum energy and maximum work

At 25 °C and 1 atm, what is the maximum amount of heat that could be extracted from the reaction of 1 mol $CaO(s)$ with $H_2O(l)$?

$$CaO(s) + H_2O(l) \longrightarrow Ca(OH)_2(s)$$

What is the maximum amount of work that could be obtained from this reaction?

Analysis:

To calculate the maximum work, we need to calculate ΔG for the reaction. Once again, because the conditions correspond to standard conditions, we need to calculate $\Delta G°$. The amount of heat obtained at constant pressure is the enthalpy change, so we need to calculate ΔH for the reaction. Because it occurs at 25 °C and 1 atm, this becomes the standard enthalpy change, $\Delta H°$. We also calculate the standard entropy change for this reaction.

Assembling the Tools:

Our tools include the concepts for calculating standard enthalpy, entropy, and free energy changes from tabulated data.

Solution:

If you worked Question 15 in the Study Guide, you already have all of the data that you need. For that question we found the following:

$$\Delta H° = -65.2 \text{ kJ}$$

$$\Delta S° = -34 \text{ J/K} \quad (T\Delta S = 10 \text{ kJ})$$

$$\Delta G° = -55 \text{ kJ (rounded)}$$

If the reaction is carried out so that no work is done, all the energy escapes as heat and the heat evolved is equal to $\Delta H°$. In other words, 65.2 kJ of heat is given off. If the reaction is carried out reversibly so that the maximum work is obtained, this work is equal to $\Delta G°$, or 55 kJ of work.

Notice that we get less energy as work than as heat. What happens to the rest of the energy? The answer is that it permits the entropy of the system to decrease. We know that for the reaction to be spontaneous, the entropy of the universe (system and surroundings taken together) must be positive, which means that overall there must be a net entropy increase and a net dispersal of energy. The 10 kJ that is *not* available for useful work amounts to the energy that must be dispersed into the surroundings to compensate for the lowering of the entropy of the system.

Is the Answer Reasonable?

We haven't really done any calculations, so the only thing to check is the reasoning in the analysis step, which seems to be sound.

Self-Test

17. Carbon monoxide is sometimes used as an industrial fuel.

    (a) What is the maximum heat obtained at 25 °C and 1 atm by burning 1 mol of CO? The

    reaction is. $2CO(g) + O_2(g) \longrightarrow 2CO_2(g)$

    _____

    (b) What is the maximum work available at 25 °C and 1 atm from the combustion of 1 mol of

    CO(g)?

    _____

18.8   Free Energy and Equilibrium

Learning Objective

Explain the connection between Gibbs free energy change and the position of equilibrium

Review

At equilibrium, the total free energy of the products equals the total free energy of the reactants and ΔG for the system is equal to zero.

$$\Delta G = 0 \quad \text{at equilibrium}$$

Because $\Delta G = 0$ at equilibrium, and because ΔG equals the maximum amount of work obtainable from a change, at equilibrium we can obtain no work from a system.

Equilibria in phase changes

For a phase change, equilibrium can occur at only one temperature. At this temperature, T,

$$T = \frac{\Delta H}{\Delta S}$$

Without much error, T can be calculated using $\Delta H°$ and $\Delta S°$ (which apply, strictly speaking, at 25 °C) because ΔH and ΔS do not change very much with temperature.

Equilibrium involving a phase change at 1 atm can only occur at one temperature. For example, pure liquid water at a pressure of 1 atm can be in equilibrium with ice only if its temperature is 0 °C. Liquid water can be in equilibrium with water vapor that has a pressure of 1 atm only when the temperature is 100 °C (the boiling point). At 25 °C and a pressure of 1 atm (and with the absence of other gases such as air), water will exist entirely as a liquid. No equilibrium exists. (See Figure 1 on the next page.)

Equilibria in homogeneous chemical systems

Nearly all homogeneous chemical reactions are able to exist in a state of equilibrium at 25 °C, and the position of equilibrium is determined by the value of $\Delta G°$. Study the free energy diagrams in Figures 18.16 and 18.17. Notice that when $\Delta G°$ is positive, the position of equilibrium lies near the reactants. On the other hand, when $\Delta G°$ is negative, the position of equilibrium lies close to the products.

When $\Delta G°$ has a reasonably large negative value (–20 kJ, or so), the position of equilibrium lies far in the direction of the products, and when the reaction occurs it will appear to go essentially to completion. On the other hand, if $\Delta G°$ has a reasonably large positive value, hardly any products will be present at equilibrium. In other words, when the reactants are mixed, they will not appear to react at all. For all practical purposes, no reaction occurs when $\Delta G°$ is positive and has a value in excess of about 20 kJ. Therefore, we can use the sign and magnitude of $\Delta G°$ as an indicator of whether or not we expect to actually observe the formation of products in a reaction.

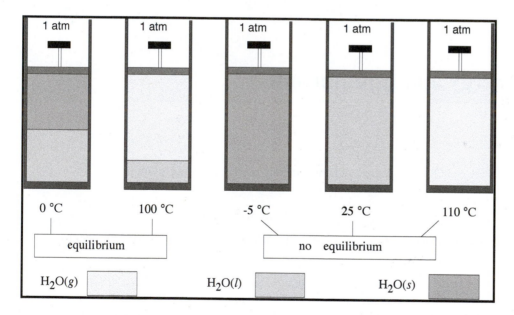

Figure 1   At a pressure of 1 atm, equilibrium between any two phases of water can only occur at a single temperature. At temperatures of –5, 25, and 110 °C, only a single phase can exist.

Summary

$\Delta G°$ large and negative	Reaction goes very nearly to completion
$\Delta G°$ large and positive	Virtually no reaction occurs. The reaction does not appear to be spontaneous
$\Delta G°$ less than about ±20 kJ	Reactants and products are both present in significant amounts at equilibrium

In the text, we've used the symbol $\Delta G_T°$ to stand for the equivalent of $\Delta G°$, but at a temperature other than 25 °C. Because ΔH and ΔS change little with temperature, we can approximate $\Delta G_T°$ by the equation

$$\Delta G_T° \approx \Delta H° - T\Delta S°$$

The sign and magnitude of $\Delta G_T°$ can be used in the same way as for $\Delta G°$ in predicting the outcome of a reaction.

Keep in mind that even though $\Delta G°$ or $\Delta G_T°$ may predict that a reaction should be "spontaneous," it tells us nothing about how rapid the reaction will be. The reaction of H_2 with O_2 has a very negative value of $\Delta G°$, so it is very "spontaneous." However, at room temperature the reaction is so slow that no reaction is observed. Thus, a change must not only be spontaneous, it must occur relatively fast for us to actually observe the change.

Example 18.6 Predicting the effect of temperature on the position of equilibrium

Consider the reaction

$$2PCl_3(g) + O_2(g) \rightleftharpoons 2POCl_3(g)$$

(a) Suppose an equilibrium mixture containing all three of the substances involved in this reaction is heated from 25 °C to 45 °C. How will the amount of PCl_3 in the mixture change?

(b) At standard state, what is the maximum amount of work that could be obtained from this reaction?

Analysis:

For part (a) there are no quantities of any reactants or products given; this problem must be asking for a reasoned, qualitative answer. Will the PCl_3 increase, decrease, or stay the same? To determine this we need to know the value and sign of $\Delta G°$ at the two temperatures. For this we need to determine the standard heat and the standard entropy change for this reaction. We will then use the Gibbs free energy equation to determine the free energy at the two different temperatures.

For part (b) the system is at standard state, so the maximum work will be $\Delta G°$.

Assembling the Tools:

We apply the tools for calculating enthalpy, entropy, and free energy changes from tabulated data.

Solution:

In Example 18.4 above, we saw that calculating the free energy using two methods gave two slightly different answers. To avoid that problem we will use the Gibbs free energy equation to determine the standard free energy change at both 298 and 318 K. We calculate ΔH°_{298} and ΔS°_{298} as we have done before. The data for this reaction are found in Appendix C.1.

$$\Delta H^\circ_{298} = [2(-1109.7 \text{ kJ})] - [2(-287.0 \text{ kJ})]$$
$$= -1645.4 \text{ kJ}$$
$$\Delta S^\circ_{298} = [2(646.5)] - [2(311.8) + 1(205)] \text{ (all J mol}^{-1} \text{ K}^{-1})$$
$$= +464 \text{ J mol}^{-1} \text{ K}^{-1}$$
$$= +0.464 \text{ kJ mol}^{-1} \text{ K}^{-1}$$

Now we calculate ΔG° at 298 K

$$\Delta G^\circ_{298} = -1645.4 \text{ kJ} - (298 \text{ K})(+0.464 \text{ kJ mol}^{-1} \text{ K}^{-1})$$
$$= -1516.1 \text{ kJ}$$

Now we calculate ΔG° at 318 K

$$\Delta G^\circ_{318} = -1645.4 \text{ kJ} - (318 \text{ K})(+0.464 \text{ kJ mol}^{-1} \text{ K}^{-1})$$
$$= -1793.0 \text{ kJ}$$

Because ΔG° is more negative at the higher temperature, the reaction proceeds more toward completion than it does at the lower temperature.

(b) The maximum work at each temperature is given by the free energy calculated at those temperatures.

Are the Answers Reasonable?

Once again the most important part of our check is to be sure we have transferred the correct values from the tables. Note that $POCl_3$ is listed for both the gas and liquid phase and we need to use data for the gas phase. In fact, the answer is just the reverse if you use the liquid phase.

Self-Test

18. At the boiling point of water, liquid and vapor are in equilibrium at a pressure of 1 atm. Use ΔH° and ΔS° for the reaction,

$$H_2O(l) \rightleftharpoons H_2O(g)$$

to calculate the boiling point of water. How does the answer compare to the actual boiling point? Does this support the statement that ΔH and ΔS are nearly temperature independent?

---

19. What would you expect to observe at 25 °C if 2 mol CO and 1 mol O_2 were mixed and allowed to react according to the following equation?

$$2CO(g) + O_2(g) \longrightarrow 2CO_2(g)$$

_____

20. What would you expect to observe at 25 °C if 2 mol of $N_2O(g)$ were mixed with 1 mol $O_2(g)$ in order to form $NO(g)$ by the reaction? $2N_2O(g) + O_2(g) \longrightarrow 4NO(g)$

_____

21. What should we expect to observe at 25 °C if we were to check a mixture of N_2O and O_2 for the presence of NO_2 in the reaction? $2N_2O(g) + 3O_2(g) \longrightarrow 4NO_2(g)$?

_____

22. Assuming that ΔH and ΔS are approximately independent of temperature, calculate the "standard" free energy change, ΔG_T°, for the reaction in Question 15 at 1 atm and 50 °C.

$$\Delta G_{323}^\circ = \underline{\hspace{5cm}}$$

Does this reaction proceed farther toward completion at this higher temperature? _____

18.9 Equilibrium constants and $\Delta G°$

Learning Objective

Describe how to calculate a Gibbs free energy change from an equilibrium constant.

Review

For a given composition in a reaction mixture, the value of ΔG for the reaction is related to the reaction quotient by Equation 18.11 in the text and repeated here.

$$\Delta G = \Delta G° + RT \ln Q \tag{18.11}$$

If you must apply this equation, remember that to calculate Q we use partial pressures (in atm) for gaseous reactions and molar concentrations if the reaction is in solution. Equation 18.11 can be used to determine where a reaction stands relative to equilibrium:

ΔG is negative.	The reaction must proceed in the forward direction to reach equilibrium.
ΔG is zero.	The reaction is at equilibrium.
ΔG is positive.	The reaction must proceed in the reverse direction to reach equilibrium.

In the text, Example 18.11 illustrates how Equation 18.12 is applied.

Thermodynamic equilibrium constants

The principal purpose of this section is to show that the equilibrium constant is related to $\Delta G°$ for the reaction. You should learn Equation 18.12 in the text which is repeated here.

$$\Delta G° = -RT \ln K \qquad (18.12)$$

In these equations, K is K_p for reactions involving gases. It is K_c for reactions in liquid solutions.

In using Equation 18.12, be sure to choose the value of R that has units that match the energy units of $\Delta G°$. If $\Delta G°$ is in kilojoules, use $R = 8.314$ J mol^{-1} K^{-1} and be sure to change kilojoules to joules so the units cancel.

In this section we also see how we can estimate the value of the thermodynamic equilibrium constant at temperatures other than 25 °C. This is done by calculating the value of $\Delta G_T°$ from $\Delta H°$ and $\Delta S°$. The appropriate equation is

$$\Delta G_T° = \Delta H° - T\Delta S°$$

Once you've obtained $\Delta G_T°$ in this way, then you use Equation 18.12 to calculate the value of K.

Example 18.7 Determining if a system is at equilibrium

The reaction

$$2C_4H_{10}(g) + 13O_2(g) \rightleftharpoons 8CO_2(g) + 10H_2O(g)$$

has $\Delta G° = -5406$ kJ at 25 °C. A certain mixture of these gases has the following partial pressures: for C_4H_{10}, 3.00×10^{-6} torr; for O_2, 12.0 torr; for CO_2, 359 torr; and for H_2O, 375 torr. Is this reaction mixture at equilibrium? If not, which way must the reaction proceed to reach equilibrium?

Analysis:

If ΔG is zero, the reaction mixture is at equilibrium; if ΔG is negative the reaction must proceed in the forward direction to reach equilibrium; and if ΔG is positive, then the reaction must proceed in the reverse direction. The core of this question is to determine ΔG. We need the value of the reaction quotient Q. This is the value of the mass action expression for the system, written using partial pressures.

Assembling the Tools:

Our tool is the equation that relates ΔG, $\Delta G°$, and Q to each other. We also need the tool for the mass action expression that uses partial pressures.

Solution:

We start by writing the mass action expression

$$\frac{P_{CO_2}^8 \, P_{H_2O}^{10}}{P_{C_4H_{10}}^2 \, P_{O_2}^{13}} \, .$$

We then need to convert the partial pressures given to us in torr to units of atmospheres. We do this by using the conversion factor (1 atm/760 torr). The pressures are for C_4H_{10}, 3.95×10^{-9} atm; for O_2, 0.0158 atm; for CO_2, 0.472 atm; and for H_2O, 0.493 atm. Substituting partial pressures expressed in atmospheres gives the value of Q.

$$Q = \frac{(0.472 \text{ atm})^8 (0.493 \text{ atm})^{10}}{(3.95 \times 10^{-9} \text{ atm})^2 (0.0158 \text{ atm})^{13}} = 3.50 \times 10^{34}$$

We now need to use $T = 298$ K and $R = 8.314$ J mol^{-1} K^{-1}, and $\Delta G° = -5406 \times 10^3$ J. These quantities are substituted into the equation above to calculate $\Delta G = -5209$ kJ. The negative value of ΔG indicates that the reaction is not in equilibrium and it will be spontaneous in the forward direction.

Is the Answer Reasonable?

It may seem surprising that the extremely large value of Q gave a small number as a natural logarithm. That is indeed the case and the result is that it is difficult to overcome the large negative value of the standard free energy. We should carefully review our math to verify this result.

Self-Test

23. The reaction $2N_2O(g) + 3O_2(g) \rightleftharpoons 4NO_2(g)$ has $\Delta G° = +0.17$ kJ. In a reaction mixture the gases involved in the reaction have the following partial pressures: N_2O, 220 torr; O_2, 120 torr; and NO_2, 458 torr. Is this reaction mixture at equilibrium? If not, in which direction must the reaction proceed to reach equilibrium?

24. The reaction $NO(g) + NO_2(g) + H_2O(g) \rightleftharpoons 2HNO_2(g)$ has $K_p = 1.56$ at 25 °C. What is $\Delta G°$ for this reaction expressed in kilojoules?

25. At 25 °C, $K_p = 4.8 \times 10^{-31}$ for the reaction $N_2(g) + O_2(g) \rightleftharpoons 2NO(g)$. What is $\Delta G°$ for this reaction expressed in kJ?

_____

26. The reaction $2N_2O(g) + 3O_2(g) \rightleftharpoons 4NO_2(g)$ has $\Delta G° = +0.17$ kJ. What is K_p for this reaction?

_____

27. Use the data in Table 18.2 to compute the value of K_p for the reaction

$$C_2H_2(g) + 2H_2(g) \rightleftharpoons C_2H_6(g)$$

_____

28. The reaction $2C_4H_{10}(g) + 13O_2(g) \rightleftharpoons 8CO_2(g) + 10H_2O(g)$ has $\Delta G° = -5406$ kJ at 25 °C. What is the value of K_p for this reaction?

_____

29. The oxidation of sulfur dioxide to sulfur trioxide by molecular oxygen,

$$2SO_2(g) + O_2(g) \rightleftharpoons 2SO_3(g)$$

has a standard heat of reaction, $\Delta H° = -196.6$ kJ and a standard entropy of reaction $\Delta S° = -189.6$ J K^{-1}. What is the value of K_p for this reaction

(a) at 25 °C?   _____

(b) at 500 °C?   _____

18.10   Bond Energies

Learning Objective

Define bond energy in terms of enthalpy.

Review

In this section, you learn how values of $\Delta H_f°$ are used to calculate bond energies. The basis for these calculations is Hess's law and the fact that the enthalpy change is a state function; that is, the same enthalpy change takes place regardless of the path followed from the reactants to the products. Study Figure 18.18. The sum of the enthalpy changes corresponding to steps 1, 2, and 3 must be equal to ΔH_f^o for the product.

In setting up an alternative path from the reactants (the elements in their standard states) to the product, we consider the following:

The conversion of the elements on the reactant side into gaseous atoms. The energy changes here are the standard heats of formation of the gaseous elements (Table 18.3). In the direction of the arrows in steps 1 and 2 in Figure 18.18, these changes are endothermic. We know this because it *always* takes energy to vaporize a solid element, and it *always* takes energy to break bonds to give atoms.

The formation of all the bonds in the product molecule. In discussing this energy change, we define the atomization energy, ΔH_{atom}, which is the energy needed to *break* all the bonds in the molecule. Whether we are forming bonds or breaking them, the amount of energy involved is the same. Only the sign of the ΔH is different, positive (endothermic) for bond breaking and negative (exothermic) for bond making. We obtain the atomization energy by adding up all the bond energies for the bonds in the molecule (Table 18.4).

Once an alternative path is established, we can use it to calculate bond energies if the value of ΔH_f^o is known, or we can calculate the value of ΔH_f^o if all the bond energies are known. In the text, this is illustrated in the calculation of the heat of formation of methyl alcohol vapor. Study the steps in the calculations and then work the Self-Test below.

Self-Test

30. Calculate the atomization energy of the molecule CH_3CN in kJ/mol. The molecule has the structure

$$
\begin{array}{c}
\mathrm{H} \\
| \\
\mathrm{H}-\mathrm{C}-\mathrm{C}\equiv\mathrm{N}{:} \\
| \\
\mathrm{H}
\end{array}
$$

———————————————

31. On a separate sheet of paper, construct a figure similar to that in Figure 18.18 in the text for the formation of $CH_3CN(g)$ from its elements. Be sure to show both the direct and alternative paths.

32. Estimate the standard heat of formation of CH_3CN vapor in kJ mol^{-1} using data in Table 18.3 in the text and your answers to Questions 30 and 31 above.

———————————————

Answers to Self-Test Questions

1. $H = E + PV$
2. 507 J
3. $\Delta H < \Delta E$; at constant pressure there would be a volume increase. Therefore, at constant pressure, the system performs some work that uses energy, which is being done on it during the volume increase. This use of energy for work will appear as less heat given off during the reaction at constant pressure.
4. No volume change would occur at constant pressure, so ΔH and ΔE are the same.
5. $\Delta E = -55.5$ kJ
6. (a) spontaneous, (b) nonspontaneous, (c) spontaneous, (d) nonspontaneous
7. (a) nonspontaneous, $\Delta H° = +438.4$ kJ (b) nonspontaneous, $\Delta H° = +41.8$ kJ
    (c) spontaneous, $\Delta H° = -26.4$ kJ
8. The melting of ice on a warm day, or the evaporation of a puddle of water.
9. When the gas molecules in one container have the entire volume of the other available to it, a state of low probability exists until the gas expands into the other container as well.
10. (a) positive, (b) positive, (c) negative, (d) negative, (e) positive
11. (a) positive, (b) negative, (c) positive, (d) positive
12. (a) ΔG is positive at all temperatures.
13. (a) $+180.2$ J/K, (b) -332.3 J/K, (c) -5.2 J/K, (d) $+227.2$ J/K
14. (a) spontaneous, (b) nonspontaneous, (c) nonspontaneous, (d) spontaneous
15. $\Delta G° = -55$ kJ
16. (a) $\Delta G° = -2651$ kJ, (b) $\Delta G° = -20.4$ kJ
17. (a) $\Delta H° = -283$ kJ/mol CO; heat evolved = 283 kJ, (b) $\Delta G° = -257.1$ kJ/mol of CO; max. work done = 257.1 kJ
18. $\Delta H° = 44.1$ kJ, $\Delta S° = 118.7$ J/K, $T_b = 372$ K = 99 °C. The actual boiling point is 100 °C = 373 K. The answer from $\Delta H°/\Delta S°$ is quite close, which supports the statement.
19. $\Delta G° = -514.2$ kJ. The reaction should go very nearly to completion.
20. $\Delta G° = +139.6$ kJ. No reaction should be observed.
21. $\Delta G° = +0.17$ kJ, 0.2 kJ when properly rounded. Substantial amounts of both N_2O and NO_2 should be present at equilibrium.
22. $\Delta G_T° = -34.3$ kJ at 50 °C (323 K). $\Delta G_T°$ is less negative than $\Delta G°$, so the reaction should not proceed as far toward completion at the higher temperature.
23. Calculated $\Delta G = -14.9$ kJ. The mixture is not at equilibrium. The reverse reaction is spontaneous, so the reaction goes to the left to reach equilibrium.
24. $\Delta G° = -1.10$ kJ
25. $\Delta G° = +173$ kJ
26. $K_p = 0.93$
27. $K_p = 2.6 \times 10^{42}$
28. $K_p = 10^{948}$ (solution is obtained by converting natural logarithms to base-10 logarithms)
29. (a) $K_p = 3.6 \times 10^{24}$ (b) $K_p = 2.4 \times 10^3$
30. $\Delta H_{atom} = 2474$ kJ/mol

31.

$$2C(g) \quad + \quad 3H(g) \quad + \quad N(g) \xrightarrow{\hspace{3cm}}$$

$$2C(s) \quad + \quad \tfrac{3}{2} H_2(g) \quad + \quad \tfrac{1}{2} N_2(g) \quad \longrightarrow \quad CH_3CN(g)$$

32. Estimated $\Delta H_f^\circ = +86$ kJ/mol (for comparison, the accepted value is +95 kJ/mol).

Tools for problem solving

In this chapter you learned to apply the following concepts as tools in solving problems. Study each one carefully so that you know what each is used for. When faced with solving a problem, recall what each tool does and consider whether it will be helpful in finding a solution.

You might want to tear these pages out to use along with solving problems in this chapter.

Converting between ΔE and ΔH (Section 18.1)
$$\Delta H = \Delta E + \Delta n_{gas} RT$$

Factors that affect the entropy (Section 18.3)
- Volume: Entropy increases with increasing volume.
- Temperature: Entropy increases with increasing temperature.
- Physical state: $S_{gas} \gg S_{liquid} > S_{solid}$. When gases are formed in a reaction, ΔS is almost always positive.
- Number of particles: Entropy increases when the number of particles increases.

Gibbs free energy (Section 18.4)
$$\Delta G = \Delta H - T\Delta S$$

ΔG as a predictor of spontaneity (Section 18.4)
- If ΔG is less than zero, the reaction is spontaneous.
- If ΔG is greater than zero, the reaction is nonspontaneous.

Standard entropies (Section 18.5)

Calculate the standard entropy change for a reaction using the following equation and standard entropy values from Table 18.1. The calculation is similar to the Hess's law calculation of $\Delta H°$ you learned in Chapter 6.

$$\Delta S° = (\text{sum of } S° \text{ of products}) - (\text{sum of } S° \text{ of reactants})$$

Calculating $\Delta G°$ from $\Delta H°$ and $\Delta S°$ (Section 18.6)

$$\Delta G° = \Delta H° - (298.15 \text{ K}) \Delta S°$$

Calculating $\Delta G°$ using $\Delta G_f°$ values (Section 18.6)

When you want $\Delta G°$ for a reaction, use $\Delta G_f°$ data from Table 18.2 and the equation

$$\Delta G° = (\text{sum of } \Delta G_f° \text{ of products}) - (\text{sum of } \Delta G_f° \text{ of reactants})$$

Using $\Delta G°$ to assess the position of equilibrium (Section 18.8)

- $\Delta G°$ is large and positive: position of equilibrium is close to reactants.
- $\Delta G°$ is large and negative: position of equilibrium is close to products.
- $\Delta G° = 0$: position of equilibrium lies about midway between reactants and products.

Calculating $\Delta G°$ at temperatures other than 25 °C (Section 18.8)

$$\Delta G_T° \approx \Delta H_{298}° - T \Delta S_{298}°$$

Relating the reaction quotient to ΔG (Section 18.9)

$$\Delta G = \Delta G° + RT \ln Q$$

- If ΔG is negative, the reaction is spontaneous in the forward direction.
- If ΔG is positive, the reaction is spontaneous in the reverse direction.
- If ΔG is zero, the reaction is at equilibrium.

Determining thermodynamic equilibrium constants (Section 18.9)

$$\Delta G° = -RT \ln K$$

Summary of Important Equations

First law of thermodynamics:

$$\Delta E = q - w$$

Definition of enthalpy:

$$H = E + PV$$

Calculating work done *by* a system:

$$w = -P\Delta V$$

Converting between ΔE and ΔH

$$\Delta H = \Delta E + \Delta n\, RT$$

Remember, Δn is the change in the number of moles of *gas*.

Calculating $\Delta S°$:

$$\Delta S° = (\text{ sum of } S° \text{ of products}) - (\text{ sum of } S° \text{ of reactants})$$

Gibbs free energy change:

$$\Delta G = \Delta H - T\Delta S$$

Calculating $\Delta G°$ for a reaction from $\Delta G_f°$:

$$\Delta G° = (\text{sum } \Delta G_f° \text{ of products}) - (\text{sum } \Delta G_f° \text{ of reactants})$$

Calculating $\Delta G°$ at a temperature other than 25°C:

$$\Delta G_T° = \Delta H_{298}° - T\,\Delta S_{298}°$$

Calculating ΔG from reaction mixture composition:

$$\Delta G = \Delta G° + RT \ln Q$$

Relating $\Delta G°$ to the equilibrium constant

$$\Delta G° = -RT \ln K$$

Notes:

Chapter 19

Electrochemistry

Some of the most useful and common practical applications of chemistry involve the use of or production of electricity. Chemical reactions that produce electrical power in batteries start our cars, run electronic calculators and portable radios, keep wristwatches running, and set proper exposures in cameras. Our lives are touched constantly by the ultimate fruits of electrolysis reactions, such as aluminum, bleach, halogenated organic molecules in plastics and insecticides, and soap. Besides all of these things, the relationship between electricity and chemical change has become an extremely useful tool in the laboratory for probing chemical systems of all kinds.

19.1   Galvanic (Voltaic) Cells

Learning Objective

Set up and use galvanic cells.

Review

Electrical devices operate by the flow of electrons; that's what electricity is. Reactions that produce or consume electrical energy are called electrochemical changes. They are oxidation–reduction reactions, and their study constitutes the field of electrochemistry. As we discuss in this chapter, the applications of electrochemistry affect our daily lives as well as our activities in the laboratory.

When a redox reaction occurs, the energy released is normally lost to the environment as heat. By separating the half–reactions and making oxidation and reduction occur in different places, that is, in different half–cells, we can cause the electron transfer to take place by way of a wire through an external electrical circuit. In this way the energy of the reaction can be harnessed. The apparatus to accomplish this is called a galvanic cell or a voltaic cell. Study the discussion in the text that describes how a galvanic cell is put together.

The overall reaction in a galvanic cell is called the cell reaction, and is obtained by using the principles of the ion–electron method (Chapter 6) to combine the half–reactions that are taking place in the individual half–cells.

The electrodes in a galvanic cell are identified by the nature of the redox processes taking place. The electrode at which oxidation occurs is the anode; the electrode where reduction occurs is the cathode. In a galvanic cell, the anode carries a slight negative charge and the cathode a slight positive charge. During operation of the cell, electrons flow from the anode to the cathode through the external electrical circuit; in the solution, cations move toward the cathode and anions move toward the anode. Keep in mind that no electrons flow through the electrolyte solutions in the cell. The movement of electrons through the wires is metallic conduction; the transport of charge by the movement of ions through the solutions is called electrolytic conduction.

For the reactions to take place in the cell, the two half–cells must be connected electrolytically—for example, by a salt bridge. The salt bridge permits ions to enter and leave the half-cells so that electrical neutrality can be maintained. In the absence of the salt bridge, no flow of electricity can occur, since, as you may recall from a physics class, a complete circuit is required.

Study the way we use standard cell notation to describe the nature of the electrodes and the electrolytes that make up the half–cells in a galvanic cell. Be sure that you understand Example 19.1 in the text and can work Practice Exercises 19.1 and 19.2 before trying the Self–Test questions given next.

Self–Test

1. The following reaction occurs spontaneously in a galvanic cell:

$$4H^+ + MnO_2 + Fe \longrightarrow Fe^{2+} + Mn^{2+} + 2H_2O$$

(a) What half–reaction occurs in the cathode compartment?

_____

(b) What electrical charge is carried by the iron electrode? _____

(c) Do electrons flow toward or away from the iron electrode? _____

(d) Using the iron half–cell as an example, explain how a salt bridge containing KNO_3 works.

_____

2. Write the standard cell notation for the galvanic cell described in the preceding question.

_____

3. What are the anode and cathode half–reactions in the galvanic cell described by the notation

$$Pb(s) \mid Pb^{2+}(aq) \mid\mid Au^{3+}(aq) \mid Au(s)$$

_____

_____

19.2   Cell Potentials

Learning Objective

Predict galvanic cell potentials.

Review

Central to the development of this section is the concept that each half–reaction has an intrinsic or innate tendency to proceed as a reduction. The magnitude of this tendency is given by the half–reaction's reduction potential, or the standard reduction potential if the concentrations of all the ions in the half–cell are 1 M, the partial pressure of any gas is 1 atm, and the temperature is 25 °C. When two half–cells compete for electrons, as they do in a galvanic cell, the one with the larger reduction potential proceeds as reduction and the other is forced to reverse and become an oxidation.

A functioning galvanic cell has a certain cell potential. This quantity is also called the potential or electromotive force produced by the cell and is measured in a unit called the volt (V). In a sense, it is a measure of the force by which the cell can push electrons through an external circuit. The standard cell potential, E°_{cell}, is the cell potential under standard conditions (1 M concentrations of ions, 1 atm pressure for any gas, 25 °C). The equation by which E°_{cell} is calculated from standard reduction potentials is

$$E^{\circ}_{cell} = \begin{pmatrix} \text{standard reduction} \\ \text{potential of the} \\ \text{substance reduced} \end{pmatrix} - \begin{pmatrix} \text{standard reduction} \\ \text{potential of the} \\ \text{substance oxidized} \end{pmatrix} \qquad \text{(19.2 in text)}$$

This is a very important and useful equation, so be sure you've learned it.

The values of standard reduction potentials are compared to that of a reference electrode called the standard hydrogen electrode, which is assigned a potential of exactly 0 V.

$$2H^{+}(aq, 1.00\ M) + 2e^{-} \rightleftharpoons H_2(g, 1\ atm) \qquad E^{\circ} = 0.00\ \text{V}$$

Be sure to study how reduction potentials are measured against the standard hydrogen electrode and review Example 19.2 in the text.

Example 19.1 Constructing a galvanic cell

Determine the materials to use to produce a galvanic cell with the largest possible potential at standard state. Each electrode must be made of a material that is part of a half–reaction and this cell must use water as a solvent.

Analysis:

Looking at the requirements, we see that the half–reaction must contain a metal, to serve as the electrode, and the selected metal should also be compatible with water. In addition, the cell potential is the difference between the two half–reactions, and the largest difference will occur by taking the qualifying half–reactions as close to the top and bottom of the table of standard reduction potentials as possible.

Assembling the Tools:

We need to use the tool that defines and identifies electrodes and the type of reaction that occurs at each.

Solution:

Starting from the top of Table 19.1, the first half–reaction we see with a usable metal is

$$Au^{3+}(aq) + 3e^- \rightleftharpoons Au(s)$$

The other half–reaction that is reasonable, at the other end of the table, is

$$Mg^{2+}(aq) + 2e^- \rightleftharpoons Mg(s)$$

The cell potential will be 1.42 V – (–2.37) = 3.79 volts.

Is the Answer Reasonable?

There is no guarantee that this galvanic cell will be practical, but it does fit the criteria provided. Checking our choices, the first six half–reactions have no metal usable as an electrode. The elements Li, Na, K, and Ca are also too reactive, and therefore incompatible, with water to use as electrodes. Our selections seem okay.

Self–Test

4. A lead half–cell was constructed using a lead electrode dipping into a 1.00 M $Pb(NO_3)_2$ solution. This was connected to a standard hydrogen electrode. A voltage of 0.13 V was measured for the cell when the positive terminal of the voltmeter was connected to the hydrogen electrode.

    (a) In the space below, sketch and label a diagram of the galvanic cell.

    (b) What substance is being reduced in the cell?

    (c) What is $E^{\circ}_{Pb^{2+}}$ for the half–cell $Pb^{2+}(aq) + 2e^- \rightleftharpoons Pb(s)$?

    _____

5. Referring to Table 19.1, to which electrode should the negative terminal of a voltmeter be connected in a cell constructed of the following half–cells?

    $$Au^{3+} + 3e^- \rightleftharpoons Au \qquad \text{and} \qquad Mg^{2+} + 2e^- \rightleftharpoons Mg$$

    _____

6. How is the volt defined in terms of SI units?_____

19.3   Utilizing Standard Reduction Potentials

Learning Objective

Using standard reduction potentials to predict spontaneous reactions.

Review

As we noted above, for a given pair of half–reactions, the one having the higher (more positive) reduction potential occurs spontaneously as reduction; the other is reversed and occurs as oxidation. After setting up the half–reactions in this way, the cell reaction is obtained by adding the reduction and oxidation half–reactions in such a way that all electrons cancel. The procedure is the same as the one you used in the ion–electron method (Section 6.2). Factors are used to adjust the coefficients so that equal numbers of electrons are gained and lost. Notice, however, that these factors are *not* used as multiplying factors when combining the half–reactions! The cell potential is obtained simply by subtracting one reduction potential from the other using Equation 19.2. For a spontaneous cell reaction, this difference has a positive algebraic sign.

When asked whether or not a given overall reaction is spontaneous (at standard state), first divide the reaction into its two half–reactions. Then find the reduction potential for each half–reaction and compute the cell potential using Equation 19.2. If the result is positive, the reaction is spontaneous; if it is negative, however, the reaction is not spontaneous in the direction written. In fact, the reaction is spontaneous in the opposite direction.

Example 19.2 Predicting a spontaneous redox reaction

What will be the spontaneous reaction at 25 °C if we add nickel and iron filings to a solution that contains 1 M $NiCl_2$ and 1 M $FeCl_2$?

Analysis:

We need to write half–reactions that contain (a) nickel metal and Ni^{2+} ions and (b) iron metal and Fe^{2+} ions. Once the half–reactions are written, the combination that gives us a balanced redox reaction *and* a positive potential, is the spontaneous reaction. We also see that all concentrations are 1 M at 25 °C, indicating standard state.

Assembling the Tools:

We need the tools for writing possible half–reactions for this experiment. A table of standard reduction potentials will be the tool to inform us which reaction will be spontaneous.

Solution:

First we write the two half–reactions as

$$Ni^{2+}(aq) + 2e^- \rightleftharpoons Ni(s) \qquad E° = -0.25 \text{ V}$$

$$Fe^{2+}(aq) + 2e^- \rightleftharpoons Fe(s) \qquad E° = -0.44 \text{ V}$$

If we write the balanced equation as $Ni^{2+}(aq) + Fe(s) \rightleftharpoons Fe^{2+}(aq) + Ni(s)$, the standard cell potential will be

$$E^\circ = -0.25\ V - (-0.44\ V) = +0.19\ V$$

Since this is a positive value, this is the spontaneous reaction at standard state.

Is the Answer Reasonable?

We need to check that our balanced equation is correct. Next we make sure that the standard reduction potential for the substance oxidized is subtracted from the standard reduction potential for the substance reduced. The latter is a very common source of error. Our process checks out and our answer is reasonable.

Self–Test

7. Suppose that the following half–reactions are used to prepare a cell.

$$ClO_3^-(aq) + 6H^+(aq) + 6e^- \rightleftharpoons Cl^-(aq) + 3H_2O \qquad E^\circ = +1.45\ V$$

$$Hg_2HPO_4(s) + H^+(aq) + 2e^- \rightleftharpoons 2Hg(l) + H_2PO_4^-(aq) \qquad E^\circ = +0.64\ V$$

   (a) Determine the net spontaneous cell reaction.

   _____

   (b) Determine the cell potential. _____

8. Without actually calculating E°_{cell}, determine the spontaneous cell reaction involving the following half–reactions.

$$BrO_3^-(aq) + 6H^+(aq) + 6e^- \rightleftharpoons Br^-(aq) + 3H_2O \qquad E^\circ = +1.44\ V$$

$$H_3AsO_4(aq) + 2H^+(aq) + 2e^- \rightleftharpoons HAsO_2(aq) + 2H_2O \qquad E^\circ = +0.58\ V$$

   _____

9. Will the following reaction occur spontaneously?

$$H_2SO_3(aq) + H_2O + Br_2(aq) \longrightarrow SO_4^{2-}(aq) + 4H^+(aq) + 2Br^-(aq)$$

   _____

19.4 E°_{cell} and ΔG°

Learning Objective

Relate standard cell potentials and standard free energies

Review

The free energy change, ΔG, and the standard free energy change, ΔG°, for a reaction, are related to the cell potential and standard cell potential, respectively, by the following equations.

$$\Delta G = -n \mathscr{F} E_{\text{cell}} \tag{19.5}$$

$$\Delta G^{\circ} = -n \mathscr{F} E^{\circ}_{\text{cell}} \tag{19.6}$$

where n is the number of electrons transferred and $\mathscr{F}$ is the Faraday constant (9.65×10^4 C/mol e^-). Since E° and E°_{cell} are in volts, and $1\,\text{V} = 1\,\text{J/C}$, ΔG and ΔG° are in units of joules.

The standard cell potential is also related to the equilibrium constant, K_c, for the reaction.

$$E^{\circ}_{\text{cell}} = \frac{RT}{n\mathscr{F}} \ln K_c \tag{19.7}$$

The value of R that must be used for this equation is 8.314 J mol^{-1} K^{-1}. As noted above, $\mathscr{F}$ equals 9.65×10^4 C/mol e^-. Study Examples 19.8 and 19.9 plus Practice Exercises 19.17 and 19.18 in the text before trying the Self–Test questions below. Remember, for Equation 19.7 above, use natural logarithms and corresponding exponentials as you employ your pocket calculator.

Example 19.3 Calculating ΔG° and K_c from standard cell potentials

What are the numerical values for ΔG° and K_c for this reaction, as written, at 25 °C?

$$\text{Ni}(s) + \text{Cd}^{2+}(aq) \rightleftharpoons \text{Cd}(s) + \text{Ni}^{2+}(aq)$$

Analysis:

We can use the value of E°_{cell} to calculate ΔG° and also K_c. The reduction potentials of Ni^{2+} and Cd^{2+} allow us to calculate the value of E°_{cell} for the reaction as written.

Assembling the Tools:

Our tools will be the two equations that relate E°_{cell} to ΔG° and K_c along with the tool that defines how to determine E°_{cell} to from standard reduction potentials.

Solution:

First we calculate $E^{\circ}_{cell} = -0.40 \text{ V} - (-0.25 \text{ V}) = -0.15 \text{ V}$ for the *reaction as written*. Using Equation 19.6,

$$\Delta G^{\circ} = -n\mathscr{F}E^{\circ}_{cell} = -(2)(96{,}495 \text{ C mol}^{-1})(-0.15 \text{ V}) = 29 \text{ kJ mol}^{-1}$$

(Remember that the units of $C \times V = $ joules (J).)

Now we use Equation 19.7,

$$E^{\circ}_{cell} = \frac{RT}{n\mathscr{F}}\ln K_c = 0.0128 \ln K_c$$

Dividing the standard cell potential by 0.0128 we obtain $\ln K_c = -11.7$.

Taking the antiln we obtain $K_c = 8.1 \times 10^{-6}$. [Note: The antiln is very sensitive and depending on how you round your numbers your result may vary from 6×10^{-6} to 9.7×10^{-6}.]

Is the Answer Reasonable?

The best check would be to convert the value of K_c to ΔG° using the equation $\Delta G^{\circ} = -RT \ln K_c$. The results agree reasonably well to suggest that our answer is correct.

Self–Test

10. Calculate ΔG° in kJ for the reaction in Question 7. _____

11. The reaction $2Al^{3+} + 3Cu \longrightarrow 3Cu^{2+} + 2Al$ has $E^{\circ}_{cell} = -2.00 \text{ V}$. What is ΔG° in kJ for this reaction?

   _____

12. The reaction $2AgBr + Pb \longrightarrow PbBr_2 + 2Ag$ has $\Delta G^{\circ} = -237 \text{ kJ}$. What is E°_{cell}?

   _____

13. What is the value of K_c for the reaction in Question 7?

_____

14. What is the value of K_c for the reaction in Question 11?

_____

15. The reaction $PbI_2(s) + Zn(s) \rightleftharpoons Pb(s) + 2I^-(aq) + Zn^{2+}(aq)$ has an equilibrium constant $K_c = 2.2 \times 10^{13}$ at 25 °C. What is the value of E°_{cell} for this reaction?

_____

19.5   Cell Potentials and Concentrations

Learning Objective

Relate the concentration of solutes to cell potentials.

Review

The effect on the cell potential caused by nonstandard concentrations of solutes is given by the Nernst equation:

$$E_{cell} = E^{\circ}_{cell} - \frac{RT}{n\mathscr{F}}\ln Q \tag{19.8}$$

where n is the number of electrons transferred and Q is the reaction quotient (the value of the mass action expression) for the reaction. Again, be careful about using natural logarithms with this equation.

Be particularly careful to follow the correct procedures for writing Q, the mass action expression. The concentrations of pure solids and liquids do not appear in the mass action expression, and many electrochemical reactions are heterogeneous (one or more of the electrode materials are solids). The concentration of water also is omitted because it is essentially a constant, too. (If you need to review this, see Section 16.7 in the text.)

Example 19.4   Writing the Nernst equation for a reaction

What is the correct form for the Nernst equation for the following reaction at 25 °C for which $E^{\circ}_{cell} = 0.38$ V?

$$3PbSO_4(s) + 2Cr(s) \longrightarrow 3Pb(s) + 3SO_4^{2-}(aq) + 2Cr^{3+}(aq)$$

Analysis:

We need to apply Equation 19.8 to this specific problem. Quantities that are specific to this reaction are the number of moles of electrons transferred, n, and the form of the mass action expression, Q. To find n, we need to determine the number of electrons gained or lost (they're the same, of course). In constructing the mass action expression, we have to be careful to omit the three solids.

Assembling the Tools:

We need the tool that defines the structure of the Nernst equation.

Solution:

The oxidation of two chromium atoms to chromium(III) ions involves a transfer of $6e^-$, so $n = 6$ for this reaction. Therefore, because $R = 8.314$ J $(\text{mol } e^-)^{-1}$ K^{-1}, $T = 298.15$ K, and $\mathscr{F} = 9.65 \times 10^4$ C $(\text{mol } e^-)^{-1}$, we can substitute into Equation 19.8 as follows.

$$E_{cell} = 0.38 \text{ V} - \frac{8.314 \text{ J (mole)}^{-1} \text{ K}^{-1} 298.15 \text{ K}}{6 \times 9.65 \times 10^4 \text{ C (mole)}^{-1}} \times \ln Q$$

The large fraction reduces to 4.28×10^{-3} J C^{-1}, but the ratio of units, J C^{-1}, is the same as volts, V. In constructing Q, we omit the concentrations of the solids. This gives the answer we seek.

$$E_{cell} = 0.38 \text{ V} - 4.28 \times 10^{-3} \text{ V} \times \ln ([SO_4^{2-}]^3[Cr^{3+}]^2)$$

Is the Answer Reasonable?

To do a check here, the quickest thing is to check that we've placed the correct quantities into the fraction that precedes the natural log term. We can also recheck the arithmetic using our calculator to confirm that this part of the solution is correct. Then, we can check that we've omitted concentration terms for solids from the mass action expression (which we have) and that the concentration terms for the ions are both in the numerator (they are) and raised to the appropriate exponents (they are).

The measurement of cell potentials provides a means for determining unknown concentrations of ions in a half–cell, as illustrated by text Example 19.10 and the example below. Notice that in this calculation we first solve for Q (the reaction quotient). Then we substitute known concentrations and solve for the unknown value.

Example 19.5 Using electrode potentials to determine concentrations

At 25 °C the following reaction was set up as a galvanic cell:

$$Ni(s) + Cd^{2+}(aq) \rightleftharpoons Cd(s) + Ni^{2+}(aq)$$

If the concentration of Ni^{2+} in one cell is 3.0×10^{-8} M, what will be the concentration of Cd^{2+} if the cell potential is measured as -0.21 volts?

Analysis:

To answer the question, we need to set up the Nernst equation and find appropriate values for all terms except $[Cd^{2+}]$. We can look up the reduction potentials of Ni^{2+} and Cd^{2+} in Table 19.1 and from them calculate the value of E°_{cell} for the reaction *as written*. Then, we use this value of E°_{cell} along with the measured cell potential to calculate Q. Then, with the given concentration of Ni^{2+} we can determine the value of $[Cd^{2+}]$.

Assembling the Tools:

This requires that we set up the Nernst equation correctly and then do the calculations carefully for the cadmium concentration, which appears to be the only unknown.

Solution:

First we write the two half–reactions as

$$Ni^{2+}(aq) + 2e^- \rightleftharpoons Ni(s) \qquad E^{\circ} = -0.25 \text{ V}$$

$$Cd^{2+}(aq) + 2e^- \rightleftharpoons Cd(s) \qquad E^{\circ} = -0.40 \text{ V}$$

If we write the balanced equation as $\quad Ni(s) + Cd^{2+}(aq) \rightleftharpoons Cd(s) + Ni^{2+}(aq)$

the standard cell potential will be $\qquad E^{\circ} = -0.40 \text{ V} - (-0.25 \text{ V}) = -0.15 \text{ V}$

The Nernst equation for this reaction is $\qquad E_{cell} = E^{\circ}_{cell} - \dfrac{RT}{n\mathscr{F}} \ln Q$

where $\qquad Q = \dfrac{[Ni^{2+}]}{[Cd^{2+}]}$

Substituting known values we get $\quad -0.21 \text{ V} = -0.15 \text{ V} - 0.0128 \ln \dfrac{[Ni^{2+}]}{[Cd^{2+}]}$

$$\ln \dfrac{[Ni^{2+}]}{[Cd^{2+}]} = 4.69$$

and we then take the anti–natural logarithm and substitute the given value for $[Ni^{2+}]$ to get

$$\dfrac{[Ni^{2+}]}{[Cd^{2+}]} = 109 = \dfrac{3.0 \times 10^{-8} \ M \ Ni^{2+}}{[Cd^{2+}]}$$

Finally, solving for $[Cd^{2+}]$ yields $2.8 \times 10^{-10} \ M \ [Cd^{2+}]$.

Is the Answer Reasonable?

We need to check that our balanced equation is correct. Next we make sure that the standard reduction potential for the substance oxidized is subtracted from the standard reduction potential for the substance reduced. Finally we check to be sure we used $n = 2$ and $R = 8.314 \text{ J mol}^{-1} \text{ K}^{-1}$ in calculating $RT/n\mathscr{F}$.

Now that we have worked an example with the Nernst equation, work Practice Exercises 19.19 and 19.20 before trying the Self-Test below.

Self–Test

16. Write the correct Nernst equation for the following reaction.

$$MnO_2(s) + 2H^+(aq) + H_3PO_2(aq) \longrightarrow Mn^{2+}(aq) + H_3PO_3(aq) + H_2O \qquad E^{\circ}_{cell} = 1.73 \text{ V}$$

_____

17. Write the correct Nernst equation for the following (unbalanced) reaction. Use data in Table 19.1.

$$Cd(s) + Cr^{3+}(aq) \longrightarrow Cd^{2+}(aq) + Cr(s)$$

_____

18. What is the cell potential for the reaction in Question 16 if $[H_3PO_2] = 0.0010 \ M$, $[Mn^{2+}] = 5.0 \times 10^{-4} \ M$, $[H_3PO_3] = 0.15 \ M$, and the pH equals 5.0?

_____

19. A chemist who wished to monitor the concentration of Cd^{2+} in the waste water leaving a chemical plant set up a galvanic cell consisting of a cadmium electrode that was immersed in a solution suspected to contain Cd^{2+}, and a silver electrode that was immersed in a 0.100 M solution of $AgNO_3$. In a particular analysis, the potential of the cell was determined to be 1.282 V. The standard cell potential for the following reaction,

$$Cd + 2Ag^+ \longrightarrow Cd^{2+} + 2Ag$$

has been accurately measured to be 1.202 V. What was the Cd^{2+} concentration in the solution that was analyzed?

_____

19.6 Electricity

Learning Objective

Understand how electricity is produced.

Review

This section describes the construction and the chemical reactions of a number of common batteries. Be sure to learn the chemical reactions that take place at the electrodes and which substances serve as cathode and anode. You should also understand the advantages and disadvantages of the various cells.

Batteries are divided into primary cells, those that cannot be recharged, and secondary cells, those that can be recharged. Automobile batteries are secondary cells since they are constantly recharged. Ordinary flashlight batteries are primary cells and are not rechargeable.

Fuel cells and lithium–ion batteries are two other sources of electricity where electron flow is generated by an electrochemical reaction. Fuel cells offer increased thermodynamic efficiency in converting the energy of chemical reactions into work because they operate under conditions approaching reversibility. Another advantage is that the fuel can be fed to them continuously, so they don't need recharging. Lithium ion batteries work by the flow of lithium ions inside the battery.

Other sources of electricity convert kinetic energy to electricity. Wind or water turning turbines in windmills or hydroelectric stations are obvious examples. Coal, oil, gas, and nuclear energy heat water to steam, which then turns turbines.

This section provides a reference for the variety of chemical sources of electricity. Working the following Self–Test will help you recall the material.

Self–Test

20. What is the cathode reaction in the lead storage battery while it is being discharged?

   _____

21. What is the cathode reaction in the lead storage battery while it is being charged?

   _____

22. A lead storage cell produces a potential of about 2 V. How can an automobile battery produce 12 V?

   _____

23. What substance serves as the anode in the common dry cell?

    _____

24. What substance serves as the anode in an alkaline battery?

    _____

25. Why doesn't the lithium–ion battery qualify as a true battery?

    _____

26. Why can a hydrometer be used to test the state of charge of a lead storage battery?

    _____

    _____

27. Write the half–reaction that takes place at the cathode during the discharge of a silver oxide battery.

    _____

28. Write equations for the cathode, anode, and net cell reaction in a nickel–metal hydride battery.

    _____

    _____

    _____

29. Write equations for the cathode, anode, and net cell reaction for a lithium–manganese dioxide battery.

    _____

    _____

    _____

30. What species is transported through the electrolyte between cathode and anode in a lithium ion battery?

    _____

31.   What function does graphite serve in a lithium ion battery?

_____

32.   What chemical reaction occurs in a hydrogen–oxygen fuel cell?

_____

33.   Why is a fuel cell more efficient at producing usable energy than a conventional system that uses combustion of the fuel?

_____

34.   What reaction enables methanol to serve as a source of hydrogen for a hydrogen–oxygen fuel cell?

_____

19.7   Electrolytic Cells

Learning Objective

Explain the nature of electrolysis.

Review

When a nonspontaneous reaction is forced to occur by the passage of electricity, the process is called *electrolysis*. An *electrolysis cell (electrolytic cell)* consists of a pair of electrodes dipping into a chemical system in which there are mobile ions (formed by melting a salt or by dissolving an electrolyte in water). *When electricity flows, oxidation–reduction reactions are forced to occur at the electrodes.*

An important thing to learn in this section is that in *any* cell, regardless of whether it is using or producing electricity, the electrode at which oxidation occurs is called the *anode* and the electrode at which reduction occurs is called the *cathode*. Thus, we name an electrode according to the chemical reaction that occurs at it, not according to its charge. In an *electrolysis cell,* an external voltage source gives the anode a positive charge and the cathode a negative charge. These are the charges that they *must* have to force oxidation and reduction to occur. The positive charge of the anode pulls electrons from substances and causes them to be oxidized, and the negative charge of the cathode pushes electrons onto other substances and causes them to be reduced.

The equation for the *cell reaction* in an electrolysis apparatus is obtained by adding the individual oxidation and reduction half–reactions that occur at the electrodes. Remember to be sure that the electrons in the cell reaction cancel. As with galvanic cells, the procedure for this is the same as in the ion–electron method.

For reactions in aqueous solution, the redox of water is possible at the electrodes. You should know the following possible electrode reactions:

Anode (oxidation of H_2O) $2H_2O(l) \longrightarrow O_2(g) + 4H^+(aq) + 4e^-$

Cathode (reduction of H_2O) $4H_2O(l) + 4e^- \longrightarrow 2H_2(g) + 4OH^-(aq)$

Standard reduction potentials can often (but not always) be used to predict electrolysis reactions. When there are two competing reduction reactions at an electrode, the half–reaction with the most positive reduction potential will tend to occur. When there are competing oxidation reactions, the half–reaction with the least positive reduction potential will tend to occur. However, there are electrode peculiarities that can sometimes alter the expected outcome. Study Example 19.11 in the text and the example below.

Example 19.6   Identifying oxidation and reduction half–reactions

When a solution of NiF_2 is electrolyzed, metallic nickel is deposited on one electrode and O_2 is produced at the other. Which substances are oxidized and reduced, and at which electrodes?

Analysis:

The electrodes are producing $Ni(s)$ and $O_2(g)$ at opposite electrodes. Since oxygen in compounds (water in this case) has a negative oxidation state, it must be oxidized to the element. Nickel on the other hand is a metal and is expected to have a positive charge in solution. Therefore it must be reduced to the elemental state. With this reasoning we can solve the problem.

Assembling the Tools:

We need the tools for writing half–reactions from the written description of what has occurred. Once the half– reactions are written we can use our tool for identifying if the substance of interest has been oxidized or reduced. Finally, our tool defining which electrode supports oxidation and which supports reduction will tell us which is the anode and which the cathode.

Solution:

First we write the two half–reactions as

$$Ni^{2+}(aq) + 2e^- \rightleftharpoons Ni(s) \qquad \text{(reduction)}$$

$$2H_2O \rightleftharpoons H^+(aq) + O_2(g) + 4e^- \qquad \text{(oxidation)}$$

Once we have identified the half–reactions, we associate the oxidation half–reaction with the anode and the reduction half–reaction with the cathode.

Are the Answers Reasonable?

We need to check that our balanced half–reaction equations are correct (look at Table 19.1 to be sure nothing has been left out of the half–reactions). The only other thing to do is to check the definition of anode and cathode (look in the glossary if needed).

Self–Test

35. What is an *electrochemical change*? _____

_____

36. What is *electrochemistry*? _____

_____

37. At which electrode (anode or cathode) would these electrolysis half–reactions occur?

   (a) $2I^-(aq) \longrightarrow I_2(aq) + 2e^-$ _____

   (b) $2Cr^{3+}(aq) + 7H_2O \longrightarrow Cr_2O_7^{2-}(aq) + 14H^+(aq) + 6e^-$ _____

   (c) $NO_3^-(aq) + 2H_2O + 3e^- \longrightarrow NO(g) + 4OH^-(aq)$ _____

38. Write the equation for the reduction of H_2O at the cathode of an electrolytic cell.

_____

39. Write the equation for the oxidation of H_2O at the anode of an electrolytic cell.

_____

40. When an aqueous solution of BaI_2 is electrolyzed, I_2 is formed at the anode and H_2 is formed at the cathode. What are the anode and cathode half–reactions?

   anode: _____

   cathode: _____

19.8   Electrolysis Stoichiometry

Learning Objective

Use electrolysis information quantitatively.

Review

The amount of electricity consumed during electrolysis is proportional to the number of moles of electrons (the number of faradays), passed through the electrolysis cell. Important relationships to remember are the following, where the coulomb (C) is the SI unit for amount of charge and the ampere (A) is the SI unit of electric current (coulombs per second or C s^{-1}).

$$1 \text{ mol } e^- = 9.65 \times 10^4 \text{ C}$$

$$1 \text{ C} = 1 \text{ A} \times \text{s}$$

$$1 \,\mathscr{F} = 1 \text{ mol } e^- = 1 \text{ faraday}$$

The value of 9.65×10^4 C $\mathscr{F}^{-1}$ is called the Faraday constant. We use the relationships above, along with balanced half–reactions or a knowledge of the number of electrons transferred according to a balanced redox equation, to relate the amount of chemical change to amperes of electrical current and to time. Study Examples 19.12 and 19.13 in the text and do Practice Exercises 19.23 to 19.26 before you work the Self–Test below.

Example 19.7 Stoichiometric calculation using electrolysis data

A solution of NaCl was electrolyzed for 45.0 min, producing Cl_2 at the anode and H_2 at the cathode. The resulting solution after electrolysis was titrated with 0.500 *M* HCl solution and required 22.3 mL of the acid to neutralize the solution. What was the current during the electrolysis?

Analysis:

The final solution is being titrated with an acid and we conclude that a base must be formed at one of the electrodes. We need to write the half–reactions for the electrodes to identify which one produces the base. That half–reaction will also give us the conversion factor between the titration results and the electrolysis time.

Assembling the Tools:

We can use the equation for calculating the coulombs of electricity using a tool developed in this section. We will need our tool for writing balanced half–reactions to obtain *n*, the number of electrons transferred, so that the current can be calculated. We will also need the titration tool to determine the moles of hydroxide ions formed by the electrolysis.

Solution:

The half–reaction for the production of hydrogen gas from water is

$$4H_2O(l) + 4e^- \longrightarrow 2H_2(g) + 4OH^-(aq) \quad \text{(reduction)}$$

and the half–reaction for the production of chlorine gas is

$$2Cl^- \longrightarrow Cl_2(g) + 2e^-$$

Therefore, the solution becomes basic as a result of the electrolysis. From the volume and concentration of the HCl solution, we can calculate the number of moles of HCl used and consequently the moles of OH^-.

$$(0.500\ M\ \text{HCl})(22.3\ \text{mL}) = 11.2\ \text{mmol HCl} = 0.0112\ \text{mol HCl} = 0.0112\ \text{mol OH}^-$$

Our reaction stoichiometry then tells us that we have used as many moles of electrons as there are moles of OH^- so we have 0.0112 moles e^-. Finally, we multiply the moles of electrons by Faraday's constant and divide by the time in seconds to obtain the amperes of current used. Together this calculation is

$$0.0112\ \text{mol } e^- \times \frac{96{,}500\ \text{C}}{\text{mol } e^-} \times \frac{1}{45\ \text{min}} \times \frac{1\ \text{min}}{60\ \text{s}} = 0.400\ \text{A}$$

Is the Answer Reasonable?

The calculated current is not unreasonably large or small. In addition we may want to calculate the moles of electrons that are used when a current of 0.400 A is allowed to flow for 45.0 minutes. We find the result is the same as we calculated from the titration.

When electrons are allowed to flow from one electrode in a galvanic cell to the other, a chemical reaction must occur. This chemical reaction will change the concentrations of the solutes in the respective cells. The following example illustrates the combination of electrolysis, to cause chemical change, combined with a calculation of a galvanic cell potential after the reaction has been stopped. We can calculate the chemical change using the stoichiometric calculations of electrochemistry. The Nernst equation is used to calculate cell potentials when the galvanic cell is not at standard state.

Example 19.8 Combining electrochemical stoichiometric and potential calculations

A galvanic cell was set up with a zinc electrode dipping into 100 mL of 1.00 M Zn^{2+} and an iron electrode dipping into 100 mL of 1.00 M Fe^{2+}. If the cell reaction

$$Zn(s) + Fe^{2+}(aq) \longrightarrow Zn^{2+}(aq) + Fe(s)$$

delivers a constant current of 0.500 A, what will be the potential of the cell after 125 minutes?

Analysis:

First we check to see if the given reaction is spontaneous as judged by the standard cell potential. Then we will calculate the coulombs of electricity used, convert that to moles of electrons with Faraday's constant, and then convert to moles of substance using conversion factors from the half–reactions. From the reaction, we see that Zn^{2+} should increase and Fe^{2+} should decrease by equal amounts. We then calculate the final concentrations and use the Nernst equation to calculate the cell potential.

Assembling the Tools:

Our tool to use is the equation to calculate the coulombs and then the moles of iron reduced and zinc oxidized. Our stoichiometric tools allow us to calculate the concentration of Fe^{2+} and Zn^{2+} left when the electrolysis is ended. The Nernst equation is then the tool used to calculate the new potential of the cell.

Solution:

Checking Table 19.1, the standard reduction potential for iron(II) is -0.44 V and for zinc it is -0.76 V. As written the standard cell potential is $+0.32$ V so we assume the reaction is spontaneous.

Next we calculate the coulombs of electrons by multiplying current by the time in seconds. This is then converted to moles of electrons using Faraday's constant, and finally the moles of electrons can be converted to the moles of Zn^{2+} produced in the reaction. All together the calculation is

$$0.500\ A \times 125\ min \times \frac{60\ s}{1\ min} \times \frac{mol\ e^-}{96,495\ C} \times \frac{1\ mol\ Zn^{2+}}{2\ mol\ e^-} = 0.0194\ mol\ Zn^{2+}$$

This number of moles is added to 100 mL (0.100 L) of solution and the molarity of Zn^{2+} increases to 1.19 M. Since the Fe^{2+} should decrease by an equal amount, its final concentration is $1.00 - 0.19 = 0.81$ M Fe^{2+}.

$$E_{cell} = 0.32\ V - \frac{8.314\ J\ (mole)^{-1}K^{-1} \times 298.15\ K}{2 \times 9.65 \times 10^4 C\ (mole)^{-1}} \times \ln Q$$

$$E_{cell} = 0.32\ V - 0.0128\ V \times \ln \frac{[Zn^{2+}]}{[Fe^{2+}]}$$

$$E_{cell} = 0.32\ V - 0.0128\ V \times \ln\ (1.19/0.81) = 0.32\ V - 0.0049\ V = 0.032\ V$$

$$E_{cell} = 0.032\ V$$

The change in cell potential is 0.0049 and when properly rounded, the cell potential does not change.

Is the Answer Reasonable?

If Q is equal to 1, the $\ln 1 = 0$ and $E_{cell} = E°_{cell}$. After calculating the change in concentration the value of Q is still close to 1 (it is 1.48) and we expect the natural log of Q to be small. We notice that $RT/n\mathscr{F}$ is also a small number (0.0128) and the product of two small numbers is expected to be very small, as we found. Therefore we conclude that the answer is reasonable.

Self–Test

41. A current of 5.00 A flows for 25.0 minutes. How many moles of electrons does this deliver?

_____

42. For how many seconds must a current of 6.00 A flow to deliver 0.225 mol of electrons?

_____

43. What current must be supplied to deliver 0.0165 mol e^- in 155 s?

_____

44. Calculate the number of moles of electrons that must pass through an electrolysis cell to produce

0.0150 mol $Cr_2O_7^{2-}$ by the reaction $2Cr^{3+} + 7H_2O \longrightarrow Cr_2O_7^{2-} + 14H^+ + 6e^-$.

_____

45. How many minutes are needed to make 0.0225 mol $Cr_2O_7^{2-}$ by the equation in Question 44 if the current is 4.00 A?

_____

46. What current will produce 25.3 g Fe in 4.00 hours by reduction of Fe^{2+} in an aqueous solution?

_____

19.9   Practical Applications of Electrolysis

Learning Objective

Describe some practical applications of electrolysis.

Review

This section also describes *electroplating* and the methods of producing some important commercial metals and chemicals by electrolysis. In studying this section, you should learn the chemical reactions involved in the various electrolytic cells and processes—the *Hall–Héroult process*, the *Downs cell*, the *diaphragm cell*, and the *mercury cell*—and the reasons why the reactions are carried out as they are.

When you feel you know the material, try the Self–Test below.

Self–Test

47. What is the purpose of electroplating? _____

_____

48. To which electrode do we connect the object to be electroplated?

_____

49. What is the name and formula of the solvent for Al_2O_3 originally used in the Hall–Héroult process?

_____

50. What is the net cell reaction in the Hall–Héroult process?

_____

51. What is the major source of magnesium? _____ What salt of magnesium is used in the electrolysis reaction that produces the free metal?

_____

52. What is the purpose of the special construction of the Downs cell?

_____

53. Why is the electrolytic refining of copper so economical?

_____

54. Give the cathode, anode, and net cell reaction for the electrolysis of brine.

anode reaction: _____

cathode reaction: _____

net reaction: _____

55. What products are formed if the brine solution is stirred while it is electrolyzed?

_____

56. Name one advantage and one disadvantage of using a diaphragm cell in the electrolysis of brine.

57. What is an advantage and a disadvantage of using a mercury cell in the electrolysis of brine?

Answers to Self–Test Questions

1. (a) $MnO_2 + 4H^+ + 2e^- \longrightarrow Mn^{2+} + 2H_2O$
   (b) negative
   (c) away
   (d) NO_3^- ions flow into the iron half–cell compartment to compensate for the charge of the Fe^{2+} ions entering the solution.

2. $Fe(s) \mid Fe^{2+}(aq) \mid\mid Mn^{2+}(aq) \mid MnO_2(s)$

3. anode: $Pb(s) \longrightarrow Pb^{2+}(aq) + 2e^-$; cathode: $Au^{3+}(aq) + 3e^- \longrightarrow Au(s)$

4. (a)

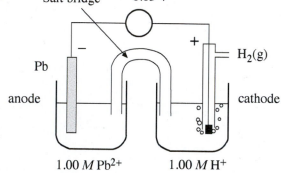

(b) H^+ (c) –0.13 V

5. magnesium

6. 1 V = 1 J/C

7. (a) $ClO_3^-(aq) + 9H^+(aq) + 3Hg_2HPO_4(s) \longrightarrow Cl^-(aq) + 3H_2O + 6Hg(l) + 3H_2PO_4^-(aq)$
   (b) $E° = 0.81$ V

8. $BrO_3^-(aq) + 3HAsO_2(aq) + 3H_2O \longrightarrow Br^-(aq) + 3H_3AsO_4(aq)$

9. yes, $E°_{cell} = +0.90$

10. $\Delta G° = -470$ kJ

11. $\Delta G° = +1158$ kcal

12. $E° = 1.23$ V

13. $K_c = 1.6 \times 10^{82}$

14. $K_c = 2 \times 10^{-203}$

15. $E°_{cell} = 0.395$ V

16. $E_{cell} = 1.73 \text{ V} - (0.0129 \text{ V}) \times \ln\left(\dfrac{[Mn^{2+}][H_3PO_3]}{[H^+]^2[H_3PO_2]}\right)$

17. $E_{cell} = -0.34 \text{ V} - (0.00429 \text{ V}) \times \ln\left(\dfrac{[Cd^{2+}]^3}{[Cr^{3+}]^2}\right)$

18. $E_{cell} = 1.47$ V

19. $2.0 \times 10^{-5} M$

20. $PbO_2(s) + 4H^+(aq) + SO_4^{2-}(aq) + 2e^- \longrightarrow PbSO_4(s) + 2H_2O$

21. $PbSO_4(s) + 2e^- \longrightarrow Pb(s) + SO_4^{2-}(aq)$

22. Six cells are connected in series, so their voltages add.

23. Zinc

24. Zinc

25. The lithium–ion battery does not employ a classical redox reaction.

26. During discharge, H_2SO_4 is used up, and the density of the electrolyte changes (decreases).

27. $Ag_2O(s) + H_2O + 2e^- \longrightarrow 2Ag(s) + 2OH^-(aq)$

28. anode: $MH(s) + OH^-(aq) \longrightarrow M(s) + H_2O + e^-$

    cathode: $NiO(H)(s) + H_2O + e^- \longrightarrow Ni(OH)_2 + OH^-(aq)$

    cell reaction: $MH(s) + NiO(OH)(s) \rightarrow Ni(OH)_2(s) + M(s)$

29. anode: $Li \longrightarrow Li^+ + e^-$

    cathode: $Mn^{IV}O_2 + Li^+ + e^- \longrightarrow Mn^{III}O_2(Li^+)$

    cell reaction: $Li + Mn^{IV}O_2 \longrightarrow Mn^{III}O_2(Li^+)$

30. Li^+ ions

31. When the battery is charged, graphite accepts Li^+ ions which slip between layers of carbon atoms.

32. $2H_2 + O_2 \rightarrow H_2O$

33. The reactions at the electrodes occur under more nearly thermodynamically reversible conditions.

34. $CH_3OH(g) + H_2O(g) \xrightarrow{\text{catalyst}} CO_2(g) + 3H_2(g)$

35. Electrochemical changes produce or are caused by electricity.

36. Electrochemistry is the study of electrochemical changes.

37. (a) anode

    (b) anode

    (c) cathode

38. $2H_2O + 2e^- \longrightarrow H_2(g) + 2OH^-(aq)$

39. $2H_2O \longrightarrow 4H^+(aq) + O_2(g) + 4e^-$

40. anode: $2I^-(aq) \longrightarrow I_2(aq) + 2e^-$

    cathode: $2H_2O + 2e^- \longrightarrow H_2(g) + 2OH^-(aq)$

41. 7.77×10^{-2} mol e^-

42. 3.62×10^3 s

43. 10.3 A

44. 0.0900 mol e^-

45. 54.3 minutes

46. 6.07 A

47. To beautify and protect metals.

48. cathode

49. cryolite, Na_3AlF_6

50. $4Al^{3+}(l) + 6O^{2-}(l) \longrightarrow 4Al(l) + 3O_2(g)$

51. the ocean; $MgCl_2$

52. To keep the Cl_2 and Na apart so they don't reform NaCl.

53. The anode mud contains precious metals whose value helps pay for the electricity that's used.

54. cathode: $2e^- + 2H_2O \longrightarrow H_2(g) + 2OH^-(aq)$

    anode: $2Cl^-(aq) \longrightarrow Cl_2(g) + 2e^-$

    net: $2Cl^-(aq) + 2H_2O \longrightarrow H_2(g) + Cl_2(g) + 2OH^-(aq)$

55. Cl^- is gradually changed to OCl^-.

56. Advantage: No OCl^- is formed by reaction of Cl_2 with OH^-.

    Disadvantage: NaOH solution is contaminated by small amounts
    of unreacted NaCl.

57. Advantage: Very pure NaOH is produced.

    Disadvantage: There is a potential for mercury pollution.

Tools for problem solving

In this chapter you learned to apply the following concepts as tools in solving problems. Study each one carefully so that you know what each is used for. When faced with solving a problem, recall what each tool does and consider whether it will be helpful in finding a solution.

You might want to tear these pages out to use along with solving problems in this chapter.

Electrode names and reactions (Section 19.1)

The cathode is the electrode at which reduction (electron gain) occurs. The anode is the electrode at which oxidation (electron loss) occurs.

Standard reduction potentials relate to E°_{cell} (Section 19.2)

$$E^{\circ}_{cell} = E^{\circ}_{reduction} - E^{\circ}_{oxidation}$$

Faraday constant (Section 19.4)

$$1\mathscr{F} = 96{,}485.340 \ \text{C/mol } e^-$$

Standard cell potentials are related to thermodynamic quantities (Section 19.4)

$$\Delta G^{\circ} = -n\mathscr{F}E^{\circ}$$

$$E^{\circ}_{cell} = \frac{RT}{n\mathscr{F}} \ln K_c$$

Nernst equation (Section 19.5)

$$E_{cell} = E^{\circ}_{cell} - \frac{RT}{n\mathscr{F}} \ln Q$$

Coulombs are related to current in amperes and time in seconds (Section 19.8)

$$C = A \times t$$

Summary of Important *Equations*

Stoichiometric relationships in electrolysis:

$$1 \text{ C} = 1 \text{ A} \times 1 \text{ s}$$

$$1 \text{ mol } e^- = 1 \ \mathscr{F} = 9.65 \times 10^4 \text{ C}$$

Using reduction potentials to calculate cell potentials:

$$E^\circ_{cell} = \begin{pmatrix} \text{standard reduction} \\ \text{potential of the} \\ \text{substance reduced} \end{pmatrix} - \begin{pmatrix} \text{standard reduction} \\ \text{potential of the} \\ \text{substance oxidized} \end{pmatrix}$$

Cell potential and free energy change:

$$\Delta G = -n \ \mathscr{F} E_{cell}$$
$$\Delta G^\circ = -n \ \mathscr{F} E^\circ_{cell}$$

Cell potential and K_c:

$$E^\circ_{cell} = \frac{RT}{n\mathscr{F}} \ln K_c$$

Nernst equation:

$$E_{cell} = E^\circ_{cell} - \frac{RT}{n\mathscr{F}} \ln Q$$

Notes:

Chapter **20**

Nuclear Reactions and Their Role in Chemistry

The unstable nuclei of many naturally occurring and synthetic isotopes present both risks and opportunities. These risks and opportunities arise from the same phenomenon, the large amount of energy involved in nuclear processes. To understand, control, and safely use nuclear reactions, we need to understand what makes some nuclei stable and others unstable, and the effect of the high–energy particles and electromagnetic waves involved.

20.1   Conservation of Mass and Energy

Learning Objective

Utilize the unified laws of mass and energy conservation.

Review

Radionuclides, isotopes that are radioactive, emit streams of particles or of electromagnetic radiation. The study of the associated energies required a rethinking of the concepts of both matter and energy, which carried Einstein to the development of a relationship that drew a distinction between the rest mass of a particle and its mass in motion. To develop the relationship between mass and energy, Einstein made a combined law, the law of conservation of mass–energy, and proposed what is now called the Einstein equation, $\Delta E = \Delta m_0 c^2$. This equation is essential to a discussion of nuclear stability because it lets us calculate nuclear binding energies (Section 20.2). It also lets us understand the huge energy yields from small quantities of "fuel" in nuclear fission (Section 20.8).

Chemists can ignore the distinction between mass and energy in all situations involving the stoichiometry of chemical reactions. A calculation in this Section using the Einstein equation illustrates how extremely small the error is when we do this for enthalpy changes in chemical reactions.

Self–Test

1.   The enthalpy of combustion of acetylene is -1.30×10^3 kJ/mol. When 1.00 mol of acetylene burns, how much mass (in nanograms) changes to energy? What is this in parts per billion of the original mass of acetylene?

———————————————

20.2   Nuclear Binding Energy

Learning Objective

Grasp the importance of the nuclear binding energy.

Review

The energy that leaves the system when nucleons come together to form a nucleus would be the energy required to break up the nucleus. This is why the energy that leaves the system is called the nuclear binding energy. The greater this binding energy is, the more stable is the nucleus. In another sense, the nuclear binding energy is the energy the nucleus does not have because some of the mass of the nucleons changed to energy and left the system as the nucleus formed. Without this energy, the nucleus is more stable than it could have been with this energy.

When binding energies per nucleon are plotted against atomic number (Figure 20.1 in the text), the curve rises rapidly from the least stable nuclei to reach a peak in the vicinity of the isotopes of atomic number 26 (iron). Then the curve drops slowly as the highest atomic numbers are approached. In other words, on strictly the grounds of net energy changes, remembering that nature tends to favor events that are exothermic, the fusion of small nuclei into larger ones should release energy. And nuclear fusion does this. Likewise, the breaking up of very large nuclei into those of intermediate atomic numbers, should also give an overall gain in nuclear stability and the release of energy. Nuclear fission does this. Many nuclei change in the direction of greater stability by less drastic events. They emit radiations that transport energy out of their nuclei.

Self–Test

2. Why do we call the nuclear binding energy the energy that a nucleus does not have?

    _____

3. How do we explain the fact that the total mass of the nucleons in helium–4 is less than the actual mass of its nucleus?

    _____

    _____

4. In the curve of Figure 20.1 in the text, what does the maximum point correspond to, a point of *high stability* or a point of *low stability* for a nucleus at or near it?

    _____

20.3 Radioactivity

Learning Objective

Understand the nature of radionuclides and their emissions.

Review

Within a nucleus, the electrostatic force causes protons to repel each other, which lessens nuclear stability. But the nuclear strong force, which causes nucleons to attract each other, acts to overcome the electrostatic force. The electrostatic force (of repulsion), however, is able to act over longer distances than the strong force, so if a nucleus does not have enough neutrons to "dilute" the electrostatic force, the nucleus is unstable. A common consequence of such instability is radioactive decay. Various radionuclides decay by one of the following modes.

Decay Mode	Change in the Nucleus	Change in	
		Mass no.	**At. no.**
alpha emission	loss of $_2^4\text{He}$ (and usually also $_0^0\gamma$)	-4	-2
beta emission	loss of $_{-1}^0e$ (and usually also $_0^0\gamma$)	none	$+1$
gamma emission	loss of $_0^0\gamma$ (1 MeV range)	none	none
positron emission	loss of $_1^0e$ (then an annihilation collision produces gamma radiation)	none	-1
neutron emission	loss of $_0^1\text{n}$	-1	none
electron capture	change of a proton into a neutron	none	-1

The energy of a radiation is usually described by some multiple of the electron–volt (eV), and the relative instability of a radionuclide is described by its half–life.

When we write nuclear equations, the sums of the mass numbers on each side of the arrow must be equal as well as the sums of the atomic numbers on each side.

The most penetrating radiations are those with neither mass nor charge (gamma and X rays) or with mass but no charge (neutrons).

Several radionuclides of high mass number do not achieve stable nuclei by one nuclear change. Additional changes occur as a radioactive disintegration series is descended to a stable isotope.

Self–Test

5. Consider the natures of the electrostatic force and the strong force in an atomic nucleus.

    (a) Which acts between both protons and neutrons? _____

    (b) Which acts only between protons? _____

    (c) Which destabilizes nuclei? _____

    (d) Which acts over the shorter distance? _____

    (e) Which is a force of attraction? _____

6. What is present in a nucleus, besides the strong force, that helps to lessen repulsions between protons?

    _____

7. Write the nuclear equations for the decay of a hypothetical isotope, $^{279}_{111}X$, by each process. (Use Z as the atomic symbol for any new nuclide that forms from each process.)

    (a) By beta and gamma emission _____

    (b) By alpha and gamma emission _____

    (c) By positron emission _____

    (d) By neutron emission _____

    (e) By electron capture and X–ray emission _____

8. State what kind of particle or photon is *emitted* when

    (a) A neutron changes to a proton. _____

    (b) An electron capture takes place. _____

    (c) The radionuclide's atomic number increases by 1. _____

    (d) The atomic number decreases by 2. _____

(e)   Two photons of gamma radiation are
        produced following decay.   _____

(f)   The mass number decreases by 1.   _____

(g)   The atomic number decreases by 1.   _____

(h)   The mass number decreases by 4.   _____

(i)   No change to a different element occurs.   _____

9.   Which is the approximate energy of gamma rays?
      (a) 0.1 MeV   (b) 1.0 MeV   (c) 10 MeV   (d) 1.0 keV   _____

10.   The radiation with the best ability to penetrate lead is

      (a)   alpha radiation   (c)   gamma radiation

      (b)   beta radiation   (d)   positron radiation

                                                                    _____

11.   Annihilation radiation photons result from the collision of an electron with

      (a)   a positron   (c)   a neutron

      (b)   another electron   (d)   a proton   _____

12.   The net effect of electron capture is the conversion of

      (a)   an electron into a proton

      (b)   a proton into a neutron

      (c)   a neutron into a proton

      (d)   a positron into an electron   _____

20.4   Band of Stability

Learning Objectives

Explore the nature of the "band of stability."

Review

The odd–even rule says that nuclear instability is prevalent among nuclides having odd numbers for either the mass number or the atomic number, and particularly when both are odd and when both make the isotope lie outside the band of stability. When one or both numbers is a magic number (2, 8, 20, 28, 50, 82, or 126), the nuclide is more stable than those nearby in the band of stability. Among the elements below atomic number 83, radionuclides with too high a neutron/proton ratio tend to be beta emitters. Those with too low a value of this ratio tend to emit positrons. Radionuclides with atomic numbers above 83 are most often alpha emitters.

Self–Test

13.   The most stable isotope of the following 4 isotopes (where we use *hypothetical* atomic symbols) is

   (a) $^{15}_{8}X$ (b) $^{131}_{53}Y$ (c) $^{16}_{8}Z$ (d) $^{32}_{15}A$ _____

14.   At which atomic number is the nuclide most likely to be both stable and have a neutron to proton ratio very nearly equal to 1?

   (a) 10   (b) 40   (c) 80   (d) 106   _____

15.   An isotope of atomic number 65 and mass number 140

   (a)   lies below the band of stability.

   (b)   lies within the band of stability.

   (c)   lies above the band of stability.

   (d)   has one of the magic numbers.   _____

16.   If a radionuclide lies above and outside the band of stability, then the ejection of what particle will move it closer to this band?

   (a)   beta particle

   (b)   gamma ray photon

   (c)   positron

   (d)   a photon of gamma emission   _____

20.5   Transmutation

Learning Objective

Explain modern transmutation methods.

Review

When *transmutation* is caused by the bombardment of nuclei with high energy particles (e.g., $_2^4He$, $_1^1p$, or $_1^2d$), generally a *compound nucleus* first forms. It then sheds its excess energy by emitting a different particle or gamma radiation. Exactly what mode of decay is taken by the compound nucleus depends only on the energy it acquired by the initial bombardment and particle capture, not on the kind of particle captured. Hundreds of isotopes, nearly all of them radioactive, and including all of the *transuranium elements* have been made this way.

Self–Test

17.   To make a compound nucleus of $_{13}^{27}Al$ from each of the following bombarding particles, what must be the target isotope? Give its symbol.

(a)   proton   _____

(b)   deuteron   _____

(c)   alpha particle   _____

18.   What particle or photon must the compound nucleus, $_{13}^{27}Al^*$, eject to change into each of the following nuclides? Give the name and symbol.

(a) $_{11}^{23}Na$ _____

(b) $_{12}^{26}Mg$ _____

(c) $_{13}^{27}Al$ _____

19.   What is the general name for elements 93 through 114?

_____

For all the elements 93 and up?   _____

20.6   Measuring Radioactivity

Learning Objective

Demonstrate a knowledge of how radioactivity is measured.

Review

The various kinds of atomic radiation are sometimes called *ionizing radiation* because they create ions (and free radicals) in their wakes, or they make phosphors scintillate (give off bursts of light). This property accounts both for the hazards of radiation and for the ease of detection. You should be able to describe in general terms how the Geiger counter, a scintillation counter, and a film dosimeter work.

In learning the units for various measurements discussed in this section, notice that the becquerel (Bq) is the SI version of the curie (Ci), and that both describe the activity of a radioactive source, not the energy of its radiations. The becquerel and the curie are thus extensive quantities—they depend on the mass of the source (as well as the half–lives of the radionuclides present).

The activity is the number of disintegrations per second and is related to the first–order rate constant (also called the decay constant) for the process

$$\text{activity} = kN$$

The half–life is related to the decay constant by

$$t_{1/2} = \frac{\ln 2}{k}$$

Be sure to study Example 20.2 in the text to see how we can calculate the activity of a sample of an isotope from its half–life. Radioactive decay is a first–order rate process that was covered in Section 13.4.

The gray (Gy) is the SI version of the rad, and both refer to the energy absorbed by a quantity of matter because of the radiation it receives, not to the activity of the source and not even solely to the actual energy associated with the radiation. Thus the rad and the gray are also extensive quantities. They depend on the duration of the exposure (as well as on the energy, usually given in some multiple of the electron volt, of the radiation itself).

The rem is always some fraction (sometimes a very large fraction) of a rad (or a gray). The exact fraction depends on the kind of radiation, because the damage to tissue varies with this factor even when different kinds of radiation have identical energies and duration of exposure. The rem is a unit for describing potential harm to humans, because it concerns effects on tissue. The SI equivalent of the rem is the sievert (Sv); 1 rem = 10^{-2} Sv.

One of the reasons radiation is dangerous to living creatures is its ability to disrupt bonds to give highly reactive species called free radicals. These are molecular fragments that have one or more unpaired electrons. The damage caused by free radicals is large because of the ability of these reactive particles to set off undesirable reactions within living cells.

Because of several radionuclides in the Earth's crust as well as cosmic radiation, we are bathed constantly in a low level of background radiation averaging close to 360 mrem per year for each person in the United States—more depending on an individual's use of medical radiation. In all applications, workers can protect themselves to a considerable extent by using dense shielding materials (e.g., lead) and by getting at some distance from the source. For every doubling of the distance, the radiation intensity drops by a factor of four, according to the inverse square law.

Self–Test

20. What is the name of a radiation detector that

    (a)   uses photographic film or plates?   _____

    (b)   contains a phosphor?   _____

    (c)   lets radiation generate ions in a gas at low pressure?   _____

21. One rd = _____ J/g

22. One Gy = 1 _____ (supply units)

23. One Bq = _____ (supply units)

24. One Ci = _____ Bq

25. 1 Gy = _____ rd

26. How are even very low doses in rems dangerous to humans?

   _____

27. What is the relationship of the rem to the rad, in general terms?

   _____

28. "Rad" stands for _____

29. "Rem" stands for _____

30. The exposure we all experience per year to background radiation is about

    (a) 3 rem (b) 36 rem (c) 360 mrem (d) 1600 μCi   _____

31. If the intensity of radiation is 100 units at a distance of 1.50 m from a source, how far away must one move to reduce the exposure to 1.00 unit?

    (a) 0.0015 m (b) 15 m (c) 12.2 m (d) 1500 m   _____

32. The half–life of cobalt–60 is 5.27 yr. Calculate the activity of a 35 mg sample of this isotope in becquerels.

20.7 Medical and Analytical Applications of Radionuclides

Learning Objective

Show an appreciation for practical applications of radionuclides.

Review

In nearly all applications, gamma emitters are the best kinds of radionuclides. This radiation is the most penetrating of all, and therefore it is the easiest to detect, and it lets the scientist use very small quantities of the radionuclide. All of its radiation serves the purpose because little if any is blocked.

In *tracer analysis*, the ability of a bodily fluid to enter a particular tissue can be traced if the fluid contains a small concentration of a radionuclide (e.g., $^{99m}_{43}Tc$ as TcO_4^-).

When a sample is bombarded by neutrons in neutron activation analysis, its various nuclei capture neutrons and become compound nuclei that then emit gamma radiation. Which frequency of gamma radiation comes out is determined by the kinds of atoms in the sample, and the intensities at these frequencies give measures of the concentrations of the atoms.

For radiological dating, pairs of radionuclides have to be identified and their relative concentrations in the sample measured. For dating very ancient rock formations, members of the pair might belong to the same radioactive disintegration series—for example, uranium–238 and lead–206—with the lighter one assumed to be produced solely by the decay of the heavier one at a rate of decay that has held constant over the millennia. For dating organic remains, the relative concentrations of carbon–14 and carbon–12 are used. As long as the living thing (e.g., a tree) lives, its level of carbon–14 is presumed to be constant. Once it dies, it no longer takes in carbon–14 and now the decay of this radionuclide at a known rate means that the age of any object made from the living thing (e.g., a wooden article) can be measured.

Self–Test

33. What method involving the use of radionuclides would be used in each situation?

    (a) Determine the existence and concentration of lead as an impurity in the fingernails of children who have eaten chips of lead–based paints.

(b)   Measure the age of the Laurentian shield of bedrock in a southern Ontario province in Canada.

_____

(c)   Measure how well blood circulates through a region of the lower leg suspected of having an early stage of gangrene with the hope of doing no more serious an amputation than absolutely necessary.

_____

34.   How does carbon–14 originate in the upper atmosphere? Write a nuclear equation.

_____

35.   In what chemical form is carbon–14 taken in by plants? Write a chemical formula.

_____

36.   For the carbon–14 method to work without any correction factors, what would have to be true about the ratio of carbon–14 to carbon–12 in all living things both today and in the past?

_____

37.   When carbon–14 dating methods were used on a sample of wood taken from a door post of an ancient archaeological site, it was found to have a specific activity of 382 Bq/g. How old was this wood sample according to calculations uncorrected for the factors that are known to cause some errors in the method? (Calculate to two significant figures.)

_____

20.8   Nuclear Fission and Fusion

Learning Objective

Develop an understanding of nuclear fission and fusion.

Review

The capture of a neutron by a uranium–235 nucleus produces an unstable, compound nucleus that breaks apart—undergoes fission. The products are isotopes of intermediate atomic number, neutrons, and energy. If the neutrons can be slowed enough by moderators (e.g., graphic or heavy water or ordinary water), some may be captured by unchanged uranium–235 nuclei and so launch a chain reaction. (Plutonium–239 is also a fissile isotope.) The difference in binding energy between uranium–235 and the product isotopes is released largely as heat that can change water to steam and so drive electrical turbines.

To operate a reactor safely, control rods can be used to capture enough neutrons to make the multiplication factor equal 1. Now the reactor is said to be critical, because exactly as many neutrons are left at the end of the fission cycle as started the cycle.

Since the concentration of fissile isotope in the fuel elements of a nuclear reactor is small, a critical mass of the isotope is not possible and an atomic bomb explosion cannot occur. Should the coolant be lost, then the reactor could melt through its containment vessel. And heat could cause a steam explosion that would rupture the vessel (as occurred at Chernobyl, Russia).

Radioactive wastes include gases, liquids, and solids. Iodine–131, cesium–137, and strontium–90 are particular problems when they get into the environment because the blood can carry them throughout the body where their radiations cause harm. Long–lived solid wastes must be kept out of human touch for centuries.

Self–Test

38. Why can nuclei capture neutrons much more easily than protons?   _____

_____

39. What is nuclear fission? _____

_____

40. The isotopes initially formed from nuclear fusion have neutron–to–proton ratios that are too *high* or too *low*?   _____

How do they adjust these ratios?   _____

_____

41. Which is generally higher, the sum of the binding energies of the nuclei produced by fission or the binding energy of the uranium–235 nucleus?

_____

42. What makes the fission of uranium–235 self–sustaining?   _____

_____

43. What is meant by "pressurized" in the pressurized water reactor?

_____

44. What is the function of each of the following in a pressurized water nuclear reactor?

    (a)   The moderator

    _____

    (b)   The cladding

    _____

    (c)   The primary coolant loop

    _____

    (d)   The secondary coolant loop

    _____

    (e)   The control rods

    _____

45. Why do each of the following radionuclide wastes pose human health problems?

    (a)   Iodine–131

    _____

    (b)   Cesium–137

    _____

    (c)   Strontium–90

    _____

Answers to Self–Test Questions

1. 14.4 ng, 0.55 ppb
2. Because it is the energy *lost* from the system when some mass changed to energy as the nucleons formed into a nucleus.
3. Some mass of the nucleons changed to energy when the nucleus formed.
4. High stability
5. (a) strong force (b) electrostatic force (c) electrostatic force (d) strong force (d) strong force
6. neutrons
7. (a) $^{279}_{111}X \longrightarrow {}^{0}_{1}e + {}^{279}_{112}Z + {}^{0}_{0}\gamma$

   (b) $^{279}_{111}X \longrightarrow {}^{4}_{2}He + {}^{279}_{109}Z + {}^{0}_{0}\gamma$

   (c) $^{279}_{111}X \longrightarrow {}^{0}_{1}e + {}^{279}_{110}Z$

   (d) $^{279}_{111}X \longrightarrow {}^{1}_{0}n + {}^{278}_{111}Z$

   (e) $^{279}_{111}X + {}^{0}_{1}e \longrightarrow {}^{279}_{110}Z + X \text{ rays}$

8. (a) beta particle (b) X ray photon (c) beta particle (d) alpha particle (e) positron (f) neutron (g) positron (h) alpha particle (i) neutron (or gamma ray photon, only)
9. b
10. c
11. a
12. b
13. c
14. a
15. a
16. a
17. (a) $^{26}_{12}Mg$ (b) $^{25}_{12}Mg$ (c) $^{23}_{11}Na$
18. (a) alpha particle (b) proton ${}^{1}_{1}p$ (c) gamma ray photon ${}^{0}_{0}\gamma$
19. actinide elements; transuranium elements
20. (a) dosimeter (b) scintillation counter (c) Geiger counter
21. 10^{-5}
22. J/kg
23. 1 disintegration/s
24. 3.7×10^{10}
25. 100
26. They generate unstable ions and radicals that initiate other chemical changes of danger to the individual.
27. The rem is a fraction of a rad, the fraction depending on how damaging a particular rad dose is in tissue.
28. radiation absorbed dose
29. radiation equivalent for man
30. c
31. b
32. 1.5×10^{12} Bq
33. (a) neutron activation analysis (b) radiological dating (c) tracer analysis
34. $^{1}_{0}n + {}^{14}_{7}N \longrightarrow {}^{15}_{7}N^* \longrightarrow {}^{14}_{6}C + {}^{1}_{1}p$

35. CO_2
36. a constant ratio of carbon–14 to carbon–12
37. 7.2×10^3 years
38. Neutrons are electrically neutral and so are not repelled by nuclei as they approach.
39. The spontaneous breaking of an unstable nucleus roughly in half.
40. Too high. They eject neutrons.
41. The sum of the binding energies of the products.
42. It produces more neutrons than needed to cause further fission events.
43. The water in the primary coolant loop is under such high pressure that even at high temperatures it is in the liquid state.
44. (a) Convert high energy neutrons into slower (thermal) neutrons.
    (b) Hold both the fuel in place and retain radioactive wastes.
    (c) Remove heat from the cladding elements as it is produced by fission.
    (d) Remove heat from the primary coolant loop and let this heat generate steam under pressure to drive the turbines.
    (e) Manage the flux of neutrons in the core so that the reactor will be critical during operation and go subcritical at shutdown.
45. (a) Concentrates in the thyroid gland where its radiation could cause thyroid cancer or other loss of thyroid function.
    (b) Transported by the blood wherever sodium goes.
    (c) Is attracted to bone tissue.

Tools for problem solving

In this chapter you learned to apply the following concepts as tools in solving problems. Study each one carefully so that you know what each is used for. When faced with solving a problem, recall what each tool does and consider whether it will be helpful in finding a solution.

You might want to tear these pages out to use along with solving problems in this chapter.

The Einstein equation (Section 20.1)

$$\Delta E = \Delta m_0 c^2 \text{ (or often just } E = mc^2)$$

Balancing nuclear equations (Section 20.3)

When you have to write and balance a nuclear equation, remember to apply the following two criteria:

1. The sums of the mass numbers on each side of the arrow must be equal.

2. The sums of the atomic numbers on each side of the arrow must be the same.

The odd–even rule (Section 20.4)

When the numbers of neutrons and protons in a nucleus are both even, the isotope is far more likely to be stable than when both numbers are odd.

Law of radioactive decay (Section 20.6)

Activity has units of disintegrations per second, or Bq. The decay constant, k, is a first–order rate constant with units of s^{-1}, and the number of atoms of the radionuclide in the sample is represented by N.

$$\text{activity} = -\frac{\Delta N}{\Delta t} = kN$$

Half–life of a radionuclide (Section 20.6)

$$t_{1/2} = \frac{\ln 2}{k}$$

The half–life is represented by $t_{1/2}$, and the decay constant by k.

Inverse square law (Section 20.6)

$$\frac{I_1}{I_2} = \frac{d_2^2}{d_1^2}$$

I represents the intensity of radiation and d the distance from the source of the radiation.

Chapter 21

Metal Complexes

21.1   Complex Ions

Learning Objective

Describe the different kinds of ligands and the rules for writing formulas for metal complexes containing them.

Review

Complex ions were discussed in Chapter 19 where the focus was on their equilibria and how their formation affects the solubilities of salts. Here we look at these substances in greater detail to find out the kinds of substances that form complexes with metals and the kinds of structures that they form.

Compounds that contain complex ions (often referred to simply as complexes) are sometimes called coordination compounds because coordinate covalent bonds generally bind the pieces that make up the complex ion. One piece is an electron pair acceptor (a metal ion at the center of the complex) and the other is an electron pair donor (Lewis base) that we call a ligand. Many ligands are anions (e.g., Cl^-, S^{2-}, CN^-, OH^-, SCN^-, $S_2O_3^{2-}$, and NO_2^-). Other ligands are electrically neutral, such as NH_3 and H_2O. All these are monodentate ligands because they "bite" or are joined to the metal ion by one coordinate covalent bond. Bidentate ligands, such as $NH_2CH_2CH_2NH_2$ or the oxalate ion ($C_2O_4^{2-}$) contain two donor atoms and are held by two coordinate covalent bonds to the metal ion. EDTA is an example of a polydentate ligand. Study some of the practical uses of EDTA.

In the formula of a complex, the acceptor atom is always written first followed by the ligands. Anionic ligands are listed first in alphabetical order, followed by neutral ligands, also in alphabetical order. When several ligands are held by the same acceptor atom, as in $[CrCl(H_2O)_5]^{2+}$, the net charge, which is the algebraic sum of the charges on the acceptor and the ligands, is placed outside the square brackets.

Complexes containing bidentate ligands, which form chelate ring structures, are significantly more stable than complexes in which the metal is surrounded by the same number of donor atoms provided by monodentate ligands. This phenomenon is called the chelate effect. The reason is because a bidentate ligand is less likely than a monodentate ligand to become completely detached from a metal ion.

Self-Test

1. The zinc ion forms a complex with four OH^- ions.

   (a) What is the formula of this complex ion? _____

   (b) Could the complex ion be isolated as a potassium salt or a chloride salt? Write the formula.

   _____

2. Fe^{3+} and SCN^- form a complex ion with a net charge of 3–. Write the formula of this complex ion.

   _____

3. Ni^{2+} and NH_3 form a complex ion with a net charge of 2+. The ratio of donor atom to ligand is 1 to 6. Write the formula for the complex.

   _____

4. A complex is formed from Co^{2+}, one ethylenediamine, 2 water molecules, and 2 chloride ions. Write the formula for the complex.

   _____

5. Sketch chelate rings formed by

   (a) ethylenediamine

   (b) oxalate ion

6. Which complex would you expect to be more stable, $[Cr(en)_3]^{3+}$ or $[Cr(NH_3)_6]^{3+}$?

   _____

21.2　Metal Complex Nomenclature

Learning Objective

Use the rules for naming metal complexes.

Review

In this section you learn the IUPAC rules for naming metal complexes. In summary, they are as follows:

Name the cation first, followed by the anion.

Name the ligands first (in alphabetical order according to the *ligand name*, not according to any number prefixes such as di or tri), followed by the metal. Write the oxidation number of the metal in parentheses using Roman numerals after the name of the metal.

Use prefixes to specify numbers of ligands (except *bis*, *tris*, etc. are used when there might be confusion if *di*, *tri*, etc. are used).

The names of anionic ligands end in *-o*.

A neutral ligand has the same name as the molecule (except H_2O is aqua and NH_3 is ammine).

Negatively charged complexes always end in the suffix *-ate*, which is appended to the name of the metal. When the suffix -ate is used, we also use the Latin stem for the metal name when the metal has a symbol not derived from the English name. (The exception is mercury, e.g., mercurate).

Self-Test

7.　Name the following complexes:

(a)　$[Cr(NH_3)_6]^{2+}$ _____

(b)　$[Cu(OH)_4]^{2-}$ _____

(c)　$[MnCl_4(NH_3)_2]^{2-}$ _____

(d)　$K_3[Co(CN)_6]$ _____

(e)　$[Ni(NH_3)_2(H_2O)_4]Cl_2$ _____

8. Write formulas for the following complexes:

   (a) dichlorobis(ethylenediamine)cobalt(III) nitrate

   _____

   (b) ammonium tetrachloro(ethylenediamine)chromate(III)

   _____

   (c) tetrabromostannate(II) ion

   _____

   (d) diaquatetrahydroxoaluminate(III) ion

   _____

   (e) diiodoargentate(I) ion

   _____

21.3 Coordination Number and Structure

Learning Objective

Show the different geometries associated with each of the common coordination numbers, and draw the structures.

Review

Coordination number is the number of ligand donor atoms that are bonded to the metal in a complex. Common coordination numbers in complexes include 2, 4, and 6. For coordination number 2, the complex usually has a linear geometry. For coordination number 4, both tetrahedral and square planar structures are observed. For coordination number 6, almost all complexes are octahedral. Such octahedral complexes are formed with monodentate ligands as well as polydentate ligands. Study Figures 21.3 and 21.4. Be sure you can sketch an octahedral complex; review Figure 21.4. You might want to go back to Section 9.1 where simplified drawing instructions are given.

Self-Test

9. What property do metal ions that form tetrahedral complexes usually have?

   _____

10. Sketch the structure of the complex $[Co(NH_3)_6]^{3+}$.

11. What is the coordination number of the metal ion in each of the following?

    (a) $[Cu(NH_3)_4]^{2+}$ _____

    (b) $[Co(NH_3)_4Cl_2]^+$ _____

    (c) $[Ni(en)_3]^{2+}$ _____

21.4   Isomers of Metal Complexes

Learning Objective

Write structural formulas for all the isomers of a given metal complex.

Review

When two or more different compounds have the same chemical formula, they are said to be *isomers* of each other. For coordination compounds, there are several ways for this to occur, but the most important one for you to learn about is the type of isomerism called *stereoisomerism*. This occurs when two or more compounds have the same formula, but differ in the way their atoms are arranged in space.

Geometric isomers of square planar complexes such as $[Pt(NH_3)_2Cl_2]$, or octahedral complexes such as $[Cr(H_2O)_4Cl_2]^+$, exist in *cis* and *trans* forms. In the *cis* isomer, the chloride ions are next to each other on *the same side* of the metal ion. In the *trans* isomer, the chloride ions are opposite each other. In general, *cis* and *trans* isomers can exist for square planar complexes with the general formula Ma_2b_2 (where M is a metal ion and a and b are monodentate ligands). *Cis* and *trans* isomers also exist for octahedral complexes with the general formula Ma_2b_4, and for octahedral complexes with the general formula MA_2b_2 (where A is a bidentate ligand and b is a monodentate ligand).

Chiral isomers occur when two structures differ only in that one is the nonsuperimposable mirror image of the other—that is, when the mirror image of one isomer looks exactly like the other isomer, but the two isomers themselves do not match exactly when one is placed over the other. Chiral isomers exist for complexes with the general formula MA_3 and cis–MA_2b_2 (where A stands for a bidentate ligand and b stands for a monodentate ligand). Study Figures 21.7 and 21.8. Chiral isomers are also known as *optical isomers* because the two isomers affect polarized light in opposite ways as illustrated in Figures 21.9 and 21.10.

Self-Test

12. Is *cis–trans* isomerism possible for tetrahedral complexes? Explain.

_____

_____

13. How many different isomers exist for the complex $[Co(en)_2Br_2]^+$?

_____

14. Is the *trans* isomer of $[Ni(C_2O_4)_2(CN)_2]^{4-}$ chiral? (Note: $C_2O_4^{2-}$ is oxalate ion, a bidentate ligand.)

_____

21.5   Bonding in Metal Complexes

Learning Objective

Describe how crystal field theory explains the properties of metal complexes

Review

The crystal field theory is used to explain the properties of complexes in which the metal ion has a partially filled *d* subshell. To understand the theory, it is necessary to know the shapes and directional properties of the *d* orbitals. Study Figure 21.13.

In an octahedral complex, we can imagine the ligands to lie along the *x*, *y*, and *z* axes of a Cartesian coordinate system with the metal ion in the center (at the origin). The negative charges of the ligands (either anions or the negative ends of ligand dipoles) point directly at the metal ion's $d_{x^2-y^2}$ and d_{z^2} orbitals, but they point between the d_{xy}, d_{xz}, and d_{yz} orbitals. Because of this, the ligands repel electrons in the $d_{x^2-y^2}$ and d_{z^2} orbitals more than they repel electrons in the other three. This raises the energies of the $d_{x^2-y^2}$ and d_{z^2} orbitals above the energies of the d_{xy}, d_{xz}, and d_{yz} orbitals. The net result is an energy level diagram like that shown in Figure 21.15. The energy difference between the two energy levels is called the crystal field splitting and is given the symbol Δ.

The magnitude of Δ depends on the nature of the ligands attached to the metal ion, the oxidation state of the metal, and the period in which the metal occurs. In general, as the oxidation state increases, other things being equal, the size of Δ becomes larger. Going down a group, Δ becomes larger, too.

In this section, the usefulness of the crystal field theory is illustrated by considering three phenomena—the stabilities of certain oxidation states of metal ions in complexes, the origin of the colors of complexes, and the magnetic properties of complexes.

For chromium(II) ion, you see that the removal of a high-energy electron, along with an increase in the magnitude of Δ that accompanies the increase in oxidation state, helps to make the oxidation of $[Cr(H_2O)_6]^{2+}$ to $[Cr(H_2O)_6]^{3+}$ energetically favorable. Stated in another way, in water, chromium(III) ion is the more stable oxidation state, because chromium(II) ion is so easily oxidized to it.

According to crystal field theory, the color of a complex arises from the absorption of a photon that has an energy equal to Δ. For transition metal complexes, this photon has a frequency that places it in the visible region of the spectrum. The color observed for the complex is the color of the light that isn't absorbed.

Some ligands always produce a large Δ, regardless of the metal ion, and some ligands always produce a small Δ. The list of ligands arranged in order of their ability to produce a large Δ is called the spectrochemical series. Your teacher will tell you whether you should memorize this series of ligands.

The amount of energy needed to cause two electrons to become paired in the same orbital is called the pairing energy, to which we have given the symbol P. For certain numbers of d electrons, there is a choice as to how the electrons are to be distributed among the higher and lower d-orbital energy levels. Pairing an electron with another in a low-energy d orbital costs energy equal to the pairing energy, but it saves an energy equal to Δ. On the other hand, placing the electron in the higher energy orbital costs an energy equal to Δ, but saves energy equal to the pairing energy. Which energy distribution prevails depends on how the magnitudes of Δ and P compare. When $\Delta > P$, pairing of electrons in the lower energy level is preferred and a low spin complex is formed; when $\Delta < P$, then spreading the electrons out as much as possible is preferred and a high spin complex is formed.

Self-Test

15. Which of the d orbitals point directly along the x, y, and z axes?

   _____

16. Which complex has the larger Δ, $[CrCl_6]^{4-}$ or $[CrCl_6]^{3-}$?

   _____

17. Which complex has the larger Δ, $Ni(CN)_4^{2-}$ or $Pt(CN)_4^{2-}$?

   _____

18. Cyanide ion produces a very large crystal field splitting. Should it be easy or difficult to oxidize $[Co(CN)_6]^{4-}$ to $[Co(CN)_6]^{3-}$? Explain your answer in terms of the populations of the d orbitals of the metal ion.

19. Which complex would be expected to absorb light of longer wavelength?

    (a) $[Ti(H_2O)_6]^{3+}$ or $[Ti(H_2O)_6]^{2+}$ _____

    (b) $[Ni(H_2O)_6]^{2+}$ or $[Ni(CN)_6]^{4-}$ _____

20. Which complex has a larger Δ, one that absorbs red light or one that absorbs blue light?

    _____

21. How many unpaired electrons would you expect to find in each of the following?

    (a) $[CrI_6]^{4-}$ _____   (c) $[Fe(H_2O)_6]^{3+}$ _____

    (b) $[Cr(CN)_6]^{4-}$ _____   (d) $[Fe(CN)_6]^{3-}$ _____

21.6   Biological Functions of Metal Ions

Learning Objective

Explain the biological function of several common metal.

Review

Most metal ions in living systems serve their function when they are found in complex ions of various kinds. The porphyrin ring system is found in hemoglobin and myoglobin, which contain Fe^{2+} that binds O_2 molecules as a ligand. A similar ring system containing Co^{2+} is found in vitamin B_{12}.

Self-Test

22. Name two biologically important complexes that contain the porphyrin structure.

    _____   _____

23. How many donor atoms does the porphyrin ring contain?   _____

Answers to Self-Test Questions

1. (a) $[Zn(OH)_4]^{2-}$
   (b) It could be isolated as a potassium salt, $K_2[Zn(OH)_2]$.
2. $[Fe(SCN)_6]^{3-}$
3. $[Ni(NH_3)_6]^{2+}$
4. $[CoCl_2(H_2O)_2(en)]$
5.

<table>
<tr><td>(a)</td><td>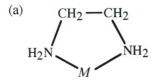</td><td>(b)</td><td>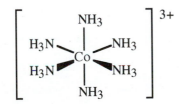</td></tr>
</table>

6. $[Cr(en)_3]^{3+}$ (chelate effect)
7. (a) hexaamminechromium(II) ion
   (b) tetrahydroxocuprate(I) ion
   (c) diamminetetrachloromanganate(II) ion
   (d) potassium hexacyanocobaltate(III)
   (e) diamminetetraaquanickel(II) chloride
8. (a) $[CoCl_2(en)_2]NO_3$
   (b) $NH_4[CrCl_4(en)]$
   (c) $[SnBr_4]^{2-}$
   (d) $[Al(OH)_4(H_2O)_2]^-$
   (e) $[AgI_2]^-$
9. They usually have filled or empty d subshells.
10.

$$\left[\begin{array}{c} NH_3 \\ H_3N - \overset{|}{\underset{|}{Co}} - NH_3 \\ H_3N \quad NH_3 \\ NH_3 \end{array} \right]^{3+}$$

11. (a) 4, (b) 6, (c) 6
12. No. For complexes of the type Ma_2b_2, only one tetrahedral structure can be constructed.
13. Three. A *trans* isomer and a pair of chiral *cis* isomers that are enantiomers.
14. No. The complex and its mirror image are superimposable.
15. $d_{x^2-y^2}$ and d_{z^2}
16. $[CrCl_6]^{3-}$, because it has chromium in the higher oxidation state.
17. $[Pt(CN)_4]^{2-}$, because Pt is below Ni in its group.

18.

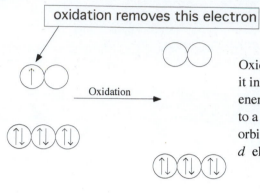

Oxidation should be easy because it involves removing a high-energy electron and it also leads to a lowering of the energy of the orbitals that hold the remaining d electrons.

$[Co(CN)_6]^{4-}$ $[Co(CN)_6]^{3-}$

19. (a) $[Ti(H_2O)_6]^{3+}$ (b) $[Ni(CN)_6]^{4-}$
20. The one that absorbs the higher energy blue light.
21. (a) 4 (b) 2 (c) 5 (d) 1
22. hemoglobin, myoglobin
23. four

Tools for problem solving

In this chapter you learned to apply the following concepts as tools in solving problems. Study each one carefully so that you know what each is used for. When faced with solving a problem, recall what each tool does and consider whether it will be helpful in finding a solution.

You might want to tear these pages out to use along with solving problems in this chapter.

Rules for writing formulas for complexes (Section 21.1)

The following rules apply whenever you have to write the formula for a complex ion:

1. The symbol for the metal ion is always given first, followed by the ligands.

2. When more than one kind of ligand is present, anionic ligands are written first (in alphabetical order), followed by neutral ligands (also in alphabetical order).

3. The charge on the complex is the algebraic sum of the charge on the metal ion and the charges on the ligands.

4. The formula is placed inside of square brackets with the charge of the complex as a superscript outside the brackets, if it is not zero.

Rules for naming complexes (Section 21.2)

Naming complexes follows rules that are an extension of the rules you learned earlier. You have to learn them and then apply them when you have to name a complex, or write a formula given the name.

Crystal field splitting pattern for octahedral complexes (Section 21.5)

Figure 21.15 forms the basis for applying the principles of crystal field theory to octahedral complexes. For a complex under consideration, set up the splitting diagram and place electrons into the d orbitals following Hund's rule. For d^4, d^5, d^6, and d^7 configurations, you may have to decide whether a high or low spin configuration is preferred.

Spectrochemical series (Section 21.5)

$$CN^- > NO_2^- > en > NH_3 > H_2O > C_2O_4^{2-} > OH^- > F^- > Cl^- > Br^- > I^-$$

Notes:

Chapter **22**

Organic Compounds, Polymers, and Biochemicals

Of the more than fifty million compounds known to chemists, approximately half of them are molecular substances, with covalent carbon–carbon bonds, that we commonly call organic compounds. Yet the study of organic compounds assumes many of the features of the study of any other field. You learn in this chapter about classifying organic compounds into families defined by functional groups. You see how such groups confer common chemical properties to all members of the same family, but that the nonfunctional, hydrocarbon groups also contribute much to physical properties.

All living processes in nature have a molecular basis, and biochemistry describes them. The complex molecules of biochemistry have functional groups like those in simpler substances and so they have similar chemical properties. Thus, this chapter is also meant to introduce you to the major kinds of biochemicals.

22.1   Organic Structures and Functional Groups

Learning Objective

Write structural formulas for organic compounds, highlighting their functional groups.

Review

The study of organic chemistry is helped by the concept of functional groups, the main idea in this section. Organic chemistry is organized around families defined by such groups, and we introduce many of the most important families here. Table 22.1 in the text lists the principal functional groups we discuss in this chapter. We suggest you take a look at them now, but you will probably find this table more useful for review later.

Substances whose molecules have oxygen or nitrogen atoms in them tend to be more polar and so more soluble in water than others of the same size. We note that some groups bear similarities in structure and therefore in chemistry to simple inorganic species. Amines are ammonia–like, for example, and alcohols are water–like.

Besides functional groups, organic molecules contain hydrocarbon–like, nonfunctional groups. Be sure to understand the value of using one symbol, R, for all such groups, no matter how many carbons are present. The fact that nonfunctional groups do not change chemically in most (if not all) of the reactions of a family is what makes such a simplifying symbol possible. R groups have few chemical reactions because they are essentially nonpolar and therefore unattractive to ionic or polar reactants.

The carbon atoms making up an R group can be in straight chains, branched chains, and rings. When rings incorporate an atom other than carbon, the ring is said to be heterocyclic.

Self–Test

1. Circle what is most likely the functional group in

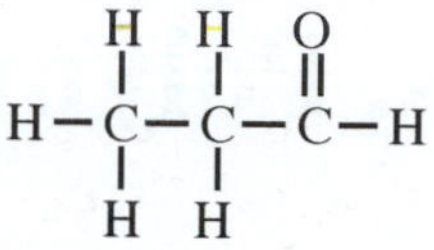

$$H-\underset{\underset{H}{|}}{\overset{\overset{H}{|}}{C}}-\underset{\underset{H}{|}}{\overset{\overset{H}{|}}{C}}-\overset{\overset{O}{||}}{C}-H$$

2. Which molecule would be more polar, CH_3-CH_3 or CH_3-NH_2?

   _____

3. What forms in the following reaction?

   $CH_3-NH_2 + H-Br \longrightarrow$ _____

4. What forms in the following reaction?

   $R-NH_2 + H-Cl \longrightarrow$ _____

5. What significance does the functional group concept have for the study of organic chemistry?

   _____

   _____

6. Study the members of the following pairs of compounds and decide if they represent *isomers*, or are *identical*, or *neither*.

   (a) $CH_3OCH_2CH_3$ and $CH_3CH_2CH_2OH$ _____

   (b)
   and
   _____

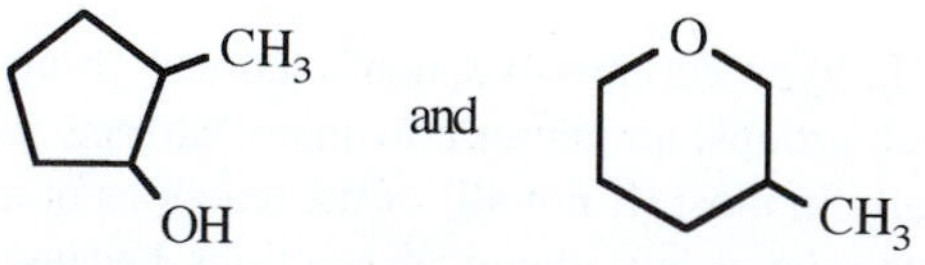

   (c) $CH_3CH_2CH_2OH$ and $HOCH_2CH_2CH_3$ _____

7. Which compound in Question 6 is a heterocyclic compound?

   _____

22.2   Hydrocarbons: Structure, Nomenclature, and Reactions

Learning Objective

Describe the nomenclature rules and major reactions of alkanes, alkenes, alkynes, and aromatic hydrocarbons.

Review

When the molecules of a compound have no double or triple bonds (no pi bonds), it is a saturated compound. Otherwise, it is unsaturated. (Compounds with double or triple bonds have the capacity to take up additional hydrogen or other atoms, much as an unsaturated solution has the capacity to take up additional solute. That's why such compounds are said to be unsaturated.) The molecules of hydrocarbons are made solely of C and H. The saturated hydrocarbons have only single bonds and are called alkanes (or cycloalkanes). Whether straight–chain, branched–chain, or ring, the saturated hydrocarbons have very few chemical properties. All hydrocarbons are hydrophobic; they do not dissolve in water.

Nomenclature

Here's how to sort out the *IUPAC rules* for all families. *Regardless of the family*, the IUPAC rules all begin with the idea of a "parent" unit—a molecular portion that is named and whose carbon skeleton is numbered. Each family has its own rule for identifying the parent. For the alkanes, the "parent" is the longest continuous sequence of carbons. In other words, a parent alkane is always, by itself, a straight–chain alkane. That's why it's important that you learn the names of the straight–chain alkanes through $C_{10}H_{22}$ (Table **22.2** in the text).

The next key idea in IUPAC nomenclature is that each family has a characteristic *name ending*. This is the same for all members of a given family. Thus "–ane" is the name ending for all alkanes and cycloalkanes, and "–ol" is the ending of all IUPAC names of alcohols.

Each family has its own IUPAC rule for selecting the *parent chain* to which substituents are attached.

Each family also has its own IUPAC rule for *numbering the parent chain;* a rule that tells us which position must be numbered 1. For the alkanes, the rule is to number from whichever end of the parent chain is nearest the first branch.

Hydrocarbon groups, called *alkyl groups*, have their own names. You should learn the names and structures for those having one to three carbons—the *methyl, ethyl, propyl,* and *isopropyl* groups.

The locations of the alkyl groups on the parent chain are identified by the numbers given to the carbons of the parent chain when the chain is properly numbered.

When you assemble all the parts of a name into one whole, remember the following rules.

1.   Each alkyl group (or other substituent) must be associated with a chain location number.

2.   Multiplier prefixes, like di– and tri–, sometimes have to be added to the names of alkyl   groups.

3.   The names of the alkyl groups are organized into the final name alphabetically, ignoring the

multiplier prefixes (di–, tri–, etc.). Thus "trimethyl" would precede "propyl" because "m"   precedes "p" in the alphabet.

4.   Two numbers in a name are always separated by a comma; e.g., 2,2–dimethylpropane.

5.   A hyphen is always used to separate a number from a word–part of a name; e.g., 2,3–dimethylbutane.

These are illustrated in Example 22.1 in the text.

Alkane chemistry

At room temperature almost nothing attacks alkanes. They do burn, and fluorine reacts violently with them. At higher temperatures, alkanes react with chlorine and bromine, and alkane molecules can be cracked, or broken down to smaller hydrocarbons.

Alkenes and alkynes

The parent of an *alkene* must be the longest chain *that includes the double bond.* The parent is numbered from whichever end gets to the double bond first, regardless of the location of alkyl branches, and the numbering proceeds through the double bond. Small alkenes often have common names such as ethane (also ethylene) or 2–butene where the double bond is between the second and third carbon. The IUPAC name is but–2–ene. Alkynes also have common names such as ethene or acetylene. Larger molecules, for instance a seven–carbon chain with the triple bond between the third and fourth carbon, is named hept–7–yne.

Many alkenes exhibit *geometric isomerism* (*cis–trans* isomerism) and have molecules that differ only in their geometry at the double bond. Notice that this isomerism in alkenes is possible only when neither end of the double bond holds identical groups. Because the structure is linear around the triple bond (*sp* bonding) *cis–trans* isomerism cannot exist around the triple bond in alkynes.

Alkene and alkyne chemistry

The carbon–carbon double bond has π electrons, making the double bond somewhat electron–rich and attractive to protons and other electron–poor species. The triple bond in alkynes is even more electron–rich. So alkenes and alkynes undergo *addition reactions* in which the reactant (or a catalyst) is able to donate H^+ to the pi bond. Thus alkenes react with hydrogen chloride and with water, provided there is an acid catalyst. When gaseous molecules of HCl add to an alkene, H^+ from HCl goes to one end of the double bond using the pi electrons to make a new C—H bond. This leaves the other end, now electron–poor, to accept the chloride ion. Overall, we have

$$\underset{/}{\overset{\backslash}{C}}=\underset{\backslash}{\overset{/}{C}} \quad + \quad H-Cl \quad \longrightarrow \quad H-\overset{|}{\underset{|}{C}}-\overset{|}{\underset{|}{C}}-Cl$$

Compare this addition with the addition of water and note that again one end of the double bond gets an H and the other end gets the rest of the adding molecule, OH in this case. Now the product is an alcohol.

$$\underset{/}{\overset{\backslash}{C}}=\underset{\backslash}{\overset{/}{C}} \quad + \quad H-O^{H} \quad \xrightarrow{\text{acid catalyst}} \quad H-\overset{|}{\underset{|}{C}}-\overset{|}{\underset{|}{C}}-O^{H}$$

Two halogens, Cl_2 and Br_2, also add to carbon–carbon double bonds. One Cl or Br goes to one end and the other Cl or Br goes to the other end of the double bond. H_2 adds similarly, but special catalysts are needed.

As a study goal, practice writing the structures of the products of the following alkenes with hydrogen chloride, water (in the presence of an acid catalyst), Cl_2, Br_2, and H_2 (assuming the special conditions).

$$H_2C{=}CH_2 \qquad CH_3CH{=}CHCH_3$$

Aromatic hydrocarbons

The typical Lewis structure for an aromatic hydrocarbon such as benzene would suggest that it is an unsaturated hydrocarbon. However, the structure below is one of a pair of resonance structures, neither of which actually exists. In molecular orbital terms the pi electrons are delocalized around the benzene ring.

The result is that aromatic hydrocarbons rarely participate in addition reactions because the benzene ring strongly resists anything that breaks up its unique delocalized pi electron network. Instead, substitution reactions occur, which leave the benzene ring system intact. Practice writing the structures of what forms when benzene reacts with concentrated sulfuric acid, nitric acid, and the two halogens, Cl_2 and Br_2, when an iron(III) halide catalyst is present.

Self–Test

8.   Write the IUPAC names of the following.

   (a) $CH_3CH_2CH_2CH_2CH_3$ _____

   (b) $CH_3\overset{\overset{\displaystyle CH_3}{|}}{C}HCH_2CH_2CH_2CH_3$ _____

   (c) $CH_3\overset{\overset{\displaystyle CH_3}{|}}{\underset{\underset{\displaystyle CH_3}{|}}{C}}CH_2CH_2CH_3$ _____

   (d)

   $$\begin{array}{l} \qquad\qquad CH_2{-}CH_3 \\ \qquad\qquad | \\ CH_2{-}CH{-}CH_2 \\ | \qquad\;\; | \\ CH_3 \quad CH_3 \end{array}$$ _____

   (e)

   $$\begin{array}{l} \qquad\qquad CH_3 \qquad CH_3 \qquad CH_2{-}CH_3 \\ \qquad\qquad | \qquad\quad\; | \qquad\qquad | \\ H_3C{-}CH_2{-}CH{-}CH_2{-}C{-}CH_2{-}CH{-}CH_2{-}CH_2{-}CH_3 \\ \qquad\qquad\qquad\qquad\quad | \\ \qquad\qquad\qquad\qquad\; CH_2{-}CH_3 \end{array}$$

_____

9. Write the structure of 2,3,3–trimethyl–4–ethylheptane.

10. Write the IUPAC names and structures of the isomers of C_5H_{12}.

11. Write the IUPAC name of the following compound.

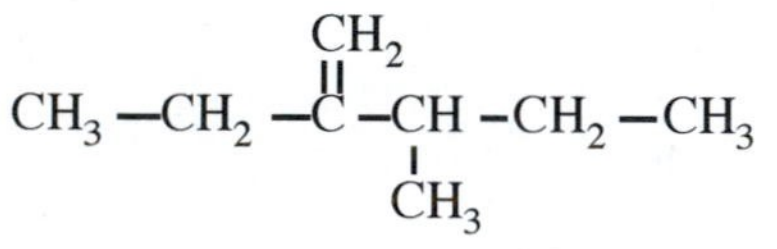

$$CH_3-CH_2-\overset{\overset{\displaystyle CH_2}{\|}}{C}-\underset{\underset{\displaystyle CH_3}{|}}{CH}-CH_2-CH_3$$

_____

12. Write the structures of the *cis* and *trans* isomers of 4–methyl–2–pentene.

13. Write the condensed structural formulas for the products of the following reactions.

    (a) The addition of hydrogen to propene _____

    (b) The addition of water to 3–hexene   _____

    (c) The addition of chlorine to 1–pentene _____

    (d) The reaction of bromine in the presence of $FeBr_3$ with benzene

    _____

14. Compare and contrast the chemical behavior toward hot concentrated sulfuric acid of cyclohexane, cyclohexene, and benzene.

   (a) Cyclohexane   _____

   (b) Cyclohexene   _____

   (c) Benzene   _____

22.3   Organic Compounds Containing Oxygen

Learning Objective

> Describe the names and typical reactions of common alcohols, ethers, aldehydes, ketones, carboxylic acids, and esters.

Review

Alcohols and ethers

To be an alcohol, the molecule must have the —OH group (or it can be written HO—) attached to a saturated carbon, one with four single bonds. (A molecule can have two or more such groups. In fact most organic compounds have more than one functional group, either alike or different.)

When you see an alcohol group, you think "This is a water–like group that confers on the molecule the following properties."

1. The —OH group helps to make the substance more soluble in water and have a higher boiling point than the corresponding hydrocarbon.

2. The molecules of the substance can be oxidized provided that the carbon bonded to the —OH group is also bonded to H.

3. Alcohols of the RCH_2OH type are oxidized, first to aldehydes and then to carboxylic acids.

4. Alcohols of the R_2CHOH type are oxidized to ketones.

5. Alcohols of the R_3COH type are not oxidized by simple loss of H_2.

6. Alcohols undergo elimination reactions, those in which the pieces of a water molecule depart from adjacent carbons and leave behind a carbon–carbon double bond.

7. Alcohols undergo substitution reactions, those in which the OH group is substituted for by Cl, Br, or I, using reactions with the corresponding concentrated HX solution (where X = Cl, Br, or I).

   For the sake of completeness, anticipating the next section, add a fifth property.

8. Alcohols react with carboxylic acids to give esters. (see below)

Ethers are compounds with a —C—O—C— functional group. A common ether, diethyl ether, was one of the first anesthetics used in 19th century surgical procedures. Ethers are only slightly polar and often more volatile than similar–sized oxygen–containing molecules. This volatility adds to their ability to burn rapidly, and at times, explode. Ethers also form peroxides, which can explode on long–term storage.

Ethers tend to be relatively unreactive, compared to other oxygen–containing organic compounds. Ethers are often used as solvents for other reactions.

Aldehydes, ketones, acids, and esters

The first task is to learn to recognize by name the functional groups that involve the carbonyl group when you see them in a complex structure. This is like being able to recognize that a wiggly blue line on a map represents a river, because functional groups are the "map signs" of organic structures. Certain chemical and physical properties are associated with each functional group "map sign."

When you learn functional groups, be sure not to tie their structures to particular positions on a page. The sequences of atoms are what count, not whether they are written left–to–right or top–to–bottom. For example, all of the following structures are esters because each has the "carbonyl–oxygen–carbon" sequence of atoms that defines the ester group.

$$\underset{\displaystyle CH_3O\overset{\displaystyle O}{\overset{\displaystyle \|}{C}}CH_3}{} \qquad \underset{\displaystyle CH_3CH_2\overset{\displaystyle O}{\overset{\displaystyle \|}{C}}OCH_2CH_3}{}$$

$$\begin{array}{c} CH_3 \\ | \\ CH_2 \\ | \\ C=O \\ | \\ O \\ | \\ CH_3 \end{array} \qquad \begin{array}{c} CH_3 \\ | \\ CH_2 \\ | \\ O \\ | \\ C=O \\ | \\ CH_3 \end{array}$$

Some important esters (e.g., those in vegetable oils or animal fats) have three ester groups per molecule.

It's important to get aldehydes and ketones straight, because they differ so much in ease of oxidation. Typically the aldehyde group can be written in a variety of ways.

$$H-\overset{\displaystyle O}{\overset{\displaystyle \|}{C}}- \quad \text{or} \quad -\overset{\displaystyle O}{\overset{\displaystyle \|}{C}}-H \quad \text{or} \quad -CH=O \quad \text{or} \quad -CHO$$

Representations of the aldehyde group

Importantly the carbonyl carbon has at least one hydrogen atom bonded to it. The remaining bond to the carbonyl carbon may be another hydrogen atom or R. The R group can be saturated or unsaturated. In ketones the carbonyl is flanked on both sides by carbon atoms from identical or different R groups.

$$-\overset{\displaystyle O}{\overset{\displaystyle \|}{C}}- \qquad\qquad CH_3-\overset{\displaystyle O}{\overset{\displaystyle \|}{C}}-CH_3$$

The ketone group.   Propanone,
(also called a keto group)   a ketone.

Carboxylic acids all have the "carbonyl–oxygen–hydrogen" sequence.

$$\overset{\displaystyle O}{\underset{\displaystyle \|}{-C}}-O-H$$

Thus all of the following structures are carboxylic acids.

You will often see the carboxylic acid group abbreviated as CO_2H or $COOH$, and sometimes it's written "backward" as HO_2C or $HOOC$.

Chemical properties of carbonyl compounds

Concerning *aldehydes*, you should learn the following.

1. The aldehyde group is one of the most easily oxidized of all functional groups, being changed by oxidation to the carboxyl group.

2. The aldehyde group adds hydrogen, catalytically, to give alcohols of the RCH_2OH type.

◆◆◆

Concerning *ketones*, there are only two properties that we studied.

1. The keto group strongly resists oxidation.

2. The keto group adds hydrogen, catalytically, to give alcohols of the R_2CHOH type.

◆◆◆

With respect to *carboxylic acids*, learn the following.

1. Carboxylic acids are weak acids (toward water), but they readily neutralize strong base, like OH^-, and form the corresponding carboxylate ions, RCO_2^-.

2. Carboxylic acids react with alcohols (acid–catalysis) to give esters.

3. The carboxyl group can be changed to the amide group by a reaction with either ammonia or an amine.

4. The carboxyl group strongly resists oxidation.

5. The carboxylate group, CO_2^-, is a good Brønsted base; when it accepts H^+, it becomes the (weakly acidic) carboxyl group, CO_2H, again.

◆◆◆

Esters undergo the following reactions.

1. The ester group is hydrolyzed (acid catalysis) to give the carboxylic acid and the alcohol from which the ester is made.

2. The ester group is also hydrolyzed by the action of aqueous alkali, like NaOH, to give the parent alcohol, only now the carboxylate ion of the parent carboxylic acid, not the free acid, forms. (This reaction is often called the *saponification* of an ester.)

Self–Test

**Questions 15–21 refer to the following numbered structures.
Some questions draw on knowledge from the preceding sections.**

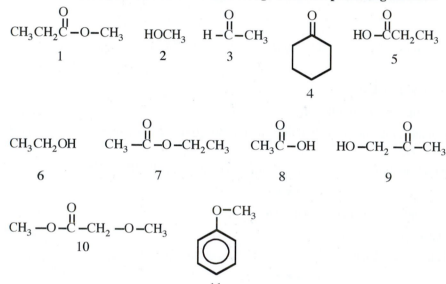

15. Which structures contain each of the following groups?

    (a)   ester group   _____

    (b)   ether group   _____

    (c)   alcohol group   _____

    (d)   carboxylic acid group   _____

    (e)   aldehyde group   _____

    (f)   ketone group   _____

16. Which compound is the most easily oxidized?   _____

17. Which compound(s) will rapidly neutralize sodium hydroxide at room temperature?

    _____

18. Two of the compounds shown will react to give compound 1. Which are they?

    _____

19. Compound 7 will react with water to give two of the compounds. Which two are they?

    _____

20. Which compound has a benzene ring?   _____

21. Which compound will react with hydrogen to give compound 6?   _____

22. Write the structure of 2,3–dimethyl–1–butanol.

    _____

23. Write the IUPAC name of the following compound.

    $$\text{CH}_3 \quad \text{OH}$$
    $$\text{CH}_3\text{CHCH}_2\text{CHCH}_3$$ _____

24. Write the structures of the products that could be made by the oxidation of each compound. If no oxidation can occur, state so.

    (a)   2–propanol   _____

    (b)   1–propanol   _____

    (c)   2–methyl–2–butanol _____

25.   Examine the following structures and then answer the following questions.

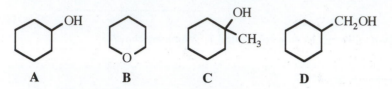

A   B   C   D

(a)   Which compound(s) cannot be oxidized?   _____

(b)   Which compound(s) cannot be dehydrated?   _____

(c)   Which compound(s) can be oxidized to a ketone?   _____

(d)   Which compound(s) can be oxidized to an aldehyde? _____

(e)   Which compound(s) can be oxidized to a carboxylic acid

(with the same number of carbons)?   _____

(f)   Which compound(s) can be dehydrated (to alkenes)?   _____

26.   Write the structure of the organic product in each situation.

(a)   OH

$\xrightarrow[\text{heat}]{H_2SO_4}$ _____

(b)   OH

$+ \; Cr_2O_7{}^{2-} \; \xrightarrow{H^+}$ _____

22.4 Organic Derivatives of Ammonia

Learning Objective

Describe the names and typical reactions of amines and amides.

Review

Amines are alkyl derivatives of ammonia and so, like ammonia, are proton acceptors or Brønsted bases. Molecules of amines can also accept and donate hydrogen bonds. Amines can be converted to amides by replacing the —OH on an organic acid with an N bonded to either: two hydrogen atoms, one hydrogen atom and an alkyl group or two alkyl groups.

The protonated forms of amines are like the ammonium ion in that they are proton donors or Brønsted acids. They can neutralize a strong base, such as OH^-. Because they are cations and interact strongly with water molecules, protonated amines are more soluble in water than the neutral amine molecules.

When considering amides, the nitrogen of an amide is not a proton acceptor (not a Brønsted base), unlike the nitrogen of an amine (or ammonia). Amides do react with water to give the parent carboxylic acid and amine (or ammonia).

Be sure to catch the structural difference between an amine and an amide. In the amides, there is always a carbonyl group, (C=O), directly attached to the nitrogen atom. The difference is important because the properties are so different. Amines are basic compounds; amides are not. Amides are broken apart by water; amines are not. Both groups are polar and both can participate in hydrogen bonds.

$$\text{Amides:} \qquad CH_3-\overset{\overset{\displaystyle O}{\|}}{C}-NH_2 \qquad CH_3-\overset{\overset{\displaystyle O}{\|}}{C}-NH-CH_3 \qquad CH_3-\overset{\overset{\displaystyle O}{\|}}{C}-\overset{\overset{\displaystyle CH_3}{|}}{N}-CH_2-CH_3$$

$$\text{Amines:} \qquad CH_3CH_2NH_2 \qquad CH_3CH_2NHCH_3 \qquad CH_3CH_2\overset{\overset{\displaystyle CH_3}{|}}{N}CH_2CH_3$$

Self–Test

27. What forms when methylamine neutralizes hydrochloric acid? Write the structure.

28. If $CH_3CH_2NH_3{}^+Cl^-$ reacts with aqueous sodium hydroxide, what forms? Write the structure.

Questions 29–34 refer to the numbered structures below.
Some of the questions require knowledge of material given earlier.

$$NH_2-CH_2-\overset{\overset{\displaystyle O}{\|}}{C}-CH_3 \qquad NH_2-CH_2CH_3 \qquad CH_3\overset{\overset{\displaystyle O}{\|}}{C}-NH_2 \qquad NH_3 \qquad CH_3NH_2$$

$$\quad\quad 1 \qquad\qquad\qquad 2 \qquad\qquad 3 \qquad\qquad 4 \qquad\quad 5$$

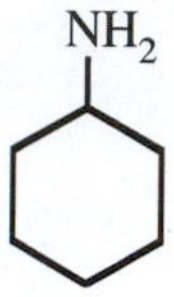

$$\underset{6}{\text{(cyclohexane with NH}_2\text{)}} \qquad CH_3\overset{\overset{\displaystyle O}{\|}}{C}-OH \qquad NH_2CH_3 \qquad CH_3NHCH_3 \qquad CH_3NH\overset{\overset{\displaystyle O}{\|}}{C}CH_3 \qquad CH_3NH_3{}^+$$

$$\qquad\qquad\qquad 7 \qquad\qquad 8 \qquad\qquad 9 \qquad\qquad 10 \qquad\qquad 11$$

29. The named groups given next are found in which structures?

    (a) the amine group   _____

    (b) an amino ketone   _____

    (c) the amide group   _____

30. Which two structures are of the same compound?   _____

31. The hydrolysis of structure 10 gives which two compounds?   _____

32. Which structures represent compounds that neutralize aqueous acids rapidly at room temperature?

    _____

33. Which structure has a carboxyl group?   _____

34. Which structure results when an amine neutralizes an acid?   _____

22.5   Organic Polymers

Learning Objective

Explain the structures, synthesis, and properties of polymers.

Review

Polymers are *macromolecules*, which are large molecules containing hundreds or even thousands of atoms. They are formed by linking together many smaller molecules called monomers. The result is long chain–like molecules composed of large numbers of identical repeating units. Not all the molecules of a given polymer are identical in size, but they do contain the same repeating unit. Study the example of polypropylene shown in this Section in the text.

Addition polymers are formed by just adding monomer units to form the chain. This is usually accomplished by a process involving *free radicals* (very reactive molecular fragments containing one or more unpaired electrons). Branching or linear chains are possible, depending on how the polymer is formed. Study the formation of polyethylene, as well as the structure of polystyrene in the text. Also, study Example 22.3 to be sure you know how to write the formula for an addition polymer.

Condensation polymers are formed by the elimination of a small molecule and the joining of monomer units. The monomer units being joined do not have to be identical; when they are different, the polymer formed is called a copolymer. Nylon is an example; learn the formula for the repeating unit of nylon 6,6.

Crosslinking occurs by the formation of bridges between adjacent polymer strands and increases the rigidity of the polymer. Rubber is formed by crosslinking the latex polymer called polyisoprene by heating it with sulfur. Polymer strands are linked by sulfur–sulfur bonds.

Crystallinity affects the properties of polymers, which become stronger as the polymer becomes more crystalline. Learn the differences between the different polyethylene polymers (LDPE, HDPE, UHMWPE).

Self–Test

35. What is the difference between an addition polymer and a condensation polymer?

_____

36. What is the function of an initiator in the creation of an addition polymer?

_____

37. Write the structural formula for the addition polymer formed from $CH_3CH_2CH=CH_2$.

38. Write the structural formula for the condensation polymer formed by elimination of CH_3OH from the following two monomer units.

$$CH_3-O-\overset{\overset{\displaystyle O}{\|}}{C}-CH_2-CH_2-CH_2-\overset{\overset{\displaystyle O}{\|}}{C}-O-CH_3 \quad \text{and} \quad HO-CH_2-CH_2-OH$$

39. What would be the repeating unit in nylon 4,4? (Sketch its structure.)

40. What is the difference between LDPE and HDPE?

_____

41. How does crosslinking affect the physical properties of a polymer?

_____

22.6   Carbohydrates, Lipids, and Proteins

Learning Objective

Name common carbohydrates, lipids, and proteins and describe their properties.

Review

This very short section is meant only to be a brief overview of *biochemistry* as well as to introduce you to the general kinds of biochemicals and how each serves in providing a living system with materials, energy, and information.

Carbohydrates

The simplest *carbohydrates*—the *monosaccharides*—involve alcohol plus aldehyde or ketone groups, so monosaccharides partake of the properties of these functional groups. Monosaccharide molecules are normally in cyclic forms that are in equilibrium with open–chain forms, and only in the open–chain forms are the aldehyde or keto groups present.

The *disaccharides* and *polysaccharides* give the monosaccharides when they react with water. These systems are made from the cyclic forms of the monosaccharides, being strung together by means of oxygen bridges. Water reacts at these bridges when disaccharides and polysaccharides are digested. We can write equations for the hydrolysis (digestion) of di– and polysaccharides as follows, and these equations will help you remember the important relationships.

For the disaccharides lactose and sucrose:

$$\text{lactose} + H_2O \xrightarrow[\text{(hydrolysis)}]{\text{digestion}} \text{galactose} + \text{glucose}$$

$$\text{sucrose} + H_2O \xrightarrow[\text{(hydrolysis)}]{\text{digestion}} \text{glucose} + \text{fructose}$$

Hydrolysis of the polysaccharides starch and cellulose is represented by:

$$\text{starch} + nH_2O \xrightarrow[\text{(hydrolysis)}]{\text{digestion}} n\text{glucose}$$

$$\text{cellulose} + nH_2O \xrightarrow[\text{(hydrolysis)}]{\text{digestion}} n\text{glucose}$$

Starch is actually a mixture of two polysaccharides, amylose and amylopectin. Cellulose, the chief constituent of the cell walls of plants, is not digestible in humans, but its acid–catalyzed hydrolysis also gives glucose as the only product.

Lipids

In lipids, the ester group is the key functional group in the *triacylglycerols*, members of the family of *lipids* that react with water during digestion to give long–chain carboxylic acids—fatty acids—and glycerol. The fatty acids often carry one or more alkene double bonds. It is not the presence of an ester group that defines the larger family of the lipids, however. To be a lipid, all that a natural product has to be is mostly hydrocarbon–like so that it tends to be far more soluble in nonpolar solvents than in water. Thus cholesterol, which has no ester group, is a lipid.

Proteins

All *proteins* consist of molecules of one or more *polypeptides*, and many proteins also include another organic molecule or a metal ion. Each polypeptide is a polymer of several (usually hundreds of) α–*amino acids*. The specific amino acids used, the number of times each is employed, and the order in which they are joined are the three factors that give each polypeptide its uniqueness.

When amino acids link together, water splits out from the carboxyl group of one amino acid and the α–amino group of the next one to give a *peptide bond*. (In a sense, a polypeptide is a condensation polymer similar in certain respects to nylon, which you studied in Chapter 13.) Polypeptides coil over much of their lengths into helices that are stabilized by hydrogen bonds. These helices usually undergo further kinking and folding. Thus each polypeptide has its own unique shape as well as unique amino acid sequence, and if this shape is lost, the protein no longer can function biologically.

Enzymes are proteins that catalyze reactions in cells. A *lock–and–key* mechanism enables an enzyme molecule to "recognize" and fit to only those substrate molecules meant for it.

Self–Test

42. What is studied in the field of biochemistry?

_____

43. What two kinds of substances are the chief sources of chemical energy for living systems?

_____

44. The biochemicals most closely involved in providing information for living things are in what family?

_____

45. What is the general name for the catalysts found in cells? _____

46. The functional group generally absent from carbohydrates is

    (a)   alcohol   (c)   aldehyde

    (b)   alkene   (d)   ketone   _____

47. Animals store glucose units as

    (a)   sucrose   (c)   starch

    (b)   glycogen   (d)   cellulose   _____

48. Sucrose digestion leads to

    (a)   glucose and fructose   (c)   glucose only

    (b)   galactose and glucose   (d)   malt sugar   _____

49. The sugar in milk is

    (a)   glucose   (c)   lactose

    (b)   maltose   (d)   sucrose   _____

50. A carbohydrate that makes up most of cotton is

    (a)   cellulose   (c)   lactose

    (b)   maltose   (d)   starch   _____

51. The digestion of lactose gives

    (a) glucose   (c) fructose and glucose

    (b) table sugar   (d) galactose and glucose   _____

52. Because of the many OH groups in glucose molecules, glucose is

    (a) insoluble in water   (c) soluble in water

    (b) nonpolar   (d) hypotonic   _____

53. The digestion of starch is an example of

    (a) oxidation   (c) neutralization

    (b) reduction   (d) hydrolysis   _____

54. Because the vegetable oils have several alkene groups per molecule, they are called

    (a) polyunsaturated   (c) polymers

    (b) polyenes   (d) polypeptides   _____

55. When triacylglycerols are digested, the reaction is the hydrolysis of

    (a) glycerol   (c) alkene groups

    (b) fatty acids   (d) ester groups   _____

56. The hydrocarbon–like portions of a phosphoglyceride are

    (a) cationic   (c) hydrophilic

    (b) hydrophobic   (d) anionic   _____

57. The surfaces of the lipid bilayer are dominated by

    (a) hydrophilic groups   (c) cholesterol

    (b) polyunsaturation   (d) nonpolar tails   _____

58. One of the services performed by proteins embedded in lipid bilayer membranes is

    (a) digestive enzyme function   (c) hormone synthesis

    (b) enzyme manufacture   (d) ion channels   _____

59. The side chain in serine is $-CH_2OH$. Therefore, serine's structure is

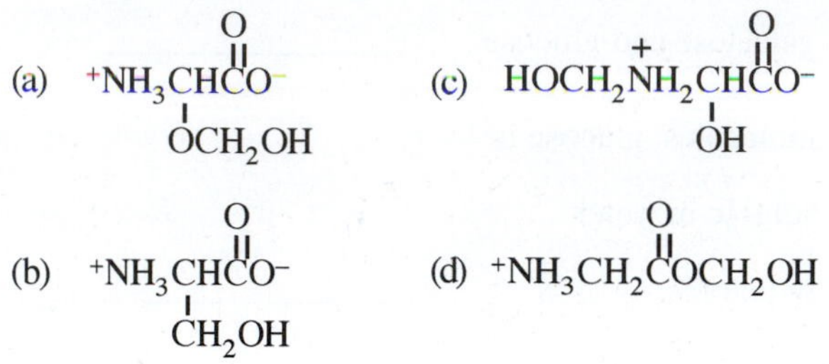

(a) $^+NH_3CH\overset{\overset{\displaystyle O}{\|}}{C}O^-$
    $|$
    OCH_2OH

(c) $HOCH_2\overset{+}{N}H_2CH\overset{\overset{\displaystyle O}{\|}}{C}O^-$
    $|$
    OH

(b) $^+NH_3CH\overset{\overset{\displaystyle O}{\|}}{C}O^-$
    $|$
    CH_2OH

(d) $^+NH_3CH_2\overset{\overset{\displaystyle O}{\|}}{C}OCH_2OH$

60. Which arrow points to the peptide bond?

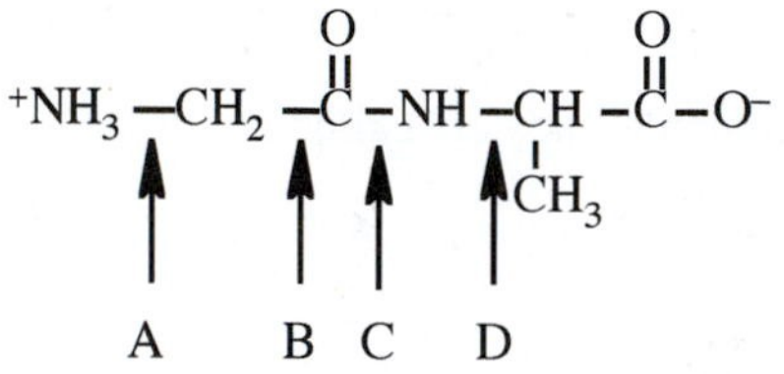

$$^+NH_3-CH_2-\overset{\overset{\displaystyle O}{\|}}{C}-NH-CH-\overset{\overset{\displaystyle O}{\|}}{C}-O^-$$
$$\uparrow \qquad \uparrow \quad \uparrow \qquad \uparrow CH_3$$
$$A \qquad B \quad C \qquad D$$

(a) A   (b) B   (c) C   (d) D

61. What is the side chain in the following compound?

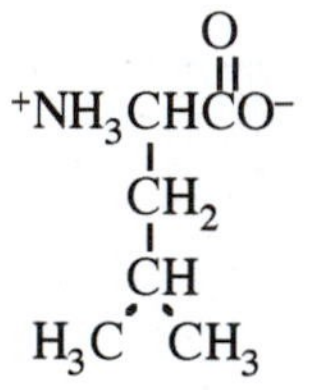

$$^+NH_3CH\overset{\overset{\displaystyle O}{\|}}{C}O^-$$
$$|$$
$$CH_2$$
$$|$$
$$CH$$
$$H_3C \quad CH_3$$

(a) $^+NH_3-$

(c) an alkyl group

(b) $-\overset{\overset{\displaystyle O}{\|}}{C}O^-$

(d) $^+NH_3CH\overset{\overset{\displaystyle O}{\|}}{C}O^-$

62. Which structure best represents the nature of the main chain or "backbone" in polypeptides?

(a) $^+NH_3CH_2\overset{\overset{O}{\|}}{C}NHCH_2\overset{\overset{O}{\|}}{C}NHCH_2\overset{\overset{O}{\|}}{C}$ — etc.

(c) $^+NH_3CH_2\overset{\overset{O}{\|}}{C}CH\overset{\overset{O}{\|}}{C}CH\overset{\overset{O}{\|}}{C}$ — etc.
$\qquad\qquad\quad\;\; |\quad\;\; |$
$\qquad\qquad -NH\;\; NH-$

(b) $^+NH_3CH_2\overset{\overset{O}{\|}}{C}O\overset{\overset{O}{\|}}{C}CH_2NHNHCH_2\overset{\overset{O}{\|}}{C}$ —etc.

(d) $^+NH_3CH_2\overset{\overset{O}{\|}}{C}ONHCH_2\overset{\overset{O}{\|}}{C}ONHCH_2\overset{\overset{O}{\|}}{C}O$ — etc.

_____

63. One of the possible dipeptides that can form from alanine (side chain = CH_3) and cysteine (side chain = $-CH_2SH$) is

(a) $^+NH_3CH\overset{\overset{O}{\|}}{C}NHCH\overset{\overset{O}{\|}}{C}O^-$
$\qquad\;\; |\qquad\quad\;\; |$
$\qquad\;\; SH\qquad\; CH_3$

(c) $^+NH_3CH_2CH_2\overset{\overset{O}{\|}}{C}NHCH\overset{\overset{O}{\|}}{C}O^-$
$\qquad\qquad\qquad\qquad\;\; |$
$\qquad\qquad\qquad\qquad CH_2SH$

(b) $^+NH_3CH\overset{\overset{O}{\|}}{C}SCH_2CH\overset{\overset{O}{\|}}{C}O^-$
$\qquad\;\; |\qquad\qquad\;\; |$
$\qquad\;\; CH_3\qquad\quad NH_2$

(d) $^+NH_3CH\overset{\overset{O}{\|}}{C}NHCH\overset{\overset{O}{\|}}{C}O^-$
$\qquad\;\; |\qquad\qquad\;\; |$
$\qquad\;\; CH_2SH\;\; CH_3$

_____

64. In protein chemistry, "native form" refers to what?

(a) a building unit of a protein

(b) the location of the hydrophobic groups in a protein molecule

(c) the shape of a protein molecule after denaturation

(d) the final shape of a protein molecule

_____

65. Enzymes are

(a) catalysts (b) B–vitamins (c) substrates (d) denatured proteins _____

22.7   Nucleic Acids, DNA, and RNA

Learning Objective

Explain how the structures of DNA and RNA enable the transmission of genetic information and the synthesis of proteins.

Review

DNA either directs the synthesis of more of itself—replication—or it directs the apparatus for making polypeptide molecules having particular sequences of their amino acid side chains. DNA is one of the two kinds of nucleic acids, and the backbone of DNA is an alternating sequence of deoxyribose–phosphate units, each one bearing a side–chain amine or base. The kind, number, sequence, and hydrogen–bonding abilities of these bases—there are four of them—determine the properties of the gene units of individual DNA double helices, according to the Crick–Watson theory. The amines come as matched base pairs, with adenine (A) pairing by hydrogen bonds to thymine (T) and guanine (G) pairing to cytosine (C).

The molecules of RNA, of which there are several types, consist of alternating sequences of ribose–phosphate units, each ribose bearing a side–chain base. The bases are the same as those in DNA except that uracil (U) replaces thymine (T). Base pairing occurs between A and U as well as between A and T.

Just prior to cell division, each of the two strands in a DNA double helix uses the pairing requirements of its side–chain bases to guide replication. Review Figure 22.10, which shows that thiamine and adenine have two hydrogen bonds between them while cytosine and guanine attract each other with three hydrogen bonds.

Between cell divisions, the sequence of bases in DNA are used to *transcribe* the genetic message to RNA. Each adjacent series of three bases on the RNA is a codon. It will eventually direct a particular amino acid unit into place during the synthesis of a polypeptide in a process commonly called *translation*. Thus the genetic code is the match–up between codons on RNA and the amino acids used to make polypeptides.

Genetic defects can arise at any stage, but those that are inherited occur as incomplete or faulty sequences of bases on DNA. Damage to DNA by chemical or photochemical processes also cause genetic defects that may have serious health effects.

Viruses consist of nucleic acids—some viruses with DNA and others with RNA—combined with proteins.

In genetic engineering, DNA material corresponding to some desired protein is inserted into a cell where it then proceeds to make the protein for which it is coded. When a cell's DNA—it might be a cell of some bacterium or a yeast—is altered by new DNA, the resulting DNA is called recombinant DNA. The technique has been used to manufacture human insulin.

Self–Test

In Questions 67 – 71, fill in the blanks with the best technical terms.

66. The force of attraction responsible for the pairing of amines in the _____ double helix is the

    _____where (use the code letters) _____

    pairs with T and _____ pairs with G.

67. The product of replication is another _____

68. The overall series of events from DNA to the RNA that directs polypeptide synthesis is called

    _____

69. After translation has occurred, the product is a(n) _____

70. A(n) _____ is a set of three nucleic acids (A,C,G,T) that represent an individual
    amino acid in a protein.

71. The information that the answer to Question 70 embodies is often called the

    _____

Answers to Self–Test Questions

1.

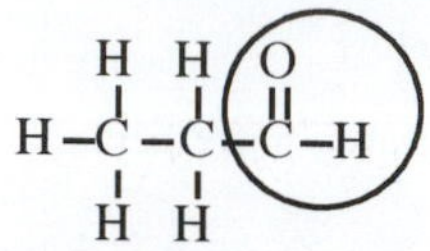

2. CH_3-NH_2

3. $CH_3-NH_3{}^+ + Br^-$

4. $R-NH_3{}^+ + Cl^-$

5. It greatly simplifies it. There are very few kinds of functional groups among the millions of organic compounds, and each kind displays mostly the same set of reactions.

6. (a) isomers, (b) isomers, (c) identical

7. The second compound of part (b).

8. (a) pentane, (b) 2–methylhexane, (c) 2,2–dimethylpentane, (d) 3–methylhexane, (e) 5,7–diethyl–3,5–dimethyldecane

9.

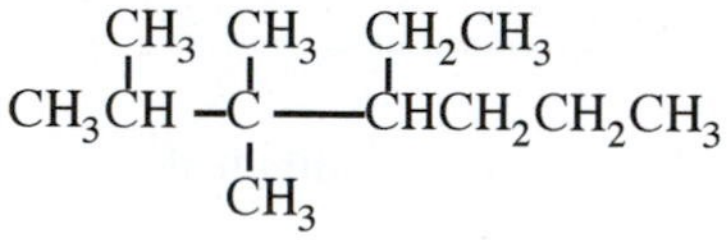

10. $CH_3CH_2CH_2CH_2CH_3$ pentane

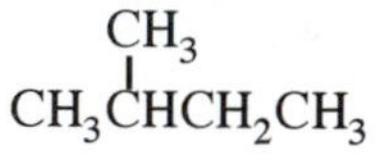

      2-methylbutane

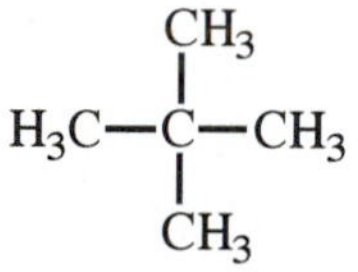

        2,2-dimethylpropane

11. 2–ethyl–3–methyl–1–pentene

12.

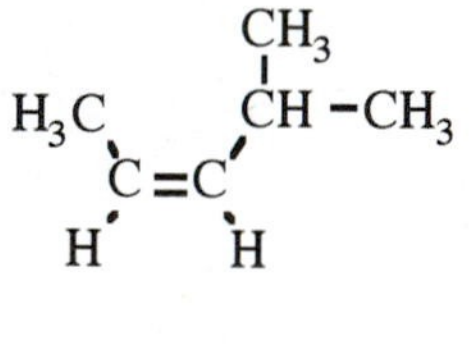

    cis-isomer 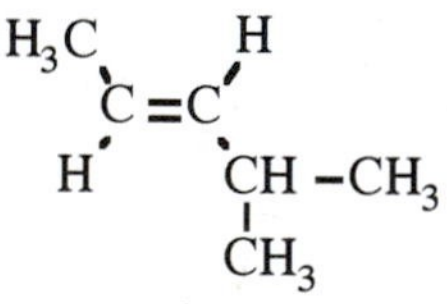   trans-isomer

13. (a) $CH_3CH_2CH_3$

    (b)

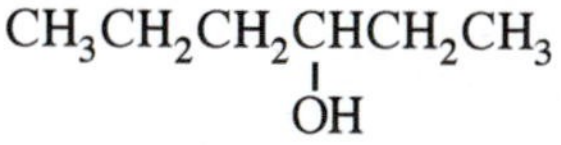

    (c)    (d) 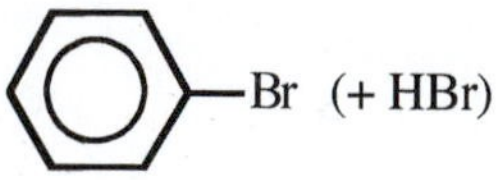 $-Br$ (+ HBr)

14. (a) No reaction

(b) —OSO₃H forms (the product of an addition reaction)

(c) —SO₃H forms (the product of a substitution reaction)

15. (a) 1, 7, 10 (b) 10, 11 (c) 2, 6, 9 (d) 5, 8 (e) 3 (f) 4, 9
16. 3
17. 5, 8
18. 2, 5
19. 6, 8
20. 11
21. 3
22.

$$CH_3\overset{\underset{\displaystyle |}{CH_3}}{CH} -\overset{\underset{\displaystyle |}{CH_3}}{CH}CH_2OH$$

23. 4–methyl–2–pentanol
24.

(a) $CH_3\overset{\overset{\displaystyle O}{||}}{C}CH_3$ (b) CH_3CH_2CHO which is further oxidized to $CH_3CH_2CO_2H$

(c) no oxidation

25. (a) B, C (b) B (c) A (d) D (e) D (f) A, C, D
26.

(a)   (b)

27. $CH_3NH_3^+Cl^-$
28. $CH_3CH_2NH_2$ (+ H_2O + NaCl)
29. (a) 1, 2, 5, 6, 8, 9 (Structure 4 is ammonia, not an amine.) (b) 1 (c) 3, 10
30. 5, 8
31. 5 (or 8) and 7
32. 1, 2, 4, 5, 6, 8, 9
33. 7
34. 11
35. In an addition polymer, monomer units are simply joined end to end. In a condensation polymer, a small molecule is eliminated when the monomer units become joined.
36. To start a free radical polymerization of monomer units. (See Section 22.5.)

37.

$$\left(\begin{array}{cc} \overset{\underset{\displaystyle |}{\underset{\displaystyle CH_2}{|}}}{\overset{\displaystyle H}{\underset{\displaystyle |}{C}}} & \overset{\displaystyle H}{\underset{\underset{\displaystyle H}{\displaystyle |}}{\underset{\displaystyle |}{C}}} \\ CH_3 & \end{array}\right)_n$$

38.

$$\left(\!\!\begin{array}{c} O \\ \| \\ O-C-CH_2-CH_2-CH_2-C-O-CH_2-CH_2 \\ \end{array}\!\!\right)_n$$

39.

$$\left(\!\!\begin{array}{c} O \qquad\qquad O \;\; H \qquad\qquad\qquad\qquad\qquad H \\ \| \qquad\qquad\;\; \| \;\; | \qquad\qquad\qquad\qquad\qquad\;\; | \\ C-CH_2-CH_2-C-N-CH_2-CH_2-CH_2-CH_2-N \\ \end{array}\!\!\right)_n$$

40. LDPE has shorter chains and more branching than HDPE.
41. Crosslinking makes a polymer stiffer, stronger, and raises its melting point.
42. The organic compounds present in living cells or that have been made from them.
43. carbohydrates and lipids
44. nucleic acids
45. enzymes
46. b
47. b
48. a
49. c
50. a
51. d
52. c
53. d
54. a
55. d
56. b
57. a
58. d
59. b
60. c
61. c
62. a
63. d
64. d
65. a
66. DNA; hydrogen bond; A; C
67. DNA double helix
68. transcription
69. polypeptide
70. codon
71. genetic code

Tools for problem solving

In this chapter you learned to apply the following concepts as tools in solving problems. Study each one carefully so that you know what each is used for. When faced with solving a problem, recall what each tool does and consider whether it will be helpful in finding a solution.

You might want to tear these pages out to use along with solving problems in this chapter.

Condensed structures (Section 22.1)

The key to writing condensed structures is keeping in mind the number of covalent bonds formed by the nonmetals in achieving an octet.

Group 4A 4 bonds	Group 5A 3 bonds	Group 6A 2 bonds
Group 7A 1 bond	hydrogen 1 bond	

In writing a condensed structure, be sure the number of hydrogens with a given element is sufficient to make up the difference between the number of non–hydrogen atoms attached and the total number of bonds required by the element in question.

Polygons as condensed structures (Section 22.1)

A carbon atom is understood at each corner; other elements in the ring are explicitly written. Edges of the polygon represent covalent bonds; double bonds are explicitly shown. The remaining bonds, as required by the covalence of the atom at a corner, are understood to hold H atoms.

Functional groups (Section 22.1)

If you can recognize a functional group in a molecule, you can place the structure into an organic family and so predict the kinds of reactions the compound can give. In particular, study Table 22.1.

IUPAC rules of nomenclature for organic compounds (Section 22.2)

For all compounds, the longest chair of carbon atoms or the longest chain with the functional group defines the parent chain for which the compound is named.

Alkanes (Section 22.2)

The chain is numbered from whichever end gives the lowest number to the first substituent. The name ending is –ane, and the prefixes that describe the number of carbon atoms in the chain are given in Table 22.2. The locations of alkyl substituents are specified by a number and a hyphen preceding the name of the alkyl group. Alkyl substituents are specified in alphabetical order.

Alkenes and alkynes (Section 22.2)

The name ending for alkenes is *–ene*, and the name ending for alkynes is *–yne*. The rules are similar to those for naming alkanes, except that the chain is numbered so that the first carbon of the double or triple bond has the lower of two possible numbers.

Alcohols (Section 22.3)

The name ending for alcohols is *–ol*. The parent chain is the longest chain that includes the alcohol group and is numbered from whichever end gives the lowest number to the –OH group.

Aldehydes and ketones (Section 22.3)

The name ending for aldehydes is *–al*. For ketones, the name ending is *–one*. The parent chain must include the carbonyl group and is numbered from whichever end gives the lowest number to the carbonyl carbon.

Carboxylic acids (Section 22.3)

The name ending is *–oic acid*.

Esters (Section 22.3) The name begins with the name of the alkyl group attached to the O atom. This is followed by a separate word taken from the name of the parent carboxylic acid, but altered by changing *–ic acid* to *–ate*.

Reactions of alkenes and alkynes (Section 22.2)
Components of a molecule add across a double bond.

$$CH_2{=}CH_2 \;+\; A{-}B \;\longrightarrow\; \underset{\underset{A}{|}}{CH_2}{-}\underset{\underset{B}{|}}{CH_2}$$

Reactions of alcohols
Oxidation of alcohols (Section 22.3)
The product(s) depend on the number of hydrogen atoms bonded to the alcohol carbon.

$$RCH_2OH \xrightarrow{\text{oxidation}} \underset{\text{aldehyde}}{R\overset{O}{\overset{\|}{C}}H} \xrightarrow[\text{oxidation}]{\text{further}} \underset{\text{carboxylic acid}}{R\overset{O}{\overset{\|}{C}}OH}$$

$$\underset{\overset{|}{\underset{}{OH}}}{R\overset{|}{C}HR'} \xrightarrow{\text{oxidation}} \underset{\text{ketone}}{R\overset{O}{\overset{\|}{C}}R'}$$

$$\underset{\overset{|}{\underset{R''}{}}}{\overset{R'}{\overset{|}{R\overset{}{C}OH}}} \xrightarrow{\text{oxidation}} \underset{\text{reaction}}{\text{no}}$$

Dehydration (Section 22.3)

$$\underset{\overset{|}{\underset{}{}}\underset{H\quad OH}{}}{RCH-CH_2} \xrightarrow[\text{heat}]{\text{acid catalyst}} RCH{=}CH_2 + H_2O$$

Substitution (Section 22.3)

$$ROH + HX(\text{conc.}) \xrightarrow{\text{heat}} RX + H_2O$$

Reactions of aldehydes and ketones: reduction (Section 22.3)

$$\underset{\text{aldehyde}}{R\overset{O}{\overset{\|}{C}}H} + H{-}H \xrightarrow[\text{catalyst}]{\text{heat, pressure}} \underset{}{R\overset{OH}{\overset{|}{C}}H_2}$$

$$\underset{\text{ketone}}{R\overset{O}{\overset{\|}{C}}R'} + H{-}H \xrightarrow[\text{catalyst}]{\text{heat, pressure}} \underset{}{R\overset{OH}{\overset{|}{C}}HR'}$$

Reactions of organic acids

Neutralization (Section 22.3)

Carboxylic acids react with bases to give salts.

$$RCO_2H \ + \ OH^- \longrightarrow \ RCO_2^- \ + \ H_2O$$

Formation of esters (Section 22.3)

Esters are formed in the reaction of a carboxylic acid with an alcohol.

$$\underset{\text{carboxylic acid}}{\overset{\overset{\displaystyle O}{\|}}{RCOH}} + \underset{\text{alcohol}}{HOR'} \underset{\text{Heat}}{\overset{H^+ \text{ catalyst}}{\rightleftharpoons}} \underset{\text{ester}}{\overset{\overset{\displaystyle O}{\|}}{RCOR'}} + H_2O$$

Saponification of esters (Section 22.3)

$$\underset{\text{ester}}{\overset{\overset{\displaystyle O}{\|}}{RCOR'}}(aq) \ + \ OH^-(aq) \overset{\text{heat}}{\longrightarrow} \underset{\text{carboxylate ion}}{\overset{\overset{\displaystyle O}{\|}}{RCO^-}}(aq) + \underset{\text{alcohol}}{R'OH}(aq)$$

Formation of amides (Section 22.4)

$$\underset{\text{carboxylic acid}}{\overset{\overset{\displaystyle O}{\|}}{RCOH}} \ + \ \underset{\text{ammonia}}{H\!-\!NH_2} \overset{\text{heat}}{\longrightarrow} \underset{\substack{\text{simple} \\ \text{amide}}}{\overset{\overset{\displaystyle O}{\|}}{RCNH_2}} + H_2O$$

Hydrolysis of amides (Section 22.4)

$$\underset{\text{simple amide}}{\overset{\overset{\displaystyle O}{\|}}{RCNH_2}} \ + \ H\!-\!OH \overset{\text{heat}}{\longrightarrow} \underset{\text{carboxylic acid}}{\overset{\overset{\displaystyle O}{\|}}{RCOH}} + NH_3$$

Addition polymerization (Section 22.5)

The addition of monomer units to a gradually growing chain by a free radical mechanism yields a long polymer chain. The repeating unit of the polymer has the formula of the monomer itself.

Condensation polymerization (Section 22.5)

Monomers are linked by the elimination of a small molecule such as H_2O or CH_3OH and the formation of a covalent bond between the remaining fragments. For copolymers, the repeating unit is that of the two monomers minus the small molecule that was eliminated.